physica status solidi (b)

ISSN 0370-1972 · SUBJECT AND AUTHOR INDEX (b) VOLUMES 103 TO 108 · 1981

ISSN 0031-8965 · SUBJECT AND AUTHOR INDEX (a) VOLUMES 63 TO 68 · 1981

physica status solidi (a)

AKADEMIE-VERLAG · BERLIN

Classification Scheme

1. Structure of Crystalline Solids
   1.1 Perfectly Periodic Structure
   1.2 Solid-State Phase Transformations
   1.3 Alloys. Metallurgy
   1.4 Microstructure (Magnetic Domains See 18; Ferroelectric Domains See 14.4.1)
   1.5 Films
   1.6 Surfaces
2. Non-Crystalline State
3. Crystal Growth
4. Bonding Properties
5. Mössbauer Spectroscopy
6. Lattice Dynamics. Phonons
7. Acoustic Properties
8. Thermal Properties
9. Diffusion
10. Defect Properties (Irradiation Defects See 11)
    10.1 Metals
    10.2 Non-Metals
11. Irradiation Effects (X-Ray Diffraction Investigations See 1 and 10)
12. Mechanical Properties (Plastic Deformations See 10)
    12.1 Metals
    12.2 Non-Metals
13. Electron States
    13.1 Band Structure
    13.2 Fermi Surfaces
    13.3 Surface and Interface States
    13.4 Impurity and Defect States
    13.5 Elementary Excitations (Phonons See 6)
        13.5.1 Excitons
        13.5.2 Plasmons
        13.5.3 Polarons
        13.5.4 Magnons
14. Electrical Properties. Transport Phenomena
    14.1 Metals. Semi-Metals
    14.2 Superconductivity. Superconducting Materials and Devices
    14.3 Semiconductors
        14.3.1 Films
        14.3.2 Surfaces and Interfaces
        14.3.3 Devices. Junctions (Contact Problems See 14.3.4)
        14.3.4 High-Field Phenomena, Space-Charge Effects, Inhomogeneities, Injected Carriers (Electroluminescence See 20.3; Junctions See 14.3.3)
    14.4 Dielectrics
        14.4.1 Ferroelectrics
15. Thermoelectric and Thermomagnetic Properties
16. Photoconductivity. Photovoltaic Effects
17. Emission of Electrons and Ions
    17.1 Field Emission Microscope Investigations
18. Magnetic Properties
    18.1 Paramagnetic Properties
    18.2 Ferromagnetic Properties
        18.2.1 Ferromagnetic Films
    18.3 Ferrimagnetic Properties
    18.4 Antiferromagnetic Properties

(Continued on cover three)

physica status solidi (b)

basic research

physica status solidi (a)

applied research

SUBJECT AND AUTHOR INDEX

Volumes 103 to 108, 1981   Volumes 63 to 68, 1981

ISSN 0370-1972/0031-8965

Edited by

E. BORCHARDT

AKADEMIE-VERLAG · BERLIN

Subscriptions and orders for this Subject and Author Index
or single copies of physica status solidi (b) and physica status solidi (a) should be directed
in the GDR:

to the Postzeitungsvertrieb or
to the Akademie-Verlag, DDR-1086 Berlin, Leipziger Straße 3 – 4;

in the other socialist countries:
to a book-shop for foreign language literature or
to the competent news-distributing agency;

in the FRG and BERLIN (WEST):
to a book-shop or
to the wholesale distributing agency Kunst und Wissen, Erich Bieber OHG, D-7000 Stuttgart 1, Wilhelmstr. 4 – 6;

in the other Western European countries:
to Kunst und Wissen Erich Bieber GmbH, CH-8008 Zürich, Dufourstr. 51;

in USA and CANADA:
to Verlag Chemie International Inc., Plaza Centre, Suite E, 1020 N.W. 6th Street, Deerfield Beach, FL 33441, USA;

in other countries:
to the international book- and journal-selling trade,
to Buchexport, Volkseigener Außenhandelsbetrieb der Deutschen Demokratischen Republik, DDR-7010 Leipzig, Postfach 160, or
to the Akademie-Verlag, DDR-1086 Berlin, Leipziger Straße 3 – 4.

Schriftleiter und verantwortlich für den Inhalt der Zeitschriften:
Professor Dr. Dr. h. c. P. Görlich, DDR-1020 Berlin, Neue Schönhauser Str. 20 bzw. DDR-6900 Jena, Schillbachstr. 24.

Herausgeber und verantwortlich für den Inhalt des Registers:
Dipl.-Phys. E. Borchardt, DDR-1020 Berlin, Neue Schönhauser Str. 20.

Verlag:
Akademie-Verlag, DDR-1086 Berlin, Leipziger Straße 3 – 4; Fernruf: 2236221 und 2236229; Telex-Nr.: 114420; Bank: Staatsbank der DDR, Berlin, Kto.-Nr.: 6836-26-20712.

Veröffentlicht unter der Lizenznummer 1310 und 1620 des Presseamtes beim Vorsitzenden des Ministerrates der Deutschen Demokratischen Republik.

Gesamtherstellung:
VEB Druckerei „Thomas Müntzer", DDR-5820 Bad Langensalza.

Bestellnummer des Registers 1068/85 R14.

Printed in the German Democratic Republic.
AN (EDV) 20 831/20 735

Preface

The fourteenth index volume contains volumes 103 to 108 of "physica status solidi (b) — basic research" as well as volumes 63 to 68 of "physica status solidi (a) — applied research", i.e. all volumes having appeared in 1981 containing 1497 papers.

The general scheme of last index volume was maintained. The basis for quoting is the subject classification given for every paper in the respective issue. Nevertheless, before taking over these subject classifications into the index volume they were once more carefully revised and in some cases improved and completed in order to get a most uniform and clear classification for all papers. The average number of quotations per paper in this index volume is nearly the same as in the last one, namely about 3.0.

Again I have to express my thanks to the readers for their interest, and I will always readily take up any critical remarks and proposals for improvements. I am especially obliged to Prof. Dr. P. Görlich, Prof. Dr. E. Gutsche, Dr. H.-J. Hänsch, Dr. H. Lange, Dr. S. Oberländer, and Dipl. Phys. K. Müller for giving me advice and unselfish help. Furthermore I have to thank Mrs. E. Schwan for her exact and reliable secretary's work at the manuscript.

Berlin, January 1982

Eva Borchardt

Classification Scheme

1. Structure of Crystalline Solids
   1.1 Perfectly Periodic Structure
   1.2 Solid-State Phase Transformations
   1.3 Alloys. Metallurgy
   1.4 Microstructure (Magnetic Domains See 18; Ferroelectric Domains See 14.4.1)
   1.5 Films
   1.6 Surfaces
2. Non-Crystalline State
3. Crystal Growth
4. Bonding Properties
5. Mössbauer Spectroscopy
6. Lattice Dynamics. Phonons
7. Acoustic Properties
8. Thermal Properties
9. Diffusion
10. Defect Properties (Irradiation Defects See 11)
    10.1 Metals
    10.2 Non-Metals
11. Irradiation Effects (X-Ray Diffraction Investigations See 1 and 10)
12. Mechanical Properties (Plastic Deformations See 10)
    12.1 Metals
    12.2 Non-Metals
13. Electron States
    13.1 Band Structure
    13.2 Fermi Surfaces
    13.3 Surface and Interface States
    13.4 Impurity and Defect States
    13.5 Elementary Excitations (Phonons See 6)
        13.5.1 Excitons
        13.5.2 Plasmons
        13.5.3 Polarons
        13.5.4 Magnons
14. Electrical Properties. Transport Phenomena
    14.1 Metals. Semi-Metals
    14.2 Superconductivity. Superconducting Materials and Devices
    14.3 Semiconductors
        14.3.1 Films
        14.3.2 Surfaces and Interfaces
        14.3.3 Devices. Junctions (Contact Problems See 14.3.4)
        14.3.4 High-Field Phenomena, Space-Charge Effects, Inhomogeneities, Injected Carriers (Electroluminescence See 20.3; Junctions See 14.3.3)
    14.4 Dielectrics
        14.4.1 Ferroelectrics
15. Thermoelectric and Thermomagnetic Properties
16. Photoconductivity. Photovoltaic Effects
17. Emission of Electrons and Ions
    17.1 Field Emission Microscope Investigations

18. Magnetic Properties
    18.1 Paramagnetic Properties
    18.2 Ferromagnetic Properties
        18.2.1 Ferromagnetic Films
    18.3 Ferrimagnetic Properties
    18.4 Antiferromagnetic Properties

19. Magnetic Resonance

20. Optical Properties
    20.1 Spectra. Optical Constants (X-Ray Spectra See 20)
    20.2 Lasers
    20.3 Luminescence (X-Ray Spectra See 20)

21. Metals. Alloys
    21.1 Transition Metals (Group Sc to Zn) and Their Alloys
        21.1.1 Fe and Fe Alloys
    21.2 Alkali Metals
    21.3 Alkali-Earth Metals
    21.4 Rare-Earth Metals
    21.5 Actinides
    21.6 Noble Metals
    21.7 Semi-Metals (Semiconducting Alloys See 22.7)

22. Semiconductors and Ionic Crystals
    22.1 Elements
        22.1.1 Germanium
        22.1.2 Silicon
        22.1.3 Group VI Elements
    22.2 III–V Compounds
        22.2.1 Arsenides
        22.2.2 Phospides
        22.2.3 Animonides
        22.2.4 Mixted Crystals
    22.3 IV–IV, II–V, and III–VI Compounds
        22.3.1 Mixed Crystals
    22.4 II–VI Compounds
        22.4.1 Sulphides
        22.4.2 Selenides
        22.4.3 Tellurides
        22.4.4 Mixed Crystals
    22.5 Halides
        22.5.1 Silver Halides
        22.5.2 Alkali Halides
        22.5.3 Alkali-Earth Halides
        22.5.4 Mixed Crystals
    22.6 Simple Oxides (CdO, ZnO See 22.4)
        22.6.1 Mixed Crystals
    22.7 Semiconducting Intermetallic Compounds
        22.7.1 Mixed Crystals
    22.8 Tertiary and Higher Compounds
        22.8.1 Salts of Oxy-Acids (Carbonates, Phosphates, Silicates Including Mica etc.)
        22.8.2 Spinels, Garnets, Ferrites
    22.9 Organic Semiconductors

23. Solidified Gases

For comments on the use of the classification scheme see p. 6.

Some Comments on the Use of the Subject and Author Index

The classification system[1]) comprises 20 broad sections each devoted to a major field of solid state physics and 3 broad sections devoted to the main groups of substances. A section may contain a number of subsections dealing with aspects of this field or special substance, respectively, which tend to receive particular attention. A paper dealing with this field will either be classified under one or more of the subsections or, if none of these are relevant, it will just appear under the general section heading (i.e. sections and subsections are equivalent in this sense). For example, a paper dealing with problems of paramagnetism would appear under 18.1 but not 18, whereas a paper on diamagnetism would be quoted under 18; a paper dealing with GaP would appear under 22.2.2, not under 22 or 22.2, whereas a paper on AlN would be quoted under 22.2, and a paper on a II–IV or a IV–VI compound would be quoted under 22.

All papers will be quoted under each relevant section. There is made a distinction between one or two principal entries (refering to the main subject matter of the paper) and secondary entries (refering to subsidiary subject matter and substance, respectively). In the principal entries the title of the paper is given in its original language and in English. In addition, every principal entry is marked by * in front of the title of the paper. In the secondary entries the title is only given in English.

Each entry gives the volume, the number of the issue, the initial and final page numbers, and the year in which the paper appeared. A paper of physica status solidi (b) or physica status solidi (a) is marked by (b) or (a) in front of the line, respectively. The letter "K" before the page numbers indicates a short note published in the offset part of the journal. Each entry is followed by references of the other entries. These references give the subject classifications and the pages in the index where the other entries appear. The references to the principal entries are given in italics.

Review articles appear, at the head of the list of entries under a given subject and are given the subtitle "Review Articles". These articles are listed in the chronological order of their publication, irrespective of being a paper of physica status solidi (b) or physica status solidi (a). The subject classification for a review article consisting of several parts is based on the article as a whole rather than on the individual parts.

The entries for original papers, short notes, and errata are made together in a single list for each subject and this follows the entries for review articles. These are again ordered according to data of publication, irrespective of being a paper of physica status solidi (b) or physica status solidi (a). Original papers are sometimes published in several parts under the same principal title (in general with special subtitles). The various parts of papers of this type are treated separately and each part is classified according to its particular contents.

In quoting errata the following rule is obeyed. If the erratum refers to a paper quoted in the preceding index, it is listed in the chronological order of its publication, otherwise the erratum is quoted together with the paper to which it refers.

---

[1]) See cover two and three, and pages 4 and 5. For the case of losing the cover by bookbinding we repeat the classification scheme on pages 4 and 5.

In using the subject index attention should be given to the following details:

subject	quoted in	remarks
representations of space groups	1 and 13.1	
Alfen waves, helicon waves	13 and 20	
hot electrons	13 and 14.3.4	
transport phenomena	14 or subsection of 14	if not clearly to be assigned to 8 or 15
galvanomagnetic effects	14 or subsection of 14	not in 18
electro-optical effects	14 and 20 (or subsection of 14 and 20)	
piezoelectricity, electrostriction	14 or subsection of 14	not in 12 or subsection of 12; different electromechanical effects are quoted in 12 or subsection of 12
ionic conduction	14.4	not in 14.3
magneto-optical effects	18 and 20 (or subsection of 18 and 20)	possibly in additional sections, e.g. Faraday effect in 13
ferromagnetic films	18.2.1	not additionally in 1.5
X-ray spectra	20	but not X-ray studies of structure
MIS and MOS structures	the section assigned to the semiconductor	not in the section assigned to the oxide or the metal
MIM structures	the section assigned to the insulator	generally not in the section assigned to the metal
epitaxial layers	the section assigned to the layer	not in the section assigned to the substrate

The author index gives the names of the authors in alphabetical order. Each name is followed by the bibliographic details of the paper and the numbers of the pages on which the paper appears in the subject index. A paper of physica status solidi (b) or physica status solidi (a) is again marked by (b) or (a) in front of the number of the volume, respectively. The principal entry is again given in italics. When several papers have been published by the same author they are arranged according to the dates of publication, irrespective of being a paper of physica status solidi (b) or physica status solidi (a).

Subject Index

1. Structure of Crystalline Solids

Original Papers and Short Notes

T. Ogawa
*Morphologies of CVD ZrC Crystals
(a) Vol. **66**, No. 1, p. 175–181 (1981).
See also *3*, p. *54*; 4, p. 59; 22.3, p. 442.

S. Chatterjee, S. Ghosh, and A. N. Basu
*Lattice Sums for Zinc Sulphide Structure Crystals
(b) Vol. **106**, No. 2, p. 693–698 (1981).
See also 4, p. 59; 22.4.1, p. 449.

J. Sivradière
*Cyclic and Dihedral Point Groups as Subgroups of the Limiting Groups
(b) Vol. **107**, No. 1, p. 117–125 (1981).
See also *13*, p. *147*.

H. G. Brühl
*Die Bestimmung der Valenzladungsdichte von Elementhalbleitern und III–V-Verbindungen mit Röntgenbeugungsmethoden (I): Grundlagen
(Determination of the Bulk Charge Density of Elemental Semiconductors and III–V Compounds by X-Ray Diffraction Methods (I): Fundamentals)
(b) Vol. **107**, No. 1, p. 289–295 (1981).
See also *13*, p. *147*; 22.1, p. 402; 22.2, p. 426.

V. Dvorak and R. Kind
*On the Low-Temperature Phase in Rb_2ZnCl_4
(b) Vol. **107**, No. 2, p. K109–K113 (1981).
See also 1.2, p. 19; 22.8, p. 490.

A. Priemuth
*Electromigration of Cu in V_2VI_3 Compounds
(a) Vol. **67**, No. 2, p. 505–510 (1981).
See also *9*, p. *91*; 14.3, p. 217; 22.7, p. 483.

G. S. Dubey and D. K. Chaturvedi
*Static and Dynamic Structure Factors of Classical Plasma in Weak and Intermediate Couplings
(b) Vol. **108**, No. 1, p. 123–129 (1981).
See also *13*, p. *149*.

M. Paić and V. Paić
Diffuse Reflectance Spectra and Polymorphism of Mercury(II) Iodide
(a) Vol. **68**, No. 2, p. 447–455 (1981).
See also 13.5.1, p. 193; *20.1*, p. *321*; *20.3*, p. *334*; 22.5, p. 460.

1.1 Perfectly Periodic Structure

Original Papers and Short Notes

P. KH. BESIRGANYAN and KH. R. DRMEYAN
*Fine Structure of X-Ray Interferograms (II): Three-Block L–L–L-Type X-Ray Interferometer with Defocusing
(a) Vol. **63**, No. 1, p. 91–95 (1981).

J. MORAWIEC and E. MIZERA
*X-Ray and TEM Investigations of PbTe Crystal
(a) Vol. **63**, No. 1, p. K1–K2 (1981).
See also 10.2, p. 106; 22, p. 391.

A. U. SHELEG, E. PELJO, and P. SUORTTI
*Structure and Thermal Motion Parameters of Tetragonal ZnP_2 between 300 and 740 K
(a) Vol. **63**, No. 2, p. 751–758 (1981).
See also 6, p. 69; 22.3, p. 440.

E. R. VANCE, D. K. AGRAWAL, and J. G. PEPIN
*Powder X-Ray Diffraction Study of Halide Sodalites
(a) Vol. **63**, No. 2, p. K189–K193 (1981).
See also *10.2*, p. *107*; 22.8, p. 485.

P. KH. BESIRGANYAN, A. K. KOCHARYAN, and K. G. TRUNI
*Diffraction of a Narrow X-Ray Beam on the Double-Block System with a Wedge-Shaped Crystal
(a) Vol. **64**, No. 2, p. 431–434 (1981).
See also 22.1.2, p. 412.

G. VAN TENDELOO and S. AMELINCKX
*Electron Diffraction and High Resolution Electron Microscopy Study of Ordering in the Gold–Manganese System (I): Lozenge Shaped Island Structures Derived from the DO_{22} Structure $Au_{3+x}Mn$
(a) Vol. **65**, No. 1, p. 73–86 (1981).
See also 1.3, p. 23; *1.4*, p. *29*; 21.1, p. 358; 21.6, p. 385.

U. PIETSCH
*Bond Charge Reconstruction for GaP from X-Ray Powder Data
(b) Vol. **105**, No. 2, p. K135–K137 (1981).
See also *4*, p. *59*; 22.2.2, p. 433.

G. VAN TENDELOO and S. AMELINCKX
*Electron Diffraction and High Resolution Electron Microscopy Study of Ordering in the Gold–Manganese System (II): Square Island Structures Derived from the Au_4Mn Structure ($Au_{4-x}Mn_1$)
(a) Vol. **65**, No. 2, p. 431–446 (1981).
See also 1.3, p. 24; *1.4*, p. *30*; 21.1, p. 359; 21.6, p. 385.

S.-L. CHANG
*Excitations and Phases of Two-Beam Dynamical Bragg Reflection of X-Rays for Infinitely Thick Crystals
(a) Vol. **65**, No. 2, p. 553–562 (1981).
See also 22.1.1, p. 405.

J. D. STEPHENSON
*An Interpretation of White (Synchrotron) X-Radiation Laue (Topograph) Patterns by Means of the Gnomonic Projection
(a) Vol. **65**, No. 2, p. 601–608 (1981).
See also 21.1.1, p. 372.

U. PIETSCH
*Bond Charge Determination from X-Ray Data for Germanium
(b) Vol. **106**, No. 1, p. K9–K11 (1981).
See also *4*, p. *59*; 22.1.1, p. 405.

W. BAŻELA and A. SZYTUŁA
Crystal and Magnetic Structure of the $NiMn_{1-t}Ti_tGe$ System
(a) Vol. **66**, No. 1, p. 45–52 (1981).
See also *18.2*, p. *271*; *18.4*, p. *283*; 22.8, p. 488.

E. A. SOZONTOV, M. V. KRUGLOV, and B. G. ZAKHAROV
*Photoemission in Bragg Diffraction of X-Rays by Bicrystals
(a) Vol. **66**, No. 1, p. 303–310 (1981).
See also *17*, p. *257*; 22.1.1, p. 406; 22.2.1, p. 429; 22.2.4, p. 438.

J. BEAUVILLAIN
*Approche quantitative des contrastes en microscopie électronique
(Quantitative Approach to Electron Microscopic Contrasts)
(a) Vol. **66**, No. 2, p. 675–682 (1981).
See also *1.4*, p. *30*; 21, p. 345.

G. L. BLERIS, J. G. ANTONOPOULOS, TH. KARAKOSTAS, and P. DELAVIGNETTE
*Grain Boundary Analysis in TEM (V): A Criterion for the Indentification of a CSL and a Characterization of an Approximate CSL
(a) Vol. **67**, No. 1, p. 249–257 (1981).
See also *1.4*, p. *31*; 22, p. 398.

S. S. SHEININ and H. S. KIM
*Absorption Coefficients in the Dynamical Theory of Electron Diffraction for the Non-Symmetrical Laue Case
(b) Vol. **107**, No. 2, p. 715–721 (1981).

N. M. OLEKHNOVICH and A. I. OLEKHNOVICH
*Dynamic Effects of Diffuse X-Ray Scattering Near Bragg Reflections
(a) Vol. **67**, No. 2, p. 427–433 (1981).
See also 22.1.2, p. 420.

S. R. SINGH, G. SINGH, and G. VAN TENDELOO
*Structure Determination of a 40 H Polytype of SiC and Its Intergrowth Structures Using Lattice Imaging Technique
(a) Vol. **67**, No. 2, p. 625–632 (1981).
See also 22.3, p. 442.

K. IGUCHI and H. HASHIMOTO
*The Flux Distribution of Electron Waves Incident on a MgO Crystal in 110 Orientation
(a) Vol. **67**, No. 2, p. 655–667 (1981).
See also 22.6, p. 479.

K. WIETESKA
*X-Ray Diffraction Investigations of High Energy α-Particle Damage in Silicon
(a) Vol. **68**, No. 1, p. 179–185 (1981).
See also *11*, p. *131*; 22.1.2, p. 422.

R. G. GOODCHILD, O. H. HUGHES, and J. C. WOOLLEY
*Crystal Structure of I III IV Se_4 Compounds
(a) Vol. **68**, No. 1, p. 239–244 (1981).
See also 22.8, p. 491.

V. I. KUSHNIR and E. V. SUVOROV
*Image Transmission through a Cylindrical Magnifying X-Ray Diffraction Lens
(a) Vol. **68**, No. 2, p. K109–K111 (1981).

L. V. LEVONIAN
*Interference Fringes of Intrabranch Scattering for X-Ray Spherical-Wave Diffraction in a Perfect Crystal
(a) Vol. **68**, No. 2, p. K199–K202 (1981).

1.2 Solid-State Phase Transformations

Review Article

Original Papers and Short Notes

J. Mertsching and H. J. Fischbeck
*The Incommensurate Peierls Phase of the Quasi-One-Dimensional Fröhlich Model with a Nearly Half-Filled Band
(b) Vol. **103**, No. 2, p. 783–795 (1981).
See also *13.1*, p. *153*.

J. Sivardière
*Lattice-Gas Model for a Simple Fluid: Condensation, Orientation, and Solidification
(b) Vol. **103**, No. 2, p. 823–829 (1981).
See also 2, p. 41.

J. Sołtys
*Order–Disorder Phase Transitions in Ternary Alloys $Cu_{3-x}Mn_xAl$
(a) Vol. **63**, No. 2, p. 401–406 (1981).
See also 1.3, p. 22; 21.1, p. 353.

J. Horváth and D. Kucharczyk
*Temperature Dependence of Lattice Parameters of $PbHPO_4$ and $PbDPO_4$ Single Crystals
(a) Vol. **63**, No. 2, p. 687–692 (1981).
See also *14.4.1*, p. *242*; 22.8.1, p. 494.

F. El-Kabbany, Y. Badr, and M. Tosson
*Infrared Study of the High-Temperature Phase Transition III–I of Na_2SO_4
(a) Vol. **63**, No. 2, p. 699–704 (1981).
See also *20.1*, p. *304*; 22.8.1, p. 494.

A. A. Tukhfatullin, V. E. Panin, V. I. Simakov, V. S. Demidenko, and A. P. Kalyanov
Electron Structure and the Nature of Co_3Ti Phase Formation
(b) Vol. **104**, No. 1, p. 69–74 (1981).
See also *13.1*, p. *153*; 21.1, p. 355.

K. Asami, T. Naka, and M. Ishiguro
Assignment of the Optical Absorption Bands in CsI:Tl and CsBr:Tl by Application of High Pressure
(b) Vol. **104**, No. 1, p. 145–149 (1981).
See also *13.4*, p. *171*; *20.1*, p. *305*; 22.5.2, p. 463.

A. Giessler, G. Schaack, and H. Bleif
Phonon Frequencies and Dynamics of the OH^- Molecule in Sodium Hydroxide
(b) Vol. **104**, No. 1, p. 151–159 (1981).
See also *6*, p. *69*; *20.1*, p. *305*; 22, p. 392.

F. Prokert
*Neutron Scattering Studies on Phase Transitions and Phonon Dispersion in $CsSrCl_3$
(b) Vol. **104**, No. 1, p. 261–265 (1981).
See also *6*, p. *69*; 22.8, p. 485.

Y. D. Yao and S. Arajs
Electrical Resistivity of Fe-Rich Fe–Pd Alloys between 77 and 1250 K
(a) Vol. **64**, No. 1, p. 95–104 (1981).
See also *14.1*, p. *199*; 21.1.1, p. 369.

E. Nebauer
*Surface Erosion during Annealing of $M^{III}X^{V}:F^{IIb}$ Systems and Its Reduction
(a) Vol. **64**, No. 1, p. 233–237 (1981).
See also *9*, p. *90*; 22.2, p. 426.

L. G. Khvostantsev and V. A. Sidorov
High-Pressure Polymorphism of Antimony. Thermoelectric Properties and Electrical Resistance Studies
(a) Vol. **64**, No. 1, p. 379–384 (1981).
See also *14.1*, p. *199*; *15*, p. *247*; 21.7, p. 387.

F. M. Mansy, N. K. Gobran, and G. Said
*On the Effect of Phase Transformation on the Steady State Creep of Cobalt
(a) Vol. **64**, No. 1, p. K81–K83 (1981).
See also *10.1*, p. *99*; 21.1, p. 356.

B. A. Lukiyanets and K. K. Danilevich-Tovstyuk
*The Influence of Layer Crystal Plate Thickness on Structural Phase Transitions
(b) Vol. **104**, No. 2, p. 485–489 (1981).
See also *6*, p. *71*.

D. Lepski and P. Burck
*Precipitate Structure Analysis by Diffraction Contrast Image Matching for a Metastable Semicoherent Carbide Phase in Molybdenum (I): Generating Inelastic Strain
(a) Vol. **64**, No. 2, p. 625–636 (1981).
See also *1.4*, p. *29*; 10.1, p. 99; 12.1, p. 135; 21, p. 340.

M. Bacmann et D. Fruchart
*Transition cristalline de $Co_3U_2O_8$ à basse température
(Crystalline Transitions of $CO_2U_2O_8$ at Low Temperatures)
(a) Vol. **64**, No. 2, p. K119–K123 (1981).
See also 22.8, p. 486.

T. Soma, H. Matsuo, and S. Kagaya
*Phase Diagrams of the Si–Sn and Ge–Sn Systems
(b) Vol. **105**, No. 1, p. 311–314 (1981).
See also *1.3*, p. *23*; 22.3, p. 441.

A. Rajewska, A. Jabłonka, D. Sikorska, and A. Sobaszek
*Neutron Critical Scattering in the Nematic Phase and at the Nematic–Isotropic Liquid Transition in PAA
(b) Vol. **105**, No. 1, p. K35–K37 (1981).
See also *2*, p. *45*; 22.9, p. 510.

V. I. Kolomytsev, V. A. Lobodyuk, and L. G. Khandros
*Shape Memory Effect and Phase Transitions in Ti–Ni Alloys. Electron Microscopy Studies of the Phase Transitions after Thermomechanical Treatment
(a) Vol. **65**, No. 1, p. 87–96 (1981).
See also 21.1, p. 358.

G. Sorge, U. Straube, and N. R. Ivanov
Elastic, Dilatational, and Electrostrictive Properties of TSCC Crystals
(a) Vol. **65**, No. 1, p. 189–196 (1981).
See also *12.2*, p. *140*; *14.4.1*, p. *243*; 22.8, p. 487.

O. Valassiades, E. K. Polychroniadis, J. Stoemenos, and N. A. Economou
On the Disordered Nature of Single Crystalline $TlBiTe_2$ and Its Influence on Its Electrical Properties
(a) Vol. **65**, No. 1, p. 215–223 (1981).
See also *10.1*, p. *100*; *14.3*, p. *213*; 22.8, p. 487.

L. A. Shebanov
*X-Ray Temperature Study of Crystallographic Characteristics of Barium Titanate
(a) Vol. **65**, No. 1, p. 321–325 (1981).
See also 14.4.1, p. 243; 22.8.1, p. 496.

V. B. Magalinskii and V. I. Zubov
*On the Constancy of the Configuration Entropy and Related Properties of Solids along the Melting Curve
(b) Vol. **105**, No. 2, p. K139–K142 (1981).
See also *8*, p. *85*; 23, p. 514.

O. Bostanjoglo
*New Phase Transitions in Arsenic Films
(a) Vol. **65**, No. 2, p. 595–600 (1981).
See also *1.5*, p. *35*; 2, p. 46; 21.7, p. 388.

S. Acharya and O. N. Srivastava
Electronic Behaviour of SnS_2 Crystals
(a) Vol. **65**, No. 2, p. 717–723 (1981).
See also *13.1*, p. *157*; *14.3*, p. *214*; 22, p. 396.

K. Terao and M. Shimizu
*Phase Transitions in Iron Alloys
(a) Vol. **65**, No. 2, p. K153–K156 (1981).
See also 6, p. 75; 8, p. 86; 21.1.1, p. 372.

N. Piślewski and L. T. H. Ferris
Investigation of the Molecular Motions in $Cd(NH_3)_6Cl_2$ by NMR Spectroscopy
(b) Vol. **106**, No. 1, p. 123–128 (1981).
See also *19*, p. *290*; 22.5, p. 459.

F. Brehat, B. Wyncke, and A. Hadni
Lattice Mode Anharmonicity in Sodium Nitrate
(b) Vol. **106**, No. 1, p. 207–214 (1981).
See also *6*, p. *76*; *20.1*, p. *313*; 22.8.1, p. 498.

T. Soma, H. Matsuo, and T. Katoh
*The Heat of Mixing in Liquid Si–Ge, Si–Sn, and Ge–Sn Mixtures
(b) Vol. **106**, No. 1, p. K5–K7 (1981).
See also *1.3*, p. *24*; 21.7, p. 388; 22.1.1, p. 405; 22.1.2, p. 415.

V. F. Degtyareva, V. E. Antonov, I. T. Belash, and E. G. Ponyatovskii
*Distortions of the F.C.C. Crystal Lattice of $Pd_{60}Cu_{40}$ Alloy Hydrogenated under High Pressure
(a) Vol. **66**, No. 1, p. 77–86 (1981).
See also 12.1, p. 136; 14.1, p. 202; 21.1, p. 360.

M. LEON, N. TERAO, and F. RUEDA
*On the Phase Transition of Chalcocite Cu_2S Thin Films
(a) Vol. **67**, No. 1, p. K11–K14 (1981).
See also *1.5*, p. *36*; 22, p. 398.

S. STOKKA and V. SAMULIONIS
Specific Heat near Two Phase Transitions in $CsBi(MoO_4)_2$ Crystals
(a) Vol. **67**, No. 1, p. K89–K92 (1981).
See also *8*, p. *87*; 22.8.1, p. 500.

V. E. SCHNEIDER and E. E. TORNAU
*The Model of the Phase Transition in Squaric Acid
(b) Vol. **107**, No. 2, p. 491–496 (1981).
See also *14.4.1*, p. *245*; 22.9, p. 512.

H. PUFF
*On the Influence of Fluctuations on the Peierls Transition of the Quasi-One-Dimensional Fröhlich Model with a Nearly Half-Filled Band
(b) Vol. **107**, No. 2, p. 543–552 (1981).
See also 13.1, p. 160.

V. DVORAK and R. KIND
On the Low-Temperature Phase in Rb_2ZnCl_4
(b) Vol. **107**, No. 2, p. K109–K113 (1981).
See also *1*, p. *10*; 22.8, p. 490.

K. S. ALEKSANDROV, A. T. ANISTRATOV, S. V. MELNIKOVA, P. V. KLEVTSOV, A. I. KRUGLIK, and V. N. VORONOV
Ferroelectric Phase Transitions in $CsLiMoO_4$ Single Crystals
(a) Vol. **67**, No. 2, p. 377–380 (1981).
See also *14.4.1*, p. *245*; 22.8.1, p. 500.

Y. UESU, J. OGAWA, N. HANAMI, B. A. STRUKOV, and J. KOBAYASHI
*X-Ray Dilatometric Study of Phase Transitions of Ferroelectric $(NH_4)_2BeF_4$
(a) Vol. **67**, No. 2, p. 693–699 (1981).
See also *14.4.1*, p. *245*; 22.8, p. 490.

F. EL-KABBANY
A Study of the Phase Transition in the System $[Na_{1-x}(NH_4)_x]_2SO_4$ by Infrared Spectroscopy
(a) Vol. **67**, No. 2, p. 729–735 (1981).
See also 6, p. 80; 20.1, p. 318; 22.8.1, p. 500.

J. MAJSZCZYK, J. RĄCZKA, and Z. CZAPLA
*A New Phase Transition in NH_4HSeO_4 Crystals
(a) Vol. **67**, No. 2, p. K123–K125 (1981).
See also 7, p. *83*; 14.4.1, p. 246; 22.8.1, p. 500.

J. WOLAK and Z. CZAPLA
*Phase Transition in $CsHSeO_4$ Single Crystal at 128 °C
(a) Vol. **67**, No. 2, p. K171–K174 (1981).
See also *14.4*, p. *240*; 22.8.1, p. 500.

M. CORTI, A. RIGAMONTI, and A. MAGISTRIS
*^{133}Cs Quadrupole Perturbed NMR Study of Jahn-Teller Phase Transitions in $CsCuCl_3$
(b) Vol. **108**, No. 1, p. 29–36 (1981).
See also *19*, p. *293*; 22.8, p. 490.

A. I. LANDA, V. E. PANIN, and M. F. ZHOROVKOV
*The Pseudopotential Method and Thermodynamics of Eutectic Systems
(b) Vol. **108**, No. 1, p. 113–122 (1981).
See also *2*, p. *50*; 8, p. 88; 13, p. 149; 21, p. 347; 21.1, p. 365.

M. RAZEGHI
*EPR Investigations of a Structural Phase Change in Lead Phosphate
(b) Vol. **108**, No. 1, p. 175–186 (1981).
See also *19*, p. *294*; 22.8.1, p. 500.

S. R. KUINDERSMA, P. R. BOUDEWIJN, and C. HAAS
Near-Infrared d–d Transitions of NiI_2, $CdI_2{:}Ni^{2+}$, and CoI_2
(b) Vol. **108**, No. 1, p. 187–194 (1981).
See also *20.1*, p. *319*; 22.5, p. 460.

T. SOMA, H. MATSUO, and M. FUNAKI
*The Electronic Theory of the Alloy Systems Composed of Alkali Metal Elements (II): Phase Diagrams
(b) Vol. **108**, No. 1, p. 221–227 (1981).
See also 1.3, p. 26; 4, p. 60; 21.2, p. 377.

J. M. PÉREZ-MATO, J. L. MAÑES, J. FERNÁNDEZ, J. ZÚÑIGA, M. J. TELLO, C. SOCÍAS, and M. A. ARRIANDIAGA
*Structural Phase Transitions in $(CH_3NH_3)_2ZnCl_4$ Crystals
(a) Vol. **68**, No. 1, p. 29–38 (1981).
See also 22.8, p. 491.

A. DA COSTA LAMAS, S.-L. CHANG, and S. CATICHA-ELLIS
*On the Use of Powder Diffractometry in the Study of Phase Transitions. Case of $NaNO_2$
(a) Vol. **68**, No. 1, p. 173–178 (1981).
See also *14.4.1*, p. *246*; 22.8.1, p. 501.

L. G. KHVOSTANTSEV and N. A. NIKOLAEV
*Polymorphism of Cerium under Hydrostatic Pressure up to 8.0 GPa
(a) Vol. **68**, No. 1, p. 193–197 (1981).
See also 21.4, p. 383.

E. I. TERUKOV, S. ROTH, G. KRABBES, and H. OPPERMANN
The Magnetic Properties of Cobalt-Doped Iron Sulphide $Fe_{1-y}Co_yS$ ($y \leqq 0.13$)
(a) Vol. **68**, No. 1, p. 233–238 (1981).
See also *18*, p. *264*; 22.8, p. 491.

G. SORGE, A. ALMEIDA, L. A. SHUVALOV, and R. M. FEDOSYUK
*Influence of the Mechanical Stress Component T_5 on the Elastic Behaviour of $KH_3(SeO_3)_2$/$KD_3(SeO_3)_2$ Crystals in the Vicinity of Their Ferroelastic Phase Transition
(a) Vol. **68**, No. 1, p. 245–251 (1981).
See also 12.2, p. 141; *14.4.1*, p. *246*; 22.8, p. 491.

E. LAZO, J. RÖESSLER, and L. A. MORAGA
*Phase Transition and Spatial Correlation of Localized States in the Falicov-Ramírez-Kimball Model
(b) Vol. **108**, No. 2, p. 341–350 (1981).
See also *13.1*, p. *161*; 22.8, p. 491.

M. SURMA, S. MACIEJEWSKI, R. KUBIAK, and M. WOŁCYRZ
Electronic Structure of the $Cd_{65}Hg_{35}$ Alloy and Its Temperature Dependence near the Phase Transition
(b) Vol. **108**, No. 2, p. 475–482 (1981).
See also *13.1*, p. *161*; 21, p. 349.

G. VÖLKEL, H.-E. MÜLLER, and W. FLOHRER
Critical Anomaly of the Electron Spin-Lattice Relaxation in X-Irradiated Rochelle Salt
(b) Vol. **108**, No. 2, p. 501–506 (1981).
See also 6, p. 81; 11, p. 131; *14.4.1*, p. *246*; 22.9, p. 513.

O. HUDÁK
*Theory of Multidomain States in the Ordered Phase of Crystals
(b) Vol. **108**, No. 2, p. 567–574 (1981).
See also *14.4.1*, p. *246*.

O. BOSTANJOGLO and G. SCHLOTZHAUER
Impulse Stimulated Crystallization of Sb Films Investigated by Time Resolved TEM
(a) Vol. **68**, No. 2, p. 555–560 (1981).
See also 2, p. 51; *3*, p. *56*; 22.1, p. 403.

C. MANOLIKAS
*Electron Microscopic Study of the Commensurate-Incommensurate Phase Transition in $Ba_2NaNb_5O_{15}$ and $Sr_2KNb_5O_{15}$
(a) Vol. **68**, No. 2, p. 653–660 (1981).
See also *1.4*, p. *32*; 22.8, p. 492.

J. STHANAPATI, D. K. DE, and A. K. PAL
*Studies on Phase Transitions in Cobalt and Ferrous Fluosilicates by Cu (II) EPR
(a) Vol. **68**, No. 2, p. K123–K127 (1981).
See also *19*, p. *295*; 22.8, p. 492.

1.3 Alloys. Metallurgy

Original Papers and Short Notes

C. Carlone, S. Jandl, and H. R. Shanks
Optical Phonons and Crystalline Symmetry of InSe
(b) Vol. **103**, No. 1, p. 123–130 (1981).
See also *6*, p. *66*; *20.1*, p. *301*; 22.3, p. 440.

S. Spooner, H. R. Child, F. H. Hsu, E. R. Vance, and J. H. Smith
*Small-Angle Neutron Scattering and Positron Annihilation in Mn–Cu Alloys
(a) Vol. **63**, No. 1, p. 31–34 (1981).
See also 21.1, p. 351.

J. Sołtys and R. Kozubski
*A Simple Model of the Order–Disorder Phase Transitions in Ternary Alloys and Its Application to Several Selected Heusler Alloys
(a) Vol. **63**, No. 1, p. 35–44 (1981).
See also 21, p. 336; 21.1, p. 351.

J. Jääskeläinen and E. Suoninen
*Structure and Ordering of Solid Solutions of Iron in Palladium
(a) Vol. **63**, No. 1, p. 241–245 (1981).
See also 21, p. 336; 21.1.1, p. 368.

J. Sołtys
Order–Disorder Phase Transitions in Ternary Alloys $Cu_{3-x}Mn_xAl$
(a) Vol. **63**, No. 2, p. 401–406 (1981).
See also *1.2*, p. *15*; 21.1, p. 353.

E. Wieser, J. Henke, M. Müller, and C. Cruz
*Connection between Structure and Magnetic Properties of a Magnetically Semi-Permanent Fe–Ni–Al–Ti Alloy
(a) Vol. **63**, No. 2, p. 487–494 (1981).
See also 5, p. 62; *18.2*, p. *268*; 21.1, p. 353; 21.1.1, p. 368.

O. Simmich
*The Influence of the Gradient Free Energy on the Movement of a Flat Interface
(a) Vol. **63**, No. 2, p. 539–545 (1981).
See also *10.1*, p. *97*; 21, p. 337.

J. J. M. Binsma, L. J. Giling, and J. Bloem
*Order–Disorder Behaviour and Tetragonal Distortion of Chalcopyrite Compounds
(a) Vol. **63**, No. 2, p. 595–603 (1981).
See also 4, p. 57; 22.8, p. 485.

J. M. Roland, X. Quillard et A. Moreau
*Etude de la phase α des alliages Cu–Al. Recherche d'un état désordonné proche de l'equilibre (Study of the α-Phase of Cu–Al Alloys. Investigations of the Near-Equilibrium Disordered State)
(a) Vol. **64**, No. 1, p. 45–51 (1981).
See also 14.1, p. 199; 21.1, p. 355.

V. Valvoda
*X-Ray Diffraction Study of Debye Temperature and Charge Distribution in Tantalum Monocarbide
(a) Vol. **64**, No. 1, p. 133 – 142 (1981).
See also 6, p. 71; 21, p. 339.

H.-U. Habermeier
*Structure and Superconducting Properties of Nb_3Ge Prepared in a UHV System
(a) Vol. **64**, No. 1, p. 195 – 206 (1981).
See also 10.1, p. 98; *14.2*, p. *205*; 21, p. 339.

W. Bażela, A. Szytuła, J. Todorović, and A. Zięba
Crystal and Magnetic Structure of the $NiMnGe_{1-x}Si_x$ System
(a) Vol. **64**, No. 1, p. 367 – 378 (1981).
See also *18.2*, p. *269*; *18.4*, p. *281*; 21.1, p. 355.

J. Wilke, V. T. Bublik, A. G. Braginskaya, A. N. Popkov, G. P. Kolchina, V. B. Osvenskii, S. P. Grishina und O. G. Stolyarov
*Untersuchung des homogenen Gebietes in unlegiertem Galliumantimonid
(Study of the Homogeneous Region in Undoped Gallium Antimonide)
(a) Vol. **64**, No. 1, p. K31 – K33 (1981).
See also 22.2.3, p. 435.

Farid A. Khwaja and M. S. K. Razmi
*A Statistical Pseudopotential Approach to the Calculation of Atomic Short-Range Order in Solid Solutions
(b) Vol. **104**, No. 2, p. 727 – 735 (1981).
See also *13*, p. *144*; 21, p. 340; 21.1, p. 356; 21.1.1, p. 370.

G. van Tendeloo, J. van Landuyt, and S. Amelinckx
*Real Space Observation of Charge Density Waves in the Incommensurate Room Temperature Phase of TaS_2
(a) Vol. **64**, No. 2, p. K105 – K106 (1981).
See also *1.4*, p. *29*; 22, p. 394.

T. Soma, H. Matsuo, and S. Kagaya
*Phase Diagrams of the Si–Sn and Ge–Sn Systems
(b) Vol. **105**, No. 1, p. 311 – 314 (1981).
See also *1.2*, p. *16*; 22.3, p. 441.

G. van Tendeloo and S. Amelinckx
*Electron Diffraction and High Resolution Electron Microscopy Study of Ordering in the Gold–Manganese System (I): Lozenge Shaped Island Structures Derived from the DO_{22} Structure $Au_{3+x}Mn$
(a) Vol. **65**, No. 1, p. 73 – 86 (1981).
See also *1.1*, p. *11*; *1.4*, p. *29*; 21.1, p. 358; 21.6, p. 385.

H. Löffler, V. Syneček, M. Simerská, G. Wendbrock, P. Bartuška, and R. Kroggel
*On the Mode of Decomposition of Al–Zn Alloys
(a) Vol. **65**, No. 1, p. 197 – 204 (1981).
See also 21, p. 342; 21.1, p. 358.

M. I. Zakharova, A. K. Khundzhua, L. Kertész, and A. Szász
*Thermal Decomposition of the B.C.C. β-Solid Solution of Titanium Alloy Containing 6.7 at% Mo, 3 at% Zr, and 1.8 at% Sn (I): Changes in Crystal and Electron Structure
(a) Vol. **65**, No. 1, p. 265–270 (1981).
See also 4, p. 58; 21.1, p. 358.

N. Kuwano, H. Mishio, M. Toki, and T. Eguchi
*Electron Microscopic Observation of Periodic Antiphase Domain Structures in the α_2-Phase of Cu–Al Alloys
(a) Vol. **65**, No. 1, p. 341–352 (1981).
Erratum
(a) Vol. **67**, No. 2, p. 737 (1981).
See also 1.4, p. 29; 21, p. 342, 21.1, p. 358.

D. Fruchart, F. Vaillant, E. Roudaut, A. Nemoz et X. G. Tessema
*Etude par diffraction neutronique de $CeRu_2D_5$
(Neutron Diffraction Study of $CeRu_2D_5$)
(a) Vol. **65**, No. 1, p. K19–K24 (1981).
See also 21.4, p. 381.

G. van Tendeloo and S. Amelinckx
Electron Diffraction and High Resolution Electron Microscopy Study of Ordering in the Gold–Manganese System (II): Square Island Structures Derived from the Au_2Mn Structure ($Au_{4-x}Mn_1$)
(a) Vol. **65**, No. 2, p. 431–446 (1981).
See also *1.1*, p. *11*; *1.4*, p. *30*; 21.1, p. 359; 21.6, p. 385.

J. Lecomte, J. P. Loup, M. Hervieu, and B. Raveau
Non-Stoichiometry and Electrical Conductivity of Strontium Niobates with Perovskite Structure (I): Defect Structure of $Sr(Sr_{1/3}Nb_{2/3})O_3$
(a) Vol. **65**, No. 2, p. 743–752 (1981).
See also *10.2*, p. *112*; *14.4*, p. *239*; 22.8.1, p. 497.

T. Soma, H. Matsuo, and T. Katoh
*The Heat of Mixing in Liquid Si-Ge, Si–Sn, and Ge–Sn Mixtures
(b) Vol. **106**, No. 1, p. K5–K7 (1981).
See also *1.2*, p. *17*; 21.7, p. 388; 22.1.1, p. 405; 22.1.2, p. 415.

J. M. Roland, X. Quillard, and S. K. Marya
*On the Kinetics of α-Phase Odering in Copper–Aluminium Alloys
(a) Vol. **66**, No. 1, p. 347–360 (1981).
See also 14.1, p. 202; 21.1, p. 360.

J. Sołtys
*X-Ray Diffraction Research of the Order–Disorder Transitions in the Ternary Heusler Alloys B_2MnAl (B = Cu, Ni, Co, Pd, Pt)
(a) Vol. **66**, No. 2, p. 485–491 (1981).
See also *1.2*, p. *18*; 21, p. 344; 21.1, p. 361; 21.6, p. 385.

M. Gasgnier and L. Névot
Analysis and Crystallographic Structures of Chromium Thin Films
(a) Vol. **66**, No. 2, p. 525–540 (1981).
See also *1.5*, p. *36*; 21.1, p. 362.

G. N. Gaidukov and J. P. Hirth
*The Influence of Thermodynamic Solution Properties on Solute Atmospheres at Dislocations
(a) Vol. **66**, No. 2, p. 601–606 (1981).
See also 8, p. 86; *10.1*, p. *102*; 21.1.1, p. 373.

T. Ishimasa, Y. Kitano, and Y. Komura
*Faulted Structure in the Fe–Cr σ-Phase
(a) Vol. **66**, No. 2, p. 703–715 (1981).
See also *10.1*, p. *102*; 21.1, p. 362; 21.1.1, p. 374.

H. V. Kiran, A. L. Shashimohan, D. K. Chakrabarty, and A. B. Biswas
*Structural and Magnetic Properties of Cooper–Nickel Ferrites
(a) Vol. **66**, No. 2, p. 743–747 (1981).
See also *18.3*, p. *279*; 22.8.2, p. 505.

M. Wołcyrz, R. Kubiak, and S. Maciejewski
X-Ray Investigation of Thermal Expansion and Atomic Thermal Vibrations of Tin, Indium, and Their Alloys
(b) Vol. **107**, No. 1, p. 245–253 (1981).
See also *6*, p. *78*; *8*, p. *87*; 21.7, p. 388.

S. F. Dubinin, Yu. A. Izyumov, S. K. Sidorov, V. N. Syromyatnikov, and S. F. Teplouchov
*The Role of the Incommensurable Atomic Structure in the Mechanism of the Martensite Transformation in the Invar Alloys
(a) Vol. **67**, No. 1, p. 75–82 (1981).
See also *1.2*, p. *18*; 21.1, p. 363; 21.1.1, p. 374.

G. van Tendeloo, F. J. A. den Broeder, S. Amelinckx, R. de Ridder, J. van Landuyt, and H. J. van Daal
*Microstructure of $Cr_{100-x}Al_x$ Alloys (10 at% $\lessapprox x \lessapprox$ 33 at%) Studied by Means of Transmission Electron Microscopy and Diffraction (I): Microstructure of the β-Phase
(a) Vol. **67**, No. 1, p. 217–232 (1981).
See also *1.4*, p. *31*; 21.1, p. 363.

F. J. A. den Broeder, G. van Tendeloo, S. Amelinckx, J. Hornstra, R. de Ridder, J. van Landuyt, and H. J. van Daal
*Microstructure of $Cr_{100-x}Al_x$ Alloys (10 at% $\lessapprox x \lessapprox$ 33 at%) Studied by Means of Transmission Electron Microscopy and Diffraction (II): Discovery of a New Phase
(a) Vol. **67**, No. 1, p. 233–248 (1981).
See also *1.4*, p. *31*; 21.1, p. 363.

F. Wagner, H. R. Wenk, C. Esling, and H. J. Bunge
*Importance of Odd Coefficients in Texture Calculations for Trigonal–Triclinic Symmetries
(a) Vol. **67**, No. 1, p. 269–285 (1981).
See also 22.8.1, p. 499.

U. De, S. Hoth, and W. Witthuhn
*Temperature and Concentration Dependence of the Electric Field Gradient in the Cadmium–Magnesium Alloy System
(b) Vol. **107**, No. 2, p. 569–577 (1981).
See also 21, p. 346; 21.3, p. 379.

S. Matthies and F. Wagner
*Study of the Ghost Phenomena in Mathematical Texture Analysis by Means of Standard Functions
(b) Vol. **107**, No. 2, p. 591–601 (1981).
See also *1.4*, p. *31*; 21.1, p. 364.

T. Soma, H. Matsuo, and Y. Kohbu
*The Electronic Theory of the Alloy Systems Composed of Alkali Metal Elements (I): Bulk Properties of Cs–K, Cs–Rb, and Rb–K Solid Solutions
(b) Vol. **107**, No. 2, p. 761–768 (1981).
See also 4, p. 60; 12.1, p. 136; 21.2, p. 377.

O. Nishikawa, M. Nachi, K. Kurihara, Y. Tsunashima, and Y. Hara
*Atom-Probe Study of Ga–Mo and Sn–Ir Interfaces
(a) Vol. **67**, No. 2, p. 461–468 (1981).
See also *1.6*, p. *39*; 21, p. 347.

E. A. Abou-Saif, M. Abd-Rabo, M. M. Ezzo, A. A. Gamil, and A. A. Mohamed
*Effect of Heat Treatment on the Structure and Mechanical Properties of Aluminium–Tin Alloys
(a) Vol. **67**, No. 2, p. K101–K105 (1981).
See also *12.1*, p. *137*; 21, p. 347.

J. G. Pepin and E. R. Vance
*Non-Confirmation of a $Cr_2Zr_3O_{10}$ Compound
(a) Vol. **67**, No. 2, p. K167–K169 (1981).
See also 22.6.1, p. 482.

T. Soma, H. Matsuo, and M. Funaki
The Electronic Theory of the Alloy Systems Composed of Alkali Metal Elements (II): Phase Diagrams
(b) Vol. **108**, No. 1, p. 221–227 (1981).
See also *1.2*, p. *20*; 4, p. 60; 21.2, p. 377.

Y. Bouazra, C. Roucau et F. Reynaud
*Simulation du contraste de précipités en théorie dynamique de la diffraction des électrons. Application aux plaquettes θ′ dans Al–4% Cu
(Simulation of Contrast of Precipitates in the Dynamical Theory of Electron Diffraction. Application to θ′ Plates in Al–4% Cu)
(a) Vol. **68**, No. 1, p. 269–280 (1981).
See also 10.1, p. 104; 21, p. 348; 21.1, p. 366.

J. K. Mukherjee and S. K. Prasad
*Classification of Computer Generated Random Binary Alloys
(a) Vol. **68**, No. 1, p. K43–K44 (1981).

D. van Dyck, D. Colaïtis, and S. Amelinckx
Diffraction from Polysynthetically Twinned Structures: The Geometrical Characteristics of the Diffraction Pattern
(a) Vol. **68**, No. 2, p. 385–398 (1981).
See also *10.2*, p. *119*; 22.8, p. 492.

D. Coliatis, D. van Dyck, and S. Amelinckx
*Electron Microscopic Study of the System mPbS–$n$$Bi_2S_3$
(a) Vol. 68, No. 2, p. 419–438 (1981).
See also 1.4, p. 32; 21.7, p. 389.

Y. Umakoshi and M. Yamaguchi
*The Stability and Energies of Planar Faults in the DO_{19} Ordered Structure
(a) Vol. 68, No. 2, p. 457–468 (1981).
See also *10.1*, p. *104*; 21.3, p. 379.

R. Ramlau and H. Löffler
*Guinier-Preston Zones in an Al–3 at%Zn Alloy
(a) Vol. 68, No. 2, p. 531–536 (1981).
See also *10.1*, p. *104*; 21, p. 349; 21.1, p. 367.

L. Varga, K. Tompa, and T. Schmidt
*Thermopower Method for Chemical Composition and Inhomogeneity Measurements in Amorphous Ni–P Samples
(a) Vol. 68, No. 2, p. 603–606 (1981).
See also 2, p. 51; *15*, p. *250*; 21.1, p. 367.

S. V. Sukhvalo, A. I. Demchenko, A. B. Shifrin, and E. M. Zub
*Orientation Transitions in Amorphous Gd–Co Films
(a) Vol. 68, No. 2, p. 637–644 (1981).
See also *2*, p. *51*; 14.1, p. 204; 18, p. 264; 21.1, p. 367; 21.4, p. 383.

F. C. Lovey, R. Rapacioli, and M. Chandrasekaran
*Precipitates with an Incommensurate Structure in β-Cu–Zn–Al
(a) Vol. 68, No. 2, p. K105–K108 (1981).
See also 21, p. 349; 21.1, p. 367.

A. Zahra and C. Y. Zahra
*Comments on "Effects of the Fluctuation of Solute Concentration in Al–10 wt%Zn Alloy on Aging at Low Temperature"
(a) Vol. 68, No. 2, p. K141–K143 (1981).
See also 8, p. 88; 21, p. 350; 21.1, p. 367.

M. Ohta, M. Yamada, T. Kanadani, and A. Sakakibara
*Reply to Comments on "Effects of the Fluctuation of Solute Concentration in Al–10 wt%Zn Alloy on Aging at Low Temperature"
(a) Vol. 68, No. 2, p. K145–K148 (1981).
See also *14.1*, p. *204*; 21, p. 350; 21.1, p. 367.

1.4 Microstructure
(Magnetic Domain See 18; Ferroelectric Domains See 14.4.1)

Original Papers and Short Notes

T. van Dijk and A. J. Burggraaf
*Grain Boundary Effects on Ionic Conductivity in Ceramic $Gd_xZr_{1-x}O_{2-(x/2)}$ Solid Solutions
(a) Vol. **63**, No. 1, p. 229–240 (1981).
See also *14.4*, p. *237*; 22.6.1, p. 481.

D. Vollet, A. F. Craievich, and R. Reich
*Sintering of Dezincified α-Brass
(a) Vol. **63**, No. 1, p. 321–328 (1981).
See also 21.1, p. 352.

C. B. Carter
Multiple Ribbons in Face-Centred Cubic Materials (III): Jogs on Double Ribbons
(a) Vol. **63**, No. 1, p. 335–344 (1981).
See also *10.1*, p. *96*; 21.1, p. 352.

M. S. Abrahams, J. L. Hutchison, and G. R. Booker
*Direct Observation of a Silicon–Sapphire Hetero-Epitaxial Interface by High Resolution Transmission Electron Microscopy
(a) Vol. **63**, No. 1, p. K3–K6 (1981).
See also 3, p. 53; 22.1.2, p. 409; 22.6, p. 474.

A. M. Dyachenko, G. V. Berenshtein, and I. E. Maronchuk
*The Internal Stresses in Semiconductors Induced by Microinclusions
(a) Vol. **63**, No. 1, p. K27–K30 (1981).
See also *10.2*, p. *106*; 22.2.1, p. 427; 22.2.4, p. 437.

V. N. Bliznyuk, M. D. Glinchuk, A. P. Pechenyi, and L. A. Litvinov
*Determination of the Interblock Misalignment of $SrTiO_3$ Single Crystals by the ESR Technique
(a) Vol. **63**, No. 1, p. K43–K46 (1981).
See also *19*, p. *286*; 22.8.1, p. 493.

G. J. Russell, B. Vincent, and J. Woods
*Precipitation Effects in Indium-Doped Zinc Selenide
(a) Vol. **63**, No. 2, p. 573–584 (1981).
See also 22.4.2, p. 452.

Lu. Ganghua and Shen GuangJun
*Identification of a $\frac{1}{6}\langle 310\rangle$ Stair-Rod Dislocation
(a) Vol. **63**, No. 2, p. 705–710 (1981).
See also *10.1*, p. *97*; 21.1, p. 354.

Y. K. Wu, J. Z. Liang, and K. H. Kuo
*Micro Inversion Domains Formed during the Crystallization of the Amorphous Alloy $Fe_{40}Ni_{40}P_{14}B_6$
(a) Vol. **64**, No. 1, p. 113–120 (1981).
See also *2*, p. *42*; 21.1, p. 355; 21.1.1, p. 369.

W. Siegel, G. Kühnel, and E. Ziegler
Electrical Properties of Grain Boundaries in n-Type and p-Type GaP
(a) Vol. **64**, No. 1, p. 249–259 (1981).
See also *14.3.2*, p. *223*; 22.2.2, p. 432.

B. Pardo, C. Pariset, et D. Renard
*Etude des franges de moirés en microscopie électronique par la théorie dynamique à *N* ondes et application aux doubles couches Sb–Bi
(Study of Moiré Fringes in Electron Diffraction by the Many-Beam Dynamic Theory and Application to Sb–Bi Double Layers)
(a) Vol. **64**, No. 1, p. 283–295 (1981).
See also 21.7, p. 387.

G. N. Gaidukov, S. K. Maksimov, and A. P. Filippov
*Effects of a Cloud of Impurity Atoms near a Dislocation on Electron Microscopic Contrast
(a) Vol. **64**, No. 2, p. 519–524 (1981).
See also *10*, p. *93*.

D. Lepski and P. Burck
*Precipitate Structure Analysis by Diffraction Contrast Image Matching for a Metastable Semicoherent Carbide Phase in Molybdenum (I): Generating Inelastic Strain
(a) Vol. **64**, No. 2, p. 625–636 (1981).
See also *1.2*, p. *16*; 10.1, p. 99; 12.1, p. 135; 21, p. 340.

G. van Tendeloo, J. van Landuyt, and S. Amelinckx
*Real Space Observation of Charge Density Waves in the Incommensurate Room Temperature Phase of TaS_2
(a) Vol. **64**, No. 2, p. K105–K106 (1981).
See also *1.3*, p. *23*; 22, p. 394.

G. van Tendeloo and S. Amelinckx
*Electron Diffraction and High Resolution Electron Microscopy Study of Ordering in the Gold–Manganese System (I): Lozenge Shaped Island Structures Derived from the DO_{22} Structure $Au_{3+x}Mn$
(a) Vol. **65**, No. 1, p. 73–86 (1981).
See also *1.1*, p. *11*; 1.3, p. 23; 21.1, p. 358; 21.6, p. 385.

B. Jenichen and R. Köhler
*Simulated X-Ray Contrast of Striations for Plane Waves in the Laue Case
(a) Vol. **65**, No. 1, p. 245–252 (1981).
See also 10.2, p. 110; 22.1.2, p. 413.

N. Kuwano, H. Mishio, M. Toki, and T. Eguchi
Electron Microscopic Observation of Periodic Antiphase Domain Structures in the α_2-Phase of Cu–Al Alloys
(a) Vol. **65**, No. 1, p. 341–352 (1981).
Erratum
(a) Vol. **67**, No. 2, p. 737 (1981).
See also *1.3*, p. *24*; 21, p. 342; 21.1, p. 358.

G. van Tendeloo, F. J. A. den Broeder, S. Amelinckx, R. de Ridder, J. van Landuyt, and H. J. van Daal

*Microstructure of $Cr_{100-x}Al_x$ Alloys (10 at% ≦ x ≦ 33 at%) Studied by Means of Transmission Electron Microscopy and Diffraction (I): Microstructure of the β-Phase
(a) Vol. **67**, No. 1, p. 217–232 (1981).
See also *1.3*, p. *25*; 21.1, p. 363.

F. J. A. den Broeder, G. van Tendeloo, S. Amelinckx, J. Hornstra, R. de Ridder, J. van Landuyt, and H. J. van Daal

*Microstructure of $Cr_{100-x}Al_x$ Alloys (10 at% ≦ x ≦ 33 at%) Studied by Means of Transmission Electron Microscopy and Diffraction (II): Discovery of a New Phase
(a) Vol. **67**, No. 1, p. 233–248 (1981).
See also *1.3*, p. *25*; 21.1, p. 363.

G. L. Bleris, J. G. Antonopoulos, Th. Karakostas, and P. Delavignette

*Grain Boundary Analysis in TEM (V): A Criterion for the Identification of a CSL and a Characterization of an Approximate CSL
(a) Vol. **67**, No. 1, p. 249–257 (1981).
See also *1.1*, p. *12*; 22, p. 398.

L. K. Fionova, A. V. Andreeva, and T. I. Zhukova

*Grain Boundary Dissociation in Aluminium of High Purity
(a) Vol. **67**, No. 1, p. K15–K19 (1981).
See also 21, p. 346.

S. Matthies and F. Wagner

*Study of the Ghost Phenomena in Mathematical Texture Analysis by Means of Standard Functions
(b) Vol. **107**, No. 2, p. 591–601 (1981).
See also *1.3*, p. *26*; 21.1, p. 364.

S. K. Maksimov

Determination of the Sign of the Displacement Vector of Defects from Electron Microscopy Fringe Contrast Images Obtained under Low Anomalous Absorption
(a) Vol. **67**, No. 2, p. 387–394 (1981).
See also *10.2*, p. *116*; 22.1.2, p. 420.

K. Kranjc, D. Kunstelj, and V. Marinković

*Parallel Moiré Effect in Electron Micrographs of Crystal Sandwiches (I): Moiré Patterns as Predicted by the Theory
(a) Vol. **67**, No. 2, p. 543–554 (1981).

D. van Dyck and R. de Ridder

*Ground States and Structures in Ordered Binary P. C. Systems with Third Neighbour Interaction
(a) Vol. **67**, No. 2, p. 603–616 (1981).
See also *4*, p. *60*.

T. Lepistö, J. Yli-Kauppila, P. Kettunen, and P. Hautojärvi

*Voids in Fatigued Copper Single Crystals
(a) Vol. **67**, No. 2, p. K93–K97 (1981).
See also *10.1*, p. *103*; 21.1, p. 365.

CH. SCHNITTLER, G. TEICHMANN, and W. KAUFMANN
*Energetic Calculations Concerning the A-Defect Structure and Formation in Silicon
(a) Vol. 68, No. 2, p. K135–K139 (1981).
See also *10.2*, p. *120*; 22.1.2, p. 424.

1.5 Films

Review Article

Original Papers and Short Notes

Y. Komem, A. Greenberg, and C. L. Bauer
*Recrystallization near Low-Angle [001] Tilt Boundaries Following Cobalt Implantation in Bicrystalline Thin Films of Gold
(a) Vol. **64**, No. 1, p. 317–324 (1981).
See also *11*, p. *123*; 21.6, p. 384.

Yu. A. Tkhorik and L. S. Khazan
*On the Possible Mechanism of Misfit Dislocation Generation
(a) Vol. **64**, No. 1, p. K55–K58 (1981).
See also *3*, p. *54*; 10, p. 93.

A. Wachniewski
Lattice Green's Function for Thin Films. The S.C. (001) Thin Film Green's Function
(b) Vol. **104**, No. 2, p. 555–562 (1981).
See also *13.5*, p. *186*.

J. H. Basson and H. Booyens
The Effect of Substrate Relaxation on the Introduction of Misfit Dislocations in Misfitting Epitaxial Layers
(a) Vol. **64**, No. 2, p. 777–786 (1981).
See also *10.1*, p. *99*; *12.1*, p. *135*; 22.2.1, p. 428; 22.2.2, p. 433; 22.2.4, p. 437.

O. Bostanjoglo
*New Phase Transitions in Arsenic Films
(a) Vol. **65**, No. 2, p. 595–600 (1981).
See also *1.2*, p. *17*; 2, p. 46; 21.7, p. 388.

S. Uthanna and P. J. Reddy
*Electrical Properties of CdSe Thin Films
(a) Vol. **65**, No. 2, p. K113–K115 (1981).
See also *14.3.1*, p. *221*; 20.1, p. 312; 22.4.2, p. 453.

L. V. Rao, B. S. Naidu, and P. J. Reddy
Electrical Conductivity and Thermopower Measurements on Silver Doped Tellurium Films
(a) Vol. **65**, No. 2, p. K135–K138 (1981).
See also *14.3.1*, p. *221*; *15*, p. *248*; 22.1.3, p. 425.

V. P. Skripov, V. P. Koverda, and V. N. Skokov
*Size Effect on Melting of Small Particles
(a) Vol. **66**, No. 1, p. 109–118 (1981).
See also 1.6, p. 38; 2, p. 47; 8, p. 86; 21, p. 344.

E. M. Trukhanov and S. I. Stenin
Initial Stage of Plastic Deformation of Substrates in Heterostructures
(a) Vol. **66**, No. 1, p. 125–131 (1981).
See also *10.2*, p. *112*; 22, p. 397; 22.1.1, p. 405; 22.1.2, p. 415; 22.6, p. 478.

E. M. Trukhanov and L. Zsoldos
*Some Features of Generation of Misfit Dislocations in Si during B Diffusion from Doped Oxide
(a) Vol. **66**, No. 1, p. 157–164 (1981).
See also 9, p. 91; *10.2*, p. *112*; 22.1.2, p. 416.

H. Müller
*Simulation of Spherulitic Crystallization in Thin Amorphous Antimony Films
(a) Vol. **66**, No. 1, p. 199–205 (1981).
See also *3*, p. *54*; 21.7, p. 388.

S. I. Stenin and A. K. Gutakovskii
*The Plastic Deformation Kinetics for Heteroepitaxial Films during the Misfit Dislocation Generation from the Growth Surface
(a) Vol. **66**, No. 1, p. 249–253 (1981).
See also 3, p. 55; *10.2*, p. *113*; 22.1.1, p. 406.

C. Donolato
An Analytical Model of SEM and STEM Charge Collection Images of Dislocations in Thin Semiconductor Layers (II): EBIC Images of Dislocations
(a) Vol. **66**, No. 2, p. 445–454 (1981).
See also *10.2*, p. *114*; *14.3.3*, p. *228*; 22.1.2, p. 417.

M. Gasgnier and L. Névot
*Analysis and Crystallographic Structures of Chromium Thin Films
(a) Vol. **66**, No. 2, p. 525–540 (1981).
See also **1.3**, p. **24**; 21.1, p. 362.

E. M. Trukhanov and S. I. Stenin
*Equilibrium Dislocation Configurations in Heterostructures
(a) Vol. **66**, No. 2, p. 591–596 (1981).
See also *10.2*, p. *115*; 22.1.1, p. 406; 22.6, p. 478.

D. M. Scott and M. A. Nicolet
*Implanted Oxygen in NiSi Formation
(a) Vol. **66**, No. 2, p. 773–778 (1981).
See also *3*, p. *55*; **11**, p. **128**; 21.1, p. 362.

R. Riesenberg
Shear Stress and Adhesion of Film Areas of Various Sizes and Shapes
(a) Vol. **66**, No. 2, p. K93–K96 (1981).
See also *10.1*, p. *102*; 21.1, p. 362.

R. Hergt and H. Pfeiffer
*Investigation of Growth Steps Due to Misorientation of Garnet LPE Films
(a) Vol. **67**, No. 1, p. 205–215 (1981).
See also *3*, p. *55*; 22.8.2, p. 505.

A. K. Gutakovskii, S. I. Stenin, and B. G. Zakharov
*The Formation of Partial Misfit Dislocations during Heteroepitaxy
(a) Vol. **67**, No. 1, p. 299–304 (1981).
See also 3, p. 55; *10.2*, p. *116*; 22.2.1, p. 430.

M. Leon, N. Terao, and F. Rueda
*On the Phase Transition of Chalcocite Cu_2S Thin Films
(a) Vol. **67**, No. 1, p. K11–K14 (1981).
See also *1.2*, p. *19*; 22, p. 398.

E. Pippel
*Pseudomorphic Growth of Palladium and Palladium-Gold Alloys on Gold
(a) Vol. **67**, No. 2, p. 443–452 (1981).
See also *3*, p. *56*; 21.6, p. 386.

V.-P. Tanninen, M. Oikkonen, and T. O. Tuomi
*X-Ray Diffraction Study of Thin Electroluminescent ZnS Films Grown by Atomic Layer Epitaxy
(a) Vol. **67**, No. 2, p. 573–583 (1981).
See also 20.3, p. 333; 22.4.1, p. 450.

H. Haller, D. Korn, and G. Zibold
Electrical Resistivity of CuCo Solid Solutions at Low Temperature
(a) Vol. **68**, No. 1, p. 135–141 (1981).
See also *14.1*, p. *203*; 21.1, p. 366; 21.6, p. 386.

W. T. McKenna and L. Ward
*The Optical Properties of Discontinuous Silver Films
(a) Vol. **68**, No. 1, p. K11–K14 (1981).
See also *20.1*, p. *320*; 21.6, p. 386.

K. Starosta
*RF Sputtering of $Al_xIn_{1-x}N$ Thin Films
(a) Vol. **68**, No. 1, p. K55–K57 (1981).
See also 22.2.4, p. 439.

K. Harnisch, S. M. Neubacher, and O. Ennemoser
*Optical Loss in Selenium Thin Films as a Measure of Crystallinity
(a) Vol. **68**, No. 1, p. K87–K91 (1981).
See also 3, p. 56; 20, p. 299; 22.1.3, p. 425.

K. Kranjc, D. Kunstelj, and V. Marinhović
Parallel Moiré Effect in Electron Micrographs of Crystal Sandwiches (II): Comparison of Theory and Experiment
(a) Vol. **68**, No. 2, p. 363–372 (1981).
See also *1.4*, p. *32*; 21, p. 349; 21.1, p. 366; 21.7, p. 389; 22.4.3, p. 455.

J. Finster and D. Schulze
*Studies of the Si/SiO_2 Interface by Angular Dependent X-Ray Photoelectron Spectroscopy
(a) Vol. **68**, No. 2, p. 505–517 (1981).
See also *17*, p. *258*; 22.1.2, p. 423; 22.6, p. 480.

1.6 Surfaces

Original Papers and Short Notes

T. G. Petrova and E. S. Syrkin
*On the Problem of Vibrational Surface States in Strongly Anisotropic Crystals with Uniaxial Long-Range Interaction
(b) Vol. **104**, No. 1, p. 375–382 (1981).
See also *6*, p. *70*.

V. G. Kohn and M. V. Kovalchuk
*On the Theory of the External Photoeffect Accompanying X-Ray Diffraction in an Ideal Crystal with Disturbed Surface Layer
(a) Vol. **64**, No. 1, p. 359–366 (1981).
See also *17*, p. *256*; 22.1.2, p. 411.

V. K. Kohn, M. V. Kovalchuk, R. M. Imamov, and E. F. Lobanovich
*The Method of Integral Characteristics in X-Ray Diffraction Studies of the Structure of the Surface Layers of Single Crystals
(a) Vol. **64**, No. 2, p. 435–442 (1981).
See also *11*, p. *124*.

S. P. Mhaskar and P. K. Naik
*Effect of Surface Relaxation on Point Defects in the Vicinity of W (100) Surfaces
(b) Vol. **105**, No. 2, p. 685–689 (1981).
See also *10.1*, p. *100*; 21, p. 343.

K. Masuda
*On the Vibrational Spectra for Atoms at B.C.C. Transition Metal Surfaces: Effects of Surface Relaxation and Reconstruction
(b) Vol. **105**, No. 2, p. K119–K124 (1981).
See also *6*, p. *75*; 21.1, p. 359.

H. Pfeiffer
*A Simple Approach to Reactive Processes on the Interface during Crystal Growth
(a) Vol. **65**, No. 2, p. 637–642 (1981).
See also *3*, p. *54*.

V. P. Skripov, V. P. Koverda, and V. N. Skokov
Size Effect on Melting of Small Particles
(a) Vol. **66**, No. 1, p. 109–118 (1981).
See also *1.5*, p. *35*; 2, p. 47; 8, p. 86; 21, p. 344.

H. Pfeiffer
*Anisotropy of Step Motion
(a) Vol. **66**, No. 1, p. 165–173 (1981).
See also *3*, p. *54*; 22.8.2, p. 504.

H. Pfeiffer and Yu. A. Baikov
*Kinetic Equation Approach to One-Component Crystal Growth. Comparison with Monte-Carlo Results
(a) Vol. **66**, No. 2, p. 521–524 (1981).
See also *3*, p. *55*.

S. P. MHASKAR and P. K. NAIK
*Helium Desorption from a Tungsten Single Crystal
(b) Vol. **107**, No. 1, p. 99–103 (1981).
See also *10.1*, p. *102*; 21, p. 345.

I. J. R. BAUMVOL
*Radiation-Enhanced Diffusion as a Coating Method for Protection against Corrosion and Wear
(a) Vol. **67**, No. 1, p. 287–298 (1981).
See also *9*, p. *91*; 11, p. 128; 21.1.1, p. 374.

K. MASUDA
*Calculation of Atomic Relaxation near the (111) 1 × 1 Surface of Covalent Semiconductors: Tight-Binding Green's Function Approach
(b) Vol. **107**, No. 2, p. 529–533 (1981).
See also *13.3*, p. *166*; 22.1, p. 402; 22.1.1, p. 407; 22.1.2, p. 419.

M. TAUT and M. SCHUBERT
Anisotropy of the Dipole Barrier of Metals
(b) Vol. **107**, No. 2, p. K139–K144 (1981).
See also *13.3*, p. *166*; 21, p. 346.

O. NISHIKAWA, M. NACHI, K. KURIHARA, Y. TSUNASHIMA, and Y. HARA
*Atom-Probe Study of Ga–Mo and Sn–Ir Interfaces
(a) Vol. **67**, No. 2, p. 461–468 (1981).
See also *1.3*, p. *26*; 21, p. 347.

H. ENDERT, E. FÖRSTER, K. GOETZ, M. P. KALASHNIKOV, J. A. MIKHAILOV, A. V. RODE, G. V. SKLIZKOV, and W.-D. ZIMMER
*Nanosecond X-Ray Topography of Crystalline Solids with a Laser-Produced Plasma Source
(a) Vol. **68**, No. 2, p. 483–488 (1981).
See also *10.2*, p. *119*; 20.2, p. 322; 22.5.3, p. 472.

2. Non-Crystalline State

Review Articles

H. BÄSSLER
Localized States and Electronic Transport in Single Component Organic Solids with Diagonal Disorder
(b) Vol. **107**, No. 1, p. 9–54 (1981).
See also *13.4*, p. *168*; 13.5.1, p. 188; *14.3*, p. *209*; 14.3.4, p. 232; 20.1, p. 301; 20.3, p. 323; 22.9, p. 507.

H. R. KHAN and K. LÜDERS
*Nuclear Magnetic Resonance on Amorphous Alloys
(b) Vol. **108**, No. 1, p. 9–18 (1981).
See also 14.2, p. 205; *19*, p. *286*; 21, p. 336; 21.1, p. 351; 21.3, p. 378.

Original Papers and Short Notes

N. N. MEDVEDEV and YU. I. NABERUKHIN
*Modelling of the Radial Distribution Function of the Hard-Sphere Liquid in a Quasi-Crystalline Model
(b) Vol. **103**, No. 1, p. 71–79 (1981).

A. BOBÁK
*Exactly Solvable Model of Amorphous Ferromagnets
(b) Vol. **103**, No. 1, p. 101–105 (1981).
See also *18.2*, p. *267*.

R. LENK
Multiple Scattering in Disordered Materials (III): Compact Representation of the Wave Number and Introduction of Local Currents
(b) Vol. **103**, No. 1, p. 253–261 (1981).
See also *13*, p. *142*.

K. N. KHANNA
*Thermodynamic Properties of Liquid Metals
(b) Vol. **103**, No. 1, p. K15–K19 (1981).
See also *8*, p. *84*; 21, p. 336; 21.2, p. 376; 21.3, p. 378; 21.4, p. 380.

S. K. DAS, R. S. TRIPATHI, and S. K. JOSHI
*Spatial Spin Correlation Function in Spin Glass
(b) Vol. **103**, No. 1, p. K31–K36 (1981).
See also *18*, p. *260*.

K. TWAROWSKI, H. K. LACHOWICZ, M. GUTOWSKI, and H. SZYMCZAK
*On the Origin of the Perpendicular Anisotropy and Magnetostriction in Amorphous RF Sputtered Gd–Co Films
(a) Vol. **63**, No. 1, p. 103–108 (1981).
See also *18.2.1*, p. *276*; 18.3, p. 278; 21.1, p. 352; 21.4, p. 380.

B. J. MULDER and S. VAN HEUSDEN
*Mechanism of the Darkening of Soda–Lime Glass by a Low-Pressure Mercury Vapour Discharge
(a) Vol. **63**, No. 1, p. 137–141 (1981).
See also 9, p. 89; 20.1, p. 302; 22, p. 391.

S. K. Trikha and S. C. Jain
*Computer Simulation of the Phase Transition in Ice
(a) Vol. **63**, No. 1, p. 163–168 (1981).
See also 22.6, p. 474.

V. K. Dolganov, N. Kroó, L. Rosta, and E. F. Sheka
Ordered and Disordered States of DOBHOP. Raman Scattering and Neutron Diffraction
(a) Vol. **63**, No. 1, p. 265–269 (1981).
See also *1.2*, p. *14*; 6, p. 67; 22.9, p. 508.

E. Nomura, K. Murata, and K. Nagami
*A Theoretical Estimate of the Voltage Dependence of the Pattern Resolution in Electron-Beam Lithography
(a) Vol. **63**, No. 1, p. 281–288 (1981).
See also *11*, p. *121*; 22.1.2, p. 408; 22.9, p. 508.

S. H. Risbud
*Amorphous Phase Separation and Transformations in the Bulk Glassy Semiconducting Alloy $CdGeAs_2$
(a) Vol. **63**, No. 1, p. 289–298 (1981).
See also 22.8, p. 484.

R. Żuberek
*Columnar Model of Magnetostriction in Amorphous Films
(a) Vol. **63**, No. 1, p. K51–K54 (1981).
See also *18.2.1*, p. *276*; 18.3, p. 278; 21.1, p. 352; 21.4, p. 380.

I. Bakonyi, L. Takács, and K. Tompa
*Dipole–Dipole Interaction and Short-Range Order in Amorphous Ni–P, Ni–Cu–P, and Ni–P–B Alloys
(b) Vol. **103**, No. 2, p. 489–497 (1981).
See also *19*, p. *287*; 21.1, p. 353.

J. Sivardière
Lattice-Gas Model for a Simple Fluid: Condensation, Orientation, and Solidification
(b) Vol. **103**, No. 2, p. 823–829 (1981).
See also *1.2*, p. *15*.

V. K. Saxena
Spin-Glass Transition in a Heisenberg System with Competing Interactions
(b) Vol. **103**, No. 2, p. K99–K104 (1981).
See also *18.2*, p. *268*; *18.4*, p. *281*.

P. John, I. M. Odeh, M. J. K. Thomas, M. J. Tricker, and J. I. B. Wilson
Silicon–Hydrogen Stretching Modes in Amorphous Silicon. A Semi-Empirical Calculation of the Infrared Band Widths
(b) Vol. **103**, No. 2, p. K141–K146 (1981).
See also *6*, p. *68*; *20.1*, p. *304*; 22.1.2, p. 409.

M. Bohl
*Electronic Structure, UV Absorption, and Dispersion of Optical Glasses
(a) Vol. **63**, No. 2, p. 529–538 (1981).
See also 13, p. 143; *20.1*, p. *304*; 22.6, p. 474.

G. LE CAER and J. M. DUBOIS
About the Asymmetries in Mössbauer Spectra of Magnetic Amorphous Transition Metal–Metalloid Alloys
(a) Vol. 64, No. 1, p. 275–282 (1981).
See also *5*, p. *63*; *18.2*, p. *269*; 21.1.1, p. 369.

G. FLEURY, A. HAMOU, C. VIGER, and C. VAUTIER
*Conductivity and Crystallization of Amorphous Selenium
(a) Vol. 64, No. 1, p. 311–316 (1981).
See also *14.3.1*, p. *220*; 22.1.3, p. 425.

P. J. SCHURER, A. H. MORRISH, and M. J. STAVN
Induced Perpendicular Anisotropy and Surface Crystallization in Amorphous $Fe_{78}B_{12}Si_{10}$ Ribbons
(a) Vol. 64, No. 1, p. 343–349 (1981).
See also 5, p. 63; *18.2.1*, p. *276*; 21.1.1, p. 369.

A. ZENTKO, M. TIMKO, P. DUHAJ, and I. ŠKORVÁNEK
The Effect of Neutron Irradiation on the Electrical Resistivity of Amorphous $Fe_{47}Ni_{25}B_{18}Si_{10}$ Alloys
(a) Vol. 64, No. 1, p. K19–K22 (1981).
See also *11*, p. *123*; *14.1*, p. *200*; 21.1.1, p. 369.

Y. C. CHEN, SHI-YU WU, and J. J. SINAI
*Excitation Spectra of Disordered Systems (I): General Theory
(b) Vol. 104, No. 2, p. 441–452 (1981).
See also *6*, p. *71*.

N. C. HALDER
*A Comparison between the Theoretical and Experimental Electronic Density of States of Liquid Metals
(b) Vol. 104, No. 2, p. 469–478 (1981).
See also *13*, p. *144*; 21, p. 340; 21.1, p. 356.

B. ESSER and I. P. ZVYAGIN
*On the Time Dependence of the Capture of Non-Equilibrium Hopping Charge Carriers
(b) Vol. 104, No. 2, p. 523–529 (1981).
See also *13.4*, p. *173*; 22, p. 393.

N. MATTERN and H. HERMANN
*Model Pair Distribution Functions for Amorphous Binary Metal–Metalloid Alloys Based on Crystal-Like Short-Range Order
(b) Vol. 104, No. 2, p. 547–554 (1981).
See also 21.1, p. 356; 21.1.1, p. 370.

P. JOHN, I. M. ODEH, M. J. K. THOMAS, M. J. TRICKER, and J. I. B. WILSON
*Studies of the Oscillator Strengths of Infrared Vibrational Modes in Glow-Discharge Hydrogenated Amorphous Silicon
(b) Vol. 104, No. 2, p. 607–612 (1981).
See also 6, p. 72; *20.1*, p. *307*; 22.1.2, p. 412.

I. Gregora, B. Velický, and M. Závětová
Refractive Index of Crystalline and Amorphous GeS
(b) Vol. **104**, No. 2, p. K95–K100 (1981).
See also *20.1*, p. *307*; 22, p. 393.

J. Šlechta
*On the Fine Structure of the Self-Consistent Continued-Fraction (SCCF) Calculation of the Density of States
(b) Vol. **104**, No. 2, p. K143–K148 (1981).
See also *13*, p. *144*.

H. C. Freyhardt, J. Goebbels, K. Lüders, and J. Reichelt
Erratum to: NMR Investigations on ^{9}Be in Amorphous Be–Nb–Zr Alloys
(b) Vol. **104**, No. 2, p. K171 (1981).
See also 14.2, p. 206; *19*, p. *288*; 21, p. 340.

S. Tyagi, J. Steinberg, A. E. Lord, Jr., and P. M. Anderson
*Acoustic Barkhausen Emission in Amorphous Metallic Alloys $Fe_{40}Ni_{40}P_{14}B_6$ and $Fe_{78}Mo_2B_{20}$
(a) Vol. **64**, No. 2, p. 443–448 (1981).
See also *18.2*, p. *269*; 21.1.1, p. 370.

L. Kraus, Z. Frait, and J. Schneider
*Ferromagnetic Resonance in Amorphous $(FeNi)_{80}P_{10}B_{10}$ Alloys (II): Resonance near the Curie Temperature
(a) Vol. **64**, No. 2, p. 449–456 (1981).
See also *18.2*, p. *269*; 19, p. 289; 21.1, p. 356; 21.1.1, p. 370.

E. M. Raspopova, M. Závětová, J. Krištofik, and B. T. Kolomiets
*The Influence of Defects Generated during Densification on the Properties of the As–Se System
(a) Vol. **64**, No. 2, p. 457–460 (1981).
See also 10.2, p. 109; 14.4, p. 238; *20.1*, p. *308*; 22, p. 393.

M. Kitao, H. Akao, T. Ishikawa, and S. Yamada
Influence of Copper Addition on Electrical and Optical Properties of Amorphous As_2Se_3
(a) Vol. **64**, No. 2, p. 493–498 (1981).
See also 13.4, p. 174; *14.3*, p. *212*; *20.1*, p. *308*; 22, p. 394.

H. Kronmüller and W. Fernengel
The Role of Internal Stresses in Amorphous Ferromagnetic Alloys
(a) Vol. **64**, No. 2, p. 593–602 (1981).
See also 10.1, p. 99; 12.1, p. 135; *18.2*, p. *270*; 21.1.1, p. 370.

P. C. M. Gubbens, A. M. van der Kraan, and K. H. J. Buschow
^{161}Dy Mössbauer Effect and Magnetic Properties of Some Amorphous and Crystalline Dy–Co Alloys
(a) Vol. **64**, No. 2, p. 657–663 (1981).
See also *5*, p. *63*; *18*, p. *261*; 21.1, p. 357; 21.4, p. 381.

D. Lukášik and P. Macko
Creation of Thermal Spikes in the GeS_{11} Glassy System by Fast Neutron Irradiation
(a) Vol. **64**, No. 2, p. 677–682 (1981).
See also *11*, p. *124*; 22, p. 394.

J. Schneider, Z. Arnold, J. Kamarad, and A. Handstein
*High Pressure Magnetic Behaviour of Amorphous Fe–P–Ge Alloys
(a) Vol. **64**, No. 2, p. K133 – K138 (1981).
See also *18.2*, p. *270*; 21.1.1, p. 370.

H. Kresse and B. Gajewska
*Dielectric Relaxation of the Liquid Crystalline S_E Phase
(a) Vol. **64**, No. 2, p. K161 – K163 (1981).
See also *14.4*, p. *239*; 22.9, p. 509.

A. R. McGurn and D. M. Esterling
*Liquid Metal Pair-Correlation Functions from Empirical Pair Potentials
(b) Vol. **105**, No. 1, p. 47 – 53 (1981).
See also *4*, p. *58*; 21.1, p. 357; 21.1.1, p. 371; 21.6, p. 385.

B. B. Sahay, P. L. Srivastava, and N. R. Mitra
*Temperature Dependence of Electrical Resistivity of Zn, Cd, Hg, Al, In, Tl, and Pb by Nonlocal Harmonic Model Potential
(b) Vol. **105**, No. 1, p. 93 – 98 (1981).
See also *14.1*, p. *200*; 21, p. 341; 21.1, p. 358.

A. F. Khokhlov, A. I. Mashin, and A. M. Satanin
The Dependence of Magnetic Ordering Temperature in Amorphous Semiconductors on Paramagnetic Centre Concentration
(b) Vol. **105**, No. 1, p. 129 – 136 (1981).
See also 11, p. 124; *18.1*, p. *266*; *18.4*, p. *282*; 19, p. 289; 22.1.1, p. 405; 22.1.2, p. 413.

H. Dersch, J. Stuke, and J. Beichler
Electron Spin Resonance of Doped Glow-Discharge Amorphous Silicon
(b) Vol. **105**, No. 1, p. 265 – 274 (1981).
See also *13.4*, p. *175*; *19*, p. *289*; 22.1.2, p. 413.

S. Tsugane, K. Morigaki, and C. Nagashima
*Anomalous Magnetoresistance in Amorphous Si–Au System
(b) Vol. **105**, No. 1, p. K5 – K8 (1981).
See also *14.3.1*, p. *220*; 21.6, p. 385; 22.1.2, p. 413.

A. Rajewska, A. Jabłonka, D. Sikorska, and A. Sobaszek
*Neutron Critical Scattering in the Nematic Phase and at the Nematic-Isotropic Liquid Transition in PAA
(b) Vol. **105**, No. 1, p. K35 – K37 (1981).
See also *1.2*, p. *16*; 22.9, p. 510.

P. Kleinert and R. Leihkauf
*On the Low-Temperature Specific Heat of Glasses
(b) Vol. **105**, No. 1, p. K67 – K70 (1981).
See also *8*, p. *85*; 22.6.1, p. 481.

L. Kaihola
*Anomolous Soft X-Ray Reflectivity of Borosilicate Glass by Successive Grazing Reflections
(a) Vol. **65**, No. 1, p. 163 – 166 (1981).
See also *20*, p. *297*; 22.6.1, p. 481.

G. Götz, H. D. Geiler, M. Wagner, K. H. Heinig, and H. Woittenneck
*Laser Annealing of Thin Buried Amorphous Layers in Silicon
(a) Vol. **65**, No. 2, p. 677–682 (1981).
See also *11*, p. *126*; 22.1.2, p. 415.

H. S. Chen, Y. Waseda, and K. T. Aust
*Structure of $Pt_{75}P_{25}$ Glass
(a) Vol. **65**, No. 2, p. 695–700 (1981).
See also 21, p. 343.

B. Rauschenbach, E. Richter, and K. Hohmuth
*Spinodal Decomposition in Amorphous Boron-Implanted Iron Films
(a) Vol. **65**, No. 2, p. K103–K105 (1981).
See also 11, p. 126; 21.1.1, p. 372.

C. E. van Dijkum and A. Lodder
*The Local Density of States of Embedded Cu and Cu(Ni) Clusters
(b) Vol. **106**, No. 1, p. 107–122 (1981).
See also *13*, p. *146*; 21.1, p. 360.

P. Valenta, K. Maier, H. Kronmüller, and K. Freitag
*Self-Diffusion of Iron in the Amorphous Alloys $Fe_{40}Ni_{40}P_{14}B_6$ and $Fe_{80}B_{20}$
(b) Vol. **106**, No. 1, p. 129–133 (1981).
See also *9*, p. *90*; 21.1.1, p. 372.

M. Giehler
The Effect of Short-Range Order on the Vibrational Spectra of 3:2 Coordinated Chalcogenide Glasses Analysis of the Vibrational Spectra by Using Pyramidal Structural
(b) Vol. **106**, No. 1, p. 193–205 (1981).
See also 4, p. 59; *6*, p. *76*; *20.1*, p. *313*; 22, p. 396.

R. J. Sayer, R. H. Prince, and W. W. Duley
*Emission Spectra of O and N Atoms and the Structure of Solid N_2/Ne Mixtures
(b) Vol. **106**, No. 1, p. 249–258 (1981).
See also *20.3*, p. *329*; 23, p. 514.

C. K. Saw and S. J. Pickart
*Aperiodicity in an Amorphous Dense Random Packed Model
(b) Vol. **106**, No. 1, p. K73–K76 (1981).
See also 21.1.1, p. 372.

V. P. Skripov, V. P. Koverda, and V. N. Skokov
Size Effect on Melting of Small Particles
(a) Vol. **66**, No. 1, p. 109–118 (1981).
See also *1.5*, p. *35*; 1.6, p. 38; 8, p. 86; 21, p. 344.

R. Klabes, J. Matthäi, M. Voelskow, G. A. Kachurin, E. V. Nidaev, and H. Bartsch
*Flash Lamp Annealing of Arsenic Implanted Silicon
(a) Vol. **66**, No. 1, p. 261–266 (1981).
See also *11*, p. *127*; 22.1.2, p. 416.

H. DERSCH, J. STUKE, and J. BEICHLER
Temperature Dependence of ESR Spectra of Doped a-Si:H
(b) Vol. **107**, No. 1, p. 307–317 (1981).
See also *13.4*, p. *180*; 14.3, p. 216; *19*, p. *292*; 22.1.2, p. 418.

J. J. SINAI, SHI-YU WU, and Y. C. CHEN
*Excitation Spectra of Disordered Systems (II): Two- and Three-Dimensional Disordered Systems
(b) Vol. **107**, No. 1, p. 367–374 (1981).
See also *6*, p. *78*.

R. L. JACOBS
Model Electronic Structure and Saturation Magnetisation of Some Substituted Fe–Si–B Metallic Glasses
(b) Vol. **107**, No. 1, p. K13–K17 (1981).
See also *13*, p. *147*; *18.2*, p. *273*; 21.1.1, p. 374.

H. VAN SWIJGENHOVEN, J. VANOPPEN, and L. M. STALS
*Vitrification of Recrystallized $Fe_{40}Ni_{38}Mo_4B_{18}$ during 5 keV Ar^+-Sputtering
(a) Vol. **67**, No. 1, p. K5–K9 (1981).
Erratum
(a) Vol. **68**, No. 2, p. K211 (1981).
See also *11*, p. *129*; 21.1, p. 363; 21.1.1, p. 375.

H. KRESSE, S. DIELE, A. WIEGELEBEN, D. DEMUS, and B. GAJEWSKA
*A New Liquid Crystalline Low Temperature Phase
(a) Vol. **67**, No. 1, p. K21–K23 (1981).
See also 14.4, p. 240; 22.9, p. 512.

A. V. SHENDRIK
*ESR of Vitreous Silica Containing Hydroxyl
(a) Vol. **67**, No. 1, p. K25–K27 (1981).
See also *19*, p. *292*; 22.6.1, p. 481.

J. KAMARÁD, J. SCHNEIDER, and Z. ARNOLD
*Pressure Dependence of Curie Temperature of Amorphous $Fe_{90}Zr_{10}$ Alloy
(a) Vol. **67**, No. 1, p. K85–K88 (1981).
See also *18.2*, p. *273*; 21.1.1, p. 375.

A. N. VTYURIN, V. P. ERMAKOV, B. I. OSTROVSKII, and V. F. SHABANOV
*Study of Optical Second Harmonic Generation in Ferroelectric Liquid Crystal
(b) Vol. **107**, No. 2, p. 397–402 (1981).
See also *14.4.1*, p. *245*; 20, p. 299; 22.9, p. 512.

MAI XUAN LI and NGUYEN MANH DUC
Low-Temperature Renormalization Group Study of the Random Spin Systems
(b) Vol. **107**, No. 2, p. 403–409 (1981).
See also *18.2*, p. *274*.

D. SEN and S. K. SARKAR
Unified Study of Metals on Pseudopotential Theory: Lead
(b) Vol. **107**, No. 2, p. 559–567 (1981).
See also *6*, p. *79*; 12.1, p. 136; *14.1*, p. *202*; 15, p. 249; 21, p. 346.

I. K. Bandrovskaya, V. Ya. Prokhorenko, V. G. Romanov, and B. I. Sokolovskii
Optical and Thermoelectric Properties of Arsenic Selenide in Glassy and Liquid States
(b) Vol. **107**, No. 2, p. K95 – K98 (1981).
See also *14.3*, p. *217*; 15, p. 249; *20.1*, p. *318*; 22, p. 398.

J. Tröltzsch and Ch. Dippmann
Thermal Measurements on Chalcogenide Glass Threshold Switching Devices
(a) Vol. **67**, No. 2, p. 527 – 531 (1981).
See also *8*, p. *88*; *14.3.3*, p. *230*; 22.8, p. 490.

Y. Aoyagi, Y. Segawa, S. Namba, T. Suhara, H. Nishihara, and H. Gamo
*Dynamic Behavior of the Photodarkening Process in As_2S_3 Chalcogenide Glass
(a) Vol. **67**, No. 2, p. 669 – 676 (1981).
See also *10.2*, p. *117*; 20, p. 299; 22, p. 399.

T. E. Jenkins and P. O'Brien
*A Room Temperature Specific Heat Anomaly in Triamantane ($C_{18}H_{24}$)
(a) Vol. **67**, No. 2, p. K161 – K162 (1981).
See also *8*, p. *88*; 22.9, p. 512.

A. I. Landa, V. E. Panin, and N. F. Zhorovkov
*The Pseudopotential Method and Thermodynamics of Eutectic Systems
(b) Vol. **108**, No. 1, p. 113 – 122 (1981).
See also *1.2*, p. *20*; 8, p. 88; 13, p. 149; 21, p. 347; 21.1, p. 365.

B. Esser
*On the Lifetime of Excited States in One-Dimensional Disordered Systems
(b) Vol. **108**, No. 1, p. 195 – 204 (1981).
See also *13.4*, p. *182*; 20.3, p. 333; 22, p. 399.

W. Grevendonk, L. van den Abeele, P. van den Keybus, B. Vanhuyse, and G. Ruymbeek
Faraday Rotation in Quartz and Oxide Glasses
(b) Vol. **108**, No. 1, p. K53 – K56 (1981).
See also 13.1, p. 160; *18*, p. *264*; *20*, p. *299*; 22.6, p. 480; 22.6.1, p. 482.

K. Handrich and J. Resch
*Nuclear Quadrupole Spin–Lattice Relaxation Time T_1 of Amorphous Solids
(b) Vol. **108**, No. 1, p. K57 – K60 (1981).
See also 6, p. 81; 13.5, p. 187; *18*, p. *264*.

J. S. Brown
*Electrical Resistivity of Liquid Ti at Its Melting Point
(b) Vol. **108**, No. 1, p. K65 – K68 (1981).
See also *14.1*, p. *203*; 21, p. 348.

T. H. O'Dell
Magnetostriction Measurements on Amorphous Ribbons by the Becker-Kersten Method
(a) Vol. **68**, No. 1, p. 221 – 226 (1981).
See also *18.2*, p. *274*; 21, p. 348.

Bhanwar Singh and K. L. Chopra
*A Novel Bi-Layer System for High Resolution Optical Imaging
(a) Vol. **68**, No. 1, p. K5 – K9 (1981).
See also *20.1*, p. *319*; 22.1.1, p. 407; 22.1.3, p. 425.

R. Göring, H. Bürger, H. Nass, and B. Schnabel
*^{11}B NMR Study on Glasses of the System TeO_2–B_2O_3
(a) Vol. 68, No. 1, p. K29 – K32 (1981).
See also *19*, p. *294*; 22.6.1, p. 482.

I. P. Zvyagin
*AC Hopping Conductivity – Analytical Theory versus Numerical Calculation
(b) Vol. 108, No. 2, p. K85 – K88 (1981).
See also *14.3*, p. *219*.

J. Bullot, P. Cordier, O. Gallais, M. Gauthier, and J. Livage
*Experimental Determination of the Disorder Energy in Amorphous V_2O_5 Layers Deposited from Gels
(a) Vol. 68, No. 2, p. 357 – 361 (1981).
See also *14.3.1*, p. *222*; 22.6, p. 480.

A. S. N. Rao, P. N. Murty, C. R. K. Murty, and T. Rs. Reddy
*EPR Study of Molecular Ordering in Liquid Crystal: NOBA
(a) Vol. 68, No. 2, p. 373 – 377 (1981).
See also *19*, p. *295*; 22.9, p. 513.

Y. J. Wang and H. Kronmüller
*Effect of Hydrogen on the Coercive Field of Metallic Glass $Co_{75}Si_{15}B_{10}$
(a) Vol. 68, No. 2, p. 499 – 504 (1981).
See also *18.2*, p. *275*; 21.1, p. 366.

O. Bostanjoglo and G. Schlotzhauer
Impulse Stimulated Crystallization of Sb Films Investigated by Time Resolved TEM
(a) Vol. 68, No. 2, p. 555 – 560 (1981).
See also 1.2, p. 21; *3*, p. *56*; 22.1, p. 403.

L. Varga, K. Tompa, and T. Schmidt
Thermopower Method for Chemical Composition and Inhomogeneity Measurements in Amorphous Ni–P Samples
(a) Vol. 68, No. 2, p. 603 – 606 (1981).
See also *1.3*, p. *27*; *15*, p. *250*; 21.1, p. 367.

S. V. Sukhvalo, A. I. Demchenko, A. B. Shifrin, and E. M. Zub
*Orientation Transitions in Amorphous Gd–Co Films
(a) Vol. 68, No. 2, p. 637 – 644 (1981).
See also *1.3*, p. *27*; 14.1, p. 204; 18, p. 264; 21.1, p. 367; 21.4, p. 383.

J. Gazsó
Determination of the Density of Localized States from the Field Effect in Wide Gap Semiconductors
(a) Vol. 68, No. 2, p. 675 – 688 (1981).
See also *13.4*, p. *184*; 14.3.1, p. 222; *14.3.4*, p. *236*; 22, p. 401.

Y. Waseda and K. T. Jacob
*Refinement of a Correlation between Isothermal Compressibility and Surface Tension of Liquid Metals
(a) Vol. 68, No. 2, p. K117 – K122 (1981).
See also *12.1*, p. *137*; 21, p. 349.

3. Crystal Growth

Review Article

E. Spenke and W. Heywang
*Twenty Five Years of Semiconductor-Grade Silicon
(a) Vol. **64**, No. 1, p. 11–44 (1981).
See also 13.4, p. 168; *14.3*, p. *209*; 14.3.3, p. 225; 22.1.2, p. 408.

Original Papers and Short Notes

T. G. Finstad
*A Xe Marker Study of the Transformation of Ni_2Si to NiSi in Thin Films
(a) Vol. **63**, No. 1, p. 223–228 (1981).
See also *1.5*, p. *34*; 21.1, p. 352.

M. S. Abrahams, J. L. Hutchison, and G. R. Booker
Direct Observation of a Silicon–Sapphire Hetero-Epitaxial Interface by High Resolution Transmission Electron Microscopy
(a) Vol. **63**, No. 1, p. K3–K6 (1981).
See also *1.4*, p. *28*; 22.1.2, p. 409; 22.6, p. 474.

K. B. Sundaram and G. K. Bhagavat
*Preparation and Properties of Indium Oxide Films
(a) Vol. **63**, No. 1, p. K15–K18 (1981).
See also 1.5, p. 34; *14.3.1*, p. *220*; 15, p. 247; 20.1, p. 303; 22.6, p. 474.

J. M. Robertson and M. W. van Tol
Luminescent Single Crystal Thin Films of Manganese Doped Barium Hexa-Aluminate
(a) Vol. **63**, No. 1, p. K59–K62 (1981).
See also *20.3*, p. *324*; 22.8.1, p. 493.

P. H. Tsien, S. C. Tsou, M. Takai, D. Röschenthaler, M. Ramin, H. Ryssel, I. Ruge, and K. Wittmaack
*Annealing of Boron-Implanted Silicon Using a CW CO_2 Laser
(a) Vol. **63**, No. 2, p. 547–555 (1981).
See also *1.5*, p. *34*; 11, p. 121; 14.3.1, p. 220; 22.1.2, p. 409.

M. G. Kalitzova, S. B. Simov, B. G. Pantchev, P. Danesh, K. I. Kirov, and A. E. Djakov
Amorphous to Crystalline State Conversion in Deposited Silicon Layers
(a) Vol. **63**, No. 2, p. 743–749 (1981).
See also *1.5*, p. *34*; *2*, p. *42*; 11, p. 122; 22.1.2, p. 410.

K. Seshan, D. K. Chakrabarty, and A. B. Biswas†
*A New Method for the Preparation of Acicular γ-Fe_2O_3
(a) Vol. **63**, No. 2, p. K123–K125 (1981).
See also *18.3*, p. *278*; 22.8.2, p. 502.

E. D. Atanasova, K. I. Kirov, and E. I. Kantardjeva
*Formation of Ultrathin Oxide Films on Silicon in RF Oxygen Plasma
(a) Vol. **64**, No. 1, p. 73–80 (1981).
See also *1.5*, p. *34*; 22.6, p. 475.

R. Hergt and H. Pfeiffer
*Anisotropy of the Displacement Rate of Garnet Growth Steps
(a) Vol. **66**, No. 1, p. 207–212 (1981).
See also 22.8.2, p. 504.

S. I. Stenin and A. K. Gutakovskii
The Plastic Deformation Kinetics for Heteroepitaxial Films during the Misfit Dislocation Generation from the Growth Surface
(a) Vol. **66**, No. 1, p. 249–253 (1981).
See also *1.5*, p. *36*; *10.2*, p. *113*; 22.1.1, p. 406.

J. F. Pouradier and G. Grimouille
*Growth of Semiconducting $CdF_2:Ho^{3+}$ Crystals and Study of Their Optical and Transport Properties
(b) Vol. **106**, No. 2, p. 657–665 (1981).
See also 13.4, p. 179; *14.3*, p. *215*; 20.1, p. 315; 22.5.3, p. 471.

H. Pfeiffer and Yu. A. Baikov
*Kinetic Equation Approach to One-Component Crystal Growth. Comparison with Monte-Carlo Results
(a) Vol. **66**, No. 2, p. 521–524 (1981).
See also *1.6*, p. *38*.

E. F. Krimmel, H. Oppolzer, H. Runge, and W. Wondrak
*Kinetics of Scanned Electron Beam Annealing of High Energy As Ion Implanted Silicon
(a) Vol. **66**, No. 2, p. 565–571 (1981).
See also *11*, p. *127*; 14.3.3, p. 228; 22.1.2, p. 417.

E. Ziegler, A. Heinrich, H. Oppermann, and G. Stöver
Electrical Properties and Non-Stoichiometry in ZnO Single Crystals
(a) Vol. **66**, No. 2, p. 635–648 (1981).
See also *14.3*, p. *216*; 22.4, p. 445.

D. M. Scott and M. A. Nicolet
*Implanted Oxygen in NiSi Formation
(a) Vol. **66**, No. 2, p. 773–778 (1981).
See also *1.5*, p. *36*; 11, p. 128; 21.1, p. 362.

R. Hergt and H. Pfeiffer
*Investigation of Growth Steps Due to Misorientation of Garnet LPE Films
(a) Vol. **67**, No. 1, p. 205–215 (1981).
See also *1.5*, p. *36*; 22.8.2, p. 505.

A. K. Gutakovskii, S. I. Stenin, and B. G. Zakharov
The Formation of Partial Misfit Dislocations during Heteroepitaxy
(a) Vol. **67**, No. 1, p. 299–304 (1981).
See also *1.5*, p. *36*; *10.2*, p. *116*; 22.2.1, p. 430.

K. Recker, F. Wallrafen, R. Büscher, and G. Lehmann
Crystal Growth, EPR, and Luminescence of $BaZnF_4:Mn$
(b) Vol. **107**, No. 2, p. 699–706 (1981).
See also *19*, p. *293*; *20.3*, p. *333*; 22.5.4, p. 473.

4. Bonding Properties

Original Papers and Short Notes

M. A. Blokhin, E. G. Orlova, and I. G. Schweitzer
*The Influence of the Environment of the Absorbing Atom on the Photoelectric Yield Spectra of Beryllium Compounds in the Soft X-Ray Region
(b) Vol. **103**, No. 1, p. 63–69 (1981).
See also *17*, p. *256*; 22.5, p. 458; 22.6, p. 474; 22.8, p. 484; 22.9, p. 507.

U. Pietsch
*X-Ray Bond Charge in GaAs and InSb
(b) Vol. **103**, No. 1, p. 93–100 (1981).
See also 22.2.1, p. 427; 22.2.3, p. 435.

R. K. Singh and P. Khare
*Study of Cohesion and Allied Properties of Silver Halide Crystals
(b) Vol. **103**, No. 1, p. 337–344 (1981).
See also 12.2, p. 138; 22.5.1, p. 461.

L. Skála
*A Successful CNDO/BW Parametrization for Calculation of the Lithium Clusters
(b) Vol. **103**, No. 1, p. K27–K29 (1981).
See also *13*, p. *142*; 21.2, p. 376.

T. P. Humphreys and G. P. Srivastava
Electronic Charge Density for Two Isoelectronic Series: Ge–GaAs–ZnSe and α-Sn–InSb–CdTe
(b) Vol. **103**, No. 1, p. K85–K90 (1981).
See also *13.1*, p. *152*; 22, p. 391.

A. Basu, S. Sengupta, and A. N. Basu
*Three-Body Interaction in the Fluorite Structure Ionic Crystals
(b) Vol. **103**, No. 2, p. 535–546 (1981).
See also *12.2*, p. *138*; 22.5.3, p. 471.

A. V. Singh and J. Shanker
*Effect of Three-Body Potential on the Dielectric Behaviour and Optic Mode Grüneisen Parameters of Alkali Halide Crystals
(b) Vol. **103**, No. 2, p. 733–738 (1981).
See also 6, p. 68; 12.2, p. 138; 14.4, p. 238; 22.5.2, p. 463.

M. M. Shukla and J. Mahanty
Compatibility of Interionic Pair Potentials and Channeling Potentials
(b) Vol. **103**, No. 2, p. K119–K121 (1981).
See also *11*, p. *121*; 13, p. 143.

J. J. M. Binsma, L. J. Giling, and J. Bloem
Order–Disorder Behaviour and Tetragonal Distortion of Chalcopyrite Compounds
(a) Vol. **63**, No. 2, p. 595–603 (1981).
See also *1.3*, p. *22*; 22.8, p. 485.

K. N. Khanna
Wigner-Seitz Radius Dependent Model Potential
(b) Vol. **105**, No. 2, p. 485–490 (1981).
See also 1, p. 9; *13*, p. *145*; 14.1, p. 200; 21, p. 342; 21.1, p. 359; 21.2, p. 376; 21.3, p. 378.

S. M. M. Rahman
*A Model Calculation on the Stability of the Glassy $Au_{0.82}Si_{0.18}$
(b) Vol. **105**, No. 2, p. K115–K118 (1981).
See also *2*, p. *46*; 21.6, p. 385; 22.1.2, p. 414.

U. Pietsch
*Bond Charge Reconstruction for GaP from X-Ray Powder Data
(b) Vol. **105**, No. 2, p. K135–K137 (1981).
See also *1.1*, p. *11*; 22.2.2, p. 433.

V. M. Talanov
Calculation of Structural Parameters of Spinels
(b) Vol. **106**, No. 1, p. 99–106 (1981).
See also *1*, p. *9*; 22.8.2, p. 504.

M. Giehler
The Effect of Short-Range Order on the Vibrational Spectra of 3:2 Coordinated Chalcogenide Glasses Analysis of the Vibrational Spectra by Using Pyramidal Structural Units
(b) Vol. **106**, No. 1, p. 193–205 (1981).
See also 2, p. 47; *6*, p. *76*; *20.1*, p. *313*; 22, p. 396.

V. K. Jain and J. Shanker
*Analysis of Interionic Potentials in II–VI Chalcogenide Crystals with NaCl Structure
(b) Vol. **106**, No. 1, p. 287–293 (1981).
See also 22.4.1, p. 448; 22.4.2, p. 453; 22.4.3, p. 454; 22.6, p. 477.

U. Pietsch
*Bond Charge Determination from X-Ray Data for Germanium
(b) Vol. **106**, No. 1, p. K9–K11 (1981).
See also *1.1*, p. *12*; 22.1.1, p. 405.

T. Ogawa
Morphologies of CVD ZrC Crystals
(a) Vol. **66**, No. 1, p. 175–181 (1981).
See also *1*, p. *10*; *3*, p. *54*; 22.3, p. 442.

S. Chatterjee, S. Ghosh, and A. N. Basu
Lattice Sums for Zinc Sulphide Structure Crystals
(b) Vol. **106**, No. 2, p. 693–698 (1981).
See also *1*, p. *10*; 22.4.1, p. 449.

I. A. Marushko
*On Quantum Theory of the Force Constant Matrix of Polyatomic Systems
(b) Vol. **106**, No. 2, p. 707–714 (1981).
See also 13, p. 146.

M. Hashimoto, Y. Ishida, R. Yamamoto, and M. Doyama
Structure and Atomic Vibration of the $\sum = 7$ Tilt Boundary in α-Titanium
(a) Vol. **68**, No. 1, p. 109–116 (1981).
See also *1.4*, p. *32*; *10.1*, p. *104*; 21.1, p. 365.

U. Pietsch
*Bond Charge Determination in Ternary Chalcopyrite Semiconductors by Means of X-Ray Structure Amplitude Investigations
(a) Vol. **68**, No. 1, p. 321–327 (1981).
See also 22.2.4, p. 439.

Y. V. G. S. Murti and T. V. Selvarajan
*An Extended Interionic Potential for Alkali Halides
(b) Vol. **108**, No. 2, p. 315–322 (1981).
See also 22.5.2, p. 469.

V. P. Gupta, A. Gupta, and V. K. Srivastava
*Electronic Properties of Gallium Chalcogenides (GaS, GaSe, and GaTe)
(b) Vol. **108**, No. 2, p. 323–327 (1981).
See also 14.4, p. 241; 22, p. 400.

K. S. Srivastava, M. Husain, P. Gupta, A. K. Srivastava, V. Kumar, K. Sinha, and S. Singh
*Shift in the Binding Energy of the Inner Electrons Due to Chemical Combination
(b) Vol. **108**, No. 2, p. 575–579 (1981).
See also *20*, p. *300*; 21, p. 349.

K. S. Sharma
*Phonon Dispersion in Barium
(b) Vol. **108**, No. 2, p. K101–K104 (1981).
See also *6*, p. *81*; 21.3, p. 379.

A. M. Ghodgaonkar and K. Ramani
*Evaluation of Bulk Modulus, Pressure Derivative of Bulk Modulus, and Cohesive Energy by Modified Morse-Kratzer Potential
(b) Vol. **108**, No. 2, p. K133–K138 (1981).
See also *12.2*, p. *141*; 22, p. 400; 22.5.1, p. 461; 22.5.3, p. 472.

V. V. Kind, V. P. Fedorov, and A. V. Korshunov
*The Manifestation of Intermolecular Hydrogen Bond in a Binary Naphthalene-β-Naphthol Mixture
(b) Vol. **108**, No. 2, p. K143–K144 (1981).
See also *20.1*, p. *321*; 22.9, p. 513.

H.-G. Brühl and U. Pietsch
*Investigation of the Bond Charge in GaP with X-Ray Data
(a) Vol. **68**, No. 2, p. 689–695 (1981).
See also *13*, p. *150*; 22.2.2, p. 434.

5. Mössbauer Spectroscopy

Original Papers and Short Notes

A. YA. DZYUBLIK
*On the Theory of Double γ-Nuclear Magnetic Resonance
(b) Vol. **104**, No. 1, p. 81–88 (1981).
See also *19*, p. *287*.

G. YA. SELYUTIN
Magnetic Contribution to the Phonon Spectrum of Fe
(b) Vol. **104**, No. 1, p. 341–344 (1981).
See also *6*, p. *70*; *18*, p. *260*; 21.1.1, p. 369.

K. K. P. SRIVASTAVA and H. C. SINGH
*Erratum to: Effect of Second-Order Spin–Orbit Coupling $\lambda' (\boldsymbol{L} \cdot \boldsymbol{S})^2$ on Electronic Spin Relaxation of Paramagnetic Fe^{2+} in $CaCO_3$ and $CdCO_3$
(b) Vol. **104**, No. 1, p. 393 (1981).
See also 18.1, p. 265; 22.8, p. 485.

F. MENIL, N. KINOMURA, L. FOURNES, J. PORTIER, and P. HAGENMULLER
Investigation of the K_2NiF_4-Type Layer Oxide Fluorides $Sr_2FeO_{3+x}F_{1-x}$ $(0 \leqq x \leqq 0.20)$
(a) Vol. **64**, No. 1, p. 261–274 (1981).
See also *14.3*, p. *211*; *18.4*, p. *281*; 22.8, p. 486.

G. LE CAER and J. M. DUBOIS
*About the Asymmetries in Mössbauer Spectra of Magnetic Amorphous Transition Metal–Metalloid Alloys
(a) Vol. **64**, No. 1, p. 275–282 (1981).
See also 2, p. 43; *18.2*, p. *269*; 21.1.1, p. 369.

J. M. GENIN, PH. BAUER, and M. J. BESNUS
*Mössbauer Study of the Temperature Dependence of the Magnetization Direction and the Hyperfine Interactions in the Laves Phase Compound $GdFe_2$
(a) Vol. **64**, No. 1, p. 325–333 (1981).
See also 18.3, p. 278; 21.1.1, p. 369.

P. J. SCHURER, A. H. MORRISH, and M. J. STAVN
Induced Perpendicular Anisotropy and Surface Crystallization in Amorphous $Fe_{78}B_{12}Si_{10}$ Ribbons
(a) Vol. **64**, No. 1, p. 343–349 (1981).
See also 2, p. 43; *18.2.1*, p. *276*; 21.1.1, p. 369.

J. J. BARA
*Dispersion Phenomena in Mössbauer Spectra
(a) Vol. **64**, No. 2, p. 509–517 (1981).

P. C. M. GUBBENS, A. M. VAN DER KRAAN, and K. H. J. BUSCHOW
*^{161}Dy Mössbauer Effect and Magnetic Properties of Some Amorphous and Crystalline Dy–Co Alloys
(a) Vol. **64**, No. 2, p. 657–663 (1981).
See also 2, p. 44; *18*, p. *261*; 21.1, p. 357; 21.4, p. 381.

P. MUTHUKUMARASAMY, T. NAGARAJAN, and A. NARAYANASAMY
*Mössbauer Study of the Ni–Cd Ferrite System
(a) Vol. **64**, No. 2, p. 747–754 (1981).
See also 18.3, p. 279; 22.8.2, p. 503.

C. M. Srivastava, S. N. Shringi, and M. V. Babu
*Mössbauer Study of the Low-Temperature Phase of Magnetite.
(a) Vol. **65**, No. 2, p. 731–735 (1981).
See also 18.3, p. 279; 22.8.2, p. 504.

A. Narayanasamy, T. Nagarajan, P. Muthukumarasamy, and N. Kimizuka
*Mössbauer Studies of a Series of New Compounds RFe_2O_4 (R = Y, Ho, Er, Tm, Yb)
(a) Vol. **66**, No. 1, p. 377–382 (1981).
See also 22.8.2, p. 504.

E. V. Zolotoyabko and V. N. Kashcheev
*The Role of Defects on Inelastic Scattering of Mössbauer Radiation near Structural Phase Transitions
(b) Vol. **106**, No. 2, p. 723–730 (1981).
See also *1.2*, p. *18*; 10.2, p. 114; 22.8.1, p. 498.

E. Wieser, A. Handstein, J. Schneider, and K. Závěta
Magnetic Structure of Amorphous $(Fe_{1-x}Mn_x)_{75}P_{15}C_{10}$ Alloys
(a) Vol. **66**, No. 2, p. 607–612 (1981).
See also 2, p. 48; *18*, p. *263*; 21.1, p. 362; 21.1.1, p. 373.

J. P. Eymery, S. B. Raju et J. F. Dinhut
*Etude par effet Mössbauer de la substitution du fer par du chrome dans les ferrites de Mn–Zn
(Mössbauer Study of the Substitution of Iron by Chromium in Mn–Zn Ferrites)
(a) Vol. **67**, No. 1, p. 89–95 (1981).
See also *18.3*, p. *280*; 22.8.1, p. 499.

R. K. Puri and L. R. Gupta
Force Constant Change of Mössbauer ^{119}Sn Impurity in Copper Host
(b) Vol. **107**, No. 2, p. 757–759 (1981).
See also *6*, p. *79*; 21.1, p. 364.

G. Albanese and A. Deriu
*Influence of Low Frequency Excitations on the Rayleigh Scattering of Mössbauer Radiation
(b) Vol. **107**, No. 2, p. K115–K117 (1981).

E. Hanga, E. Tataru, M. Morariu, L. Diamandescu, and N. Popescu-Pogrion
*Magnetic Properties of Lithium Silicate Glass Containing Manganese Ferrite
(a) Vol. **67**, No. 2, p. 725–728 (1981).
See also *18.1*, p. *266*; 22.6, p. 480.

P. K. Acharyya, A. K. Tewari, S. K. Roy, and D. L. Bhattacharya
*Effect of Anharmonicity on the Second-Order Doppler Shift for a Vacancy-Trapped Mössbauer Impurity Atom in Metallic Solid Solution
(b) Vol. **108**, No. 1, p. 211–220 (1981).
See also 10.1, p. 103; 21 p. 347.

V. G. Baryshevskii and V. V. Skadorov
*Mössbauer Radiation Diffraction in Deformed Crystals
(b) Vol. **108**, No. 2, p. 631–643 (1981).

6. Lattice Dynamics. Phonons

Review Article

Original Papers and Short Notes

D. N. Talwar, M. Vandevyver, K. Kunc, and M. Zigone
*IR Absorption and Raman Scattering Spectra in Mixed Cuprous Halides
(b) Vol. **103**, No. 1, p. 381–389 (1981).
See also *20.1*, p. *302*; 22.5.4, p. 473.

Yu. P. Tsyashchenko and G. E. Krasnyanskii
*Studies on Anharmonic Damping Mechanisms of High-Frequency Phonons in Crystals Containing Complex Ions
(b) Vol. **103**, No. 1, p. 391–395 (1981).
See also *20.1*, p. *302*; 22.8.1, p. 493.

N. Savvides and H. J. Goldsmid
*On the Boundary Scattering of Phonons in Silicon
(b) Vol. **103**, No. 1, p. K13–K14 (1981).
See also *8*, p. *84*; 22.1.2, p. 408.

D. Houde and D. Cheeke
*Acoustic Mode Softening in $K_2Pt(CN)_4Br_{0.3} \cdot 3\,H_2O$ (KCP)
(b) Vol. **103**, No. 1, p. K21–K25 (1981).
See also *12.2*, p. *138*; 22.8, p. 484.

U. A. Jayasooriya, G. J. Kearley, S. F. A. Kettle, and H. J. Lauter
*A Comparison of Inelastic Neutron Scattering and Vibrational Spectral Data for Phase V of Ammonium Nitrate
(a) Vol. **63**, No. 1, p. 169–177 (1981).
See also *20.1*, p. *303*; 22.8.1, p. 493.

V. K. Dolganov, N. Kroó, L. Rosta, and E. F. Sheka
Ordered and Disordered States of DOBHOP. Raman Scattering and Neutron Diffraction
(a) Vol. **63**, No. 1, p. 265–269 (1981).
See also *1.2*, p. *14*; 2, p. 41; 22.9, p. 508.

A. A. Bahgat
*Correlation between Einstein and Debye Models of Lattice Vibrations for Mössbauer Fraction
(a) Vol. **63**, No. 1, p. K39–K41 (1981).
See also *5*, p. *62*.

A. Czachor and A. Holas
*Propagation of Displacement Pulses in Harmonic Crystals
(b) Vol. **103**, No. 2, p. 471–475 (1981).
See also 7, p. 82.

S. Ghosh, A. N. Basu, and S. Sengupta
*Extended Deformable Shell Model Calculation of the Lattice Statics and Dynamics of LiH and LiD Crystals
(b) Vol. **103**, No. 2, p. 595–603 (1981).
See also *12.2*, p. *138*; 22, p. 391.

T. Skettrup
*Urbach's Rule and Phase Fluctuations of the Transmitted Light
(b) Vol. **103**, No. 2, p. 613–621 (1981).
See also 13.5.1, p. 188; *20.1*, p. *303*; 22.4, p. 445.

T. Soma, Y. Saito, and H. Matsuo
*Phonon Dispersion Curves near the Covalent–Metallic Transition Pressure of GaAs, GaSb, and InSb
(b) Vol. **103**, No. 2, p. K173–K177 (1981).
See also 8, p. 84; 22.2.1, p. 427; 22.2.3, p. 435.

C. K. Sarkar
Erratum to: On the Cyclotron Resonance Linewidth in Semiconductors at Low Temperatures
(b) Vol. **103**, No. 2, p. K179 (1981).
See also *19*, p. *287*; 22, p. 392.

A. U. Sheleg, E. Peljo, and P. Suortti
Structure and Thermal Motion Parameters of Tetragonal ZnP_2 between 300 and 740 K
(a) Vol. **63**, No. 2, p. 751–758 (1981).
See also *1.1*, p. *11*; 22.3, p. 440.

N. N. Sirota and T. E. Zhabko
X-Ray Study of the Anisotropy of Thermal Properties in Titanium
(a) Vol. **63**, No. 2, p. K211–K215 (1981).
See also *8*, p. *84*; 21.1, p. 355.

P. Bogusławski
*The Theory of the Electron Spin–Lattice Relaxation in Zincblende Semiconductors
(b) Vol. **104**, No. 1, p. 89–95 (1981).
See also *13.1*, p. *153*; 22.2 p. 426; 22.4, p. 445.

J. Sosnowski, S. Bednarski, W. Bührer, A. Czachor, and E. Maliszewski
*Phonon Dispersion Relations in the $Bi_{0.95}Sb_{0.05}$ Alloy
(b) Vol. **104**, No. 1, p. 97–102 (1981).
See also 21.7, p. 387.

A. Giessler, G. Schaack, and H. Bleif
*Phonon Frequencies and Dynamics of the OH^- Molecule in Sodium Hydroxide
(b) Vol. **104**, No. 1, p. 151–159 (1981).
See also 1.2, p. 15; *20.1*, p. *305*; 22, p. 392.

H. Teichler
*Theory of Multi-Phonon Transitions beyond the Condon Approximation
(b) Vol. **104**, No. 1, p. 239–251 (1981).
See also *10.2*, p. *108*; 22, p. 392.

U. Pietsch and K. Unger
The Temperature Dependence and Dynamical Properties of the Bond Charge in GaAs and InSb
(b) Vol. **104**, No. 1, p. 253–260 (1981).
See also *4*, p. *58*; *13*, p. *143*; 22.2.1, p. 427; 22.2.3, p. 435.

F. Prokert
*Neutron Scattering Studies on Phase Transitions and Phonon Dispersion in $CsSrCl_3$
(b) Vol. **104**, No. 1, p. 261–265 (1981).
See also *1.2*, p. *15*; 22.8, p. 485.

E. L. ALBUQUERQUE and N. C. CHAO
*On the Propagation of Elastic Surface Waves in Piezoelectric Crystals
(b) Vol. **104**, No. 1, p. K11–K14 (1981).
See also 7, p. 82.

B. R. SOOD
*On the Possibility of Having Phasers
(a) Vol. **64**, No. 1, p. 81–84 (1981).
See also *20.2*, p. *322*.

V. VALVODA
X-Ray Diffraction Study of Debye Temperature and Charge Distribution in Tantalum Monocarbide
(a) Vol. **64**, No. 1, p. 133–142 (1981).
See also *1.3*, p. *23* ; 21, p. 339.

Y. C. CHEN, SHI-YU WU, and J. J. SINAI
*Excitation Spectra of Disordered Systems (I): General Theory
(b) Vol. **104**, No. 2, p. 441–452 (1981).
See also *2*, p. *43*.

B. A. LUKIYANETS and K. K. DANILEVICH-TOVSTYUK
*The Influence of Layer Crystal Plate Thickness on Structural Phase Transitions
(b) Vol. **104**, No. 2, p. 485–489 (1981).
See also *1.2*, p. *16*.

E. S. KISELYOVA
*The Phenomenon of Self-Induced Transparency in the Range of the Proper Lattice Absorption
(b) Vol. **104**, No. 2, p. 497–505 (1981).
See also *20.1*, p. *306*.

R. TARANKO, E. TARANKO, and K. I. WYSOKIŃSKI
Effect of Electron–Phonon Interaction on the Optical Conductivity of Alloys
(b) Vol. **104**, No. 2, p. 507–512 (1981).
See also 13, p. 144; *20.1*, p. *306*; 21, p. 340.

E. JAHNE, P. KLEINERT, B. KH. BAIRAMOV, and V. V. TOPOROV
*Impurity Vibrational Properties of Ga in InP Crystals
(b) Vol. **104**, No. 2, p. 531–539 (1981).
See also 10.2, p. 109; *20.1*, p. *307*; 22.2.2, p. 432.

F. BROUERS and O. L. T. DE MENEZES
*Electron–Phonon Interaction in Mixed Valence Systems
(b) Vol. **104**, No. 2, p. 541–546 (1981).
See also *13*, p. *144*; 21.4, p. 381.

E. ANASTASSAKIS and M. CARDONA
*Internal Strains Raman-Active Optical Phonons
(b) Vol. **104**, No. 2, p. 589–600 (1981).
See also *12.2*, p. *139*; 22.1, p. 402; 22.5.3, p. 471; 22.6, p. 475.

B. Borzecka, S. F. Sagnowski, and S. Hodorowicz
Proton Magnetic Resonance Study on Molecular Dynamics in $[Zn(H_2O)_6](ClO_4)_2$
(a) Vol. **64**, No. 2, p. 557–564 (1981).
See also *19*, p. *289*; 22.8, p. 486.

W. D. Wagner and R. Claus
*Phase Matching for Difference Frequency Generation in the View of Spontaneous Raman Scattering by Hot Polaritons
(a) Vol. **64**, No. 2 p. 647–656 (1981).
See also *20.1*, p. *308*; 22.2.2, p. 432; 22.8.1, p. 496.

P. D. Pathak and R. J. Desai
Thermal Properties of Some H.C.P. Metals (II): Debye-Waller Factors of Zinc, Cadmium, and Magnesium
(a) Vol. **64**, No. 2, p. 741–745 (1981).
See also *8*, p. *85*; 21, p. 341; 21.1, p. 357; 21.3, p. 378.

S. K. Sarkar, D. Sen, and S. Sengupta
*A Study in the Homology of Lattice Mechanical Properties of Alkali Metals
(b) Vol. **105**, No. 1, p. 99–105 (1981).
See also 21.2, p. 376.

K. Masuda
*Lattice Vibrations around a Vacancy in B.C.C. Transition Metals
(b) Vol. **105**, No. 1, p. 107–113 (1981).
See also *10.1*, p. *100*; 21, p. 341; 21.1.1, p. 371.

K. K. P. Srivastava and H. C. Singh
*Paramagnetic Spin Relaxation Due to Spin–Spin Interaction Assisted by Spin–Phonon Interaction or Vice Versa (II)
(b) Vol. **105**, No. 1, p. 115–120 (1981).
See also 13, p. 144; *18.1*, p. *265*.

D. Fekete
A Systematic Application of Gyarmati's Wave Theory of Thermodynamics to Thermal Waves in Solids
(b) Vol. **105**, No. 1, p. 161–174 (1981).
See also *8*, p. *85*.

S. P. Sanyal and R. K. Singh
Study of Anharmonic Properties of Alkaline Earth Oxides
(b) Vol. **105**, No. 1, p. 193–199 (1981).
See also *4*, p. *58*; *12.2*, p. *140*; 22.6, p. 476.

K. Kambas, J. Spyridelis, and M. Balkanski
*Far Infrared and Raman Optical Study of α- and β-In_2S_3 Compounds
(b) Vol. **105**, No. 1, p. 291–296 (1981).
See also *20.1*, p. *309*; 22.3, p. 441.

V. S. Gorelik, B. S. Umarov, and L. G. Reznik
*The Peculiarity of the A_1-Type Polariton in Lithium-Niobate Crystal
(b) Vol. **105**, No. 2, p. K73–K75 (1981).
See also *20.1*, p. *311*; 22.8.1, p. 497.

V. V. Artamonov, M. Ya. Valakh, and M. P. Lisitsa
*Interplayer Interaction in Phonon Spectra of ε-GaSe Crystals
(b) Vol. **105**, No. 2, p. K103–K106 (1981).
See also *20.1*, p. *311*; 22.3, p. 441.

P. Bussemer and A. Taubert
*To the Asymmetry of the Absorption Spectrum at Localized Centres
(b) Vol. **105**, No. 2, p. K107–K110 (1981).
See also *13.4*, p. *177*; *20.1*, p. *311*.

I. V. Bodnar, L. V. Golubev, V. G. Plotnichenko, and E. A. Smolyaninova
*Raman Scattering in $CuGaSe_2$
(b) Vol. **105**, No. 2, p. K111–K114 (1981).
See also *20.1*, p. *311*; 22.8, p. 487.

K. Masuda
*On the Vibrational Spectra for Atoms at B.C.C. Transition Metal Surfaces: Effects of Surface Relaxation and Reconstruction
(b) Vol. **105**, No. 2, p. K119–K124 (1981).
See also *1.6*, p. *38*; 21.1, p. 359.

V. P. Dudkevich, V. A. Bukreev, Vl. M. Mukhortov, Yu. I. Golovko, Yu. G. Sindeev, V. M Mukhortov, and E. G. Fesenko
Internal Size Effect in Condensed $BaTiO_3$ Ferroelectric Films
(a) Vol. **65**, No. 2, p. 463–467 (1981).
See also *14.4.1*, p. *244*; 22.8.1, p. 497.

K. Bohmhammel, P. Deus, and H. A. Schneider
*Specific Heat, Debye Temperature, and Related Properties of Compound Semiconductors $A^{II}B^{IV}C_2^{V}$
(a) Vol. **65**, No. 2, p. 563–569 (1981).
See also *8*, p. *85*; 22.8, p. 487.

K. Terao and M. Shimizu
Phase Transitions in Iron Alloys
(a) Vol. **65**, No. 2, p. K153–K156 (1981).
See also *1.2*, p. *17*; 8, p. 86; 21.1.1, p. 372.

G. Antonioli, D. Bianchi, and P. Franzosi
*Intrinsic Urbach Rule and Electron–Phonon Interaction in GaAs and Related III–V Compounds
(b) Vol. **106**, No. 1, p. 79–84 (1981).
See also *20.1*, p. *312*; 22.2, p. 426; 22.2.1, p. 429.

D. Jumeau
*Studies by UV and IR Absorption of Particular Motions of NO_2^- in Single Crystals of $NaNO_2$
(b) Vol. **106**, No. 1, p. 149–156 (1981).
See also 14.4.1, p. 244; *20.1*, p. *312*; 22.8.1, p. 498.

R. Ramji Rao and A. Rajput
*Lattice Dynamics and Some Related Thermal Properties of Body-Centred Cubic Metals
(b) Vol. **106**, No. 1, p. 393–406 (1981).
See also 12.1, p. 136; 21, p. 344; 21.1.1, p. 372; 21.2, p. 376.

M. M. Beg, X. Mahmood, N. Ahmad, J. Aslam, Q. H. Khan, and N. M. Butt
*Concentration Dependence of the Debye Temperature for the $K_xRb_{x-1}F$ System
(b) Vol. **106**, No. 1, p. K43–K46 (1981).
See also 22.5.4, p. 473.

N. M. Gasanly, B. M. Dzhavadov, A. S. Ragimov, V. I. Tagirov, and R. E. Guseinov
*Long-Wave Optical Phonons in In_6S_7 Layer Crystals
(b) Vol. **106**, No. 1, p. K47–K51 (1981).
See also *20.1*, p. *313*; 22.3, p. 442.

A. Król, W. Nazarewicz, L. Gluziński, and M. J. Kozielski
*Infrared Studies of Defects in ZnS Crystals Double-Doped with Li and Al, Ga, and In
(b) Vol. **106**, No. 2, p. 489–498 (1981).
See also 10.2, p. 114; *20.1*, p. *314*; 22.4.1, p. 449.

G. C. Barker and J. W. Tucker
*Orbit–Lattice Interaction Hamiltonian for Iron-Group Ions in Non-Cubic Sites
(b) Vol. **106**, No. 2, p. 529–536 (1981).
See also *13.4*, p. *179*; 22.6, p. 478.

M. D. Tiwari
*Specific Heat of Ar–Xe Solid Solutions at Low Temperatures
(b) Vol. **106**, No. 2, p. 569–573 (1981).
See also *8*, p. *86*; 23, p. 514.

V. D. Iskra
Phonon Assisted Interband Optical Transitions in Semiconductors Substaining a Coulomb Random Field
(b) Vol. **106**, No. 2, p. 627–633 (1981).
See also *13.1*, p. *159*; *20.1*, p. *315*; 22, p. 397.

M. J. Gelten and L. A. Bosch
*A New Method to Determine the Coupled Mode Parameters of a Plasmon–Multiphonon System
(b) Vol. **106**, No. 2, p. 635–645 (1981).
See also *13.5.2*, p. *195*; 20.1, p. 315; 22, p. 397.

S. Mochizuki and M. Satoh
Infrared Spectra of Single-Crystal NiO at High Temperatures
(b) Vol. **106**, No. 2, p. 667–674 (1981).
See also *20.1*, p. *315*; 22.6, p. 478.

P. Kleinert
*Impurity Effects on the Adatom Phonon Density of States of a Mixed Diatomic Linear Chain
(b) Vol. **106**, No. 2, p. K107–K109 (1981).
See also *10*, p. *94*.

A. P. Zhernov
Temperature Dependence of the Righi-Leduc Coefficients for Na and K Metals
(b) Vol. **107**, No. 2, p. 411–421 (1981).
See also *8*, p. *87*; *14.1*, p. *202* ; 21.2, p. 377.

D. K. Sinha and M. Gupta
Nonlinear Parametric Interaction in a Piezosemiconducting Medium
(b) Vol. **107**, No. 2, p. 469–478 (1981).
See also *7*, p. *83*; 22, p. 398.

Doan Nhat Quang, B. Esser, and R. Keiper
*On the Theory of Resonance Raman Scattering in Heavily Doped Degenerate Semiconductors
(b) Vol. **107**, No. 2, p. 535–542 (1981).
See also *20.1*, p. *317*; 22, p. 398.

D. Sen and S. K. Sarkar
*Unified Study of Metals on Pseudopotential Theory: Lead
(b) Vol. **107**, No. 2, p. 559–567 (1981).
See also 2, p. 49; 12.1, p. 136; *14.1*, p. *202*; 15, p. 249; 21, p. 346.

F. Brehat, B. Wyncke, and A. Hadni
*Far-Infrared Reflectivity Spectra of KH_2PO_4 between 300 and 400 K
(b) Vol. **107**, No. 2, p. 723–727 (1981).
See also 14.4.1, p. 245; *20.1*, p. *317*; 22.8.1, p. 500.

G. Debiais
Optical Function of a Thin Film of Mercury Sulfide
(b) Vol. **107**, No. 2, p. 729–736 (1981).
See also *20.1*, p. *317*; 22.4.1, p. 450.

F. A. Oliveira, M. G. Cottam, and D. R. Tilley
*Green Functions, Power Spectra, and Raman Scattering of Thin Film Polaritons
(b) Vol. **107**, No. 2, p. 737–747 (1981).
See also 13.5, p. 187; *20*, p. *299*; 22.5.2, p. 468.

R. K. Puri and L. R. Gupta
*Force Constant Change of Mössbauer ^{119}Sn Impurity in Copper Host
(b) Vol. **107**, No. 2, p. 757–759 (1981).
See also *5*, p. *65*; 21.1, p. 364.

L. Weiss and A. Y. Rumyantsev
*Phonon Dispersion of Cr_3Si
(b) Vol. **107**, No. 2, p. K75–K78 (1981).
See also 21.1, p. 364.

O. A. C. Nunes
*LO Phonon Instability in the Absorption of Radiation by Free Carriers in Semiconductors
(b) Vol. **107**, No. 2, p. K83–K85 (1981).
See also 20.1, p. 312.

B. D. Indu and R. P. Gairola
*Phonon Heat Conductivity of Solids
(b) Vol. **107**, No. 2, p. K99–K103 (1981).
See also *8*, p. *87*.

A. M. Aliev, M. A. Nizametdinova, and V. Ya. Steinshreiber
Lattice Dynamics and Specific Heat of Thallium Monoselenide
(b) Vol. **107**, No. 2, p. K151–K154 (1981).
See also *8*, p. *88*; 22.4.2, p. 453.

S. R. Kuindersma
*A Far-Infrared Study of the Lattice Vibrations of NiI_2 and CoI_2
(b) Vol. **107**, No. 2, p. K163–K166 (1981).
See also *20.1*, p. *318*; 22.5, p. 460.

F. El-Kabbany
A Study of the Phase Transition in the System $[Na_{1-x}(NH_4)_x]_2SO_4$ by Infrared Spectroscopy
(a) Vol. **67**, No. 2, p. 729–735 (1981).
See also *1.2*, p. *19*; 20.1, p. 318; 22.8.1, p. 500.

T. G. Kerimova, F. R. Adzalova, R. Kh. Nani, and V. Ya. Shteinshreiber
*Interaction of Free Current Carriers with Longitudinal Optical Phonons in $CdIn_2Se_4$
(a) Vol. **67**, No. 2, p. K143–K146 (1981).
See also 14.3, p. 217; *20.1*, p. *318*; 22.8, p. 490.

E. Anachkova, H. Ratajczak, and I. Savatinova
*Water Vibrations in Ferroelectric $K_4Fe(CN_6) \cdot 3\,H_2O$
(b) Vol. **108**, No. 1, p. 65–70 (1981).
See also 14.4.1, p. 246; *20.1*, p. *318*; 22.8, p. 490.

U. Nowak, W. Richter, and G. Sachs
*LO-Phonon–Plasmon Dispersion in GaAs. Hydrodynamical Theory and Experimental Results
(b) Vol. **108**, No. 1, p. 131–143 (1981).
See also *13.5.2*, p. *195*; 20.1, p. 318; 22.2.1, p. 430.

Z. V. Popović and H. J. Stolz
*Infrared and Raman Spectra of Germanium Dichalcogenides (II): $GeSe_2$
(b) Vol. **108**, No. 1, p. 153–163 (1981).
See also *20.1*, p. *319*; 22, p. 399.

I. Gorczyca
*Calculations of Deformation-Potential Constants in InSb
(b) Vol. **108**, No. 1, p. 205–209 (1981).
See also *13*, p. *149*; 22.2.3, p. 436.

T. Chattopadhyay and H. G. von Schnering
*Temperature Dependence of the Infrared Absorption Spectrum of P_4S_3
(b) Vol. **108**, No. 1, p. 241–244 (1981).
See also *20.1*, p. *319*; 22, p. 399.

K. Handrich and J. Resch
Nuclear Quadrupole Spin–Lattice Relaxation Time T_1 of Amorphous Solids
(b) Vol. **108**, No. 1, p. K57–K60 (1981).
See also *2*, p. *50*; 13.5, p. 187; *18*, p. *264*.

M. Kobayashi and I. Yokota
*Collective Motions in a Metal-Doped Molten Salt
(b) Vol. **108**, No. 2, p. 351–355 (1981).
See also *13.5.2*, p. *195*; 22.5.1, p. 461.

H. Wipf, M. V. Klein, and W. S. Williams
*Vacancy-Induced and Two-Phonon Raman Scattering in ZrC_x, NbC_x, HfC_x, and TaC_x
(b) Vol. **108**, No. 2, p. 489–500 (1981).
See also *20.1*, p. *320*; 22.3, p. 443.

G. Völkel, H.-E. Müller, and W. Flohrer
Critical Anomaly of the Electron Spin–Lattice Relaxation in X-Irradiated Rochelle Salt
(b) Vol. **108**, No. 2, p. 501–506 (1981).
See also 1.2, p. 21; 11, p. 131; *14.4.1*, p. *246*; 22.9, p. 513.

S. Asano et N. Yamashita
*Effet du champ magnétique sur la luminescence de l'ion Pb^{2+} dans les luminophores CaO, CaS, CaSe et MgS
(Effect of a Magnetic Field on the Luminescence of the Pb^{2+} Ion in CaO, CaS, CaSe, and MgS Phosphors)
(b) Vol. **108**, No. 2, p. 549–558 (1981).
See also 13.4, p. 184; 18, p. 264; *20.3*, p. *334*; 22.4.1, p. 450; 22.4.2, p. 453; 22.6, p. 480.

K. S. Sharma
*Phonon Dispersion in Barium
(b) Vol. 108, No. 2, p. K101–K104 (1981).
See also *4*, p. *61*; 21.3, p. 379.

F. Brehat, M. El Sherif, and B. Wyncke
*Overdamped Modes in Ammonium Dihydrogen Phosphate at 300 K
(b) Vol. **108**, No. 2, p. K105–K108 (1981).
See also *20.1*, p. *320*; 22.8.1, p. 501.

R. S. Singh, H. C. Gupta, and B. B. Tripathi
*Phonon Anomalies in Transition Metals
(b) Vol. **108**, No. 2, p. K113–K116 (1981).
See also 21.1, p. 366.

7. Acoustic Properties

Original Papers and Short Notes

A. Czachor and A. Holas
Propagation of Displacement Pulses in Harmonic Crystals
(b) Vol. **103**, No. 2, p. 471–475 (1981).
See also *6*, p. *67*.

E. K. Naimi
*Acoustoelectric and Acoustomagnetic Effects in Superconductors
(b) Vol. **103**, No. 2, p. K109–K113 (1981).
See also *14.2*, p. *205*.

V. D. Krevchik, R. A. Muminov, and A. Ya. Yafasov
*Influence of Ultrasound on Ionic Diffusion Process in Semiconductors
(a) Vol. **63**, No. 2, p. K159–K162 (1981).
See also *9*, p. *89*; 22.1.2, p. 410.

E. L. Albuquerque and N. C. Chao
On the Propagation of Elastic Surface Waves in Piezoelectric Crystals
(b) Vol. **104**, No. 1, p. K11–K14 (1981).
See also *6*, p. *71*.

O. A. Grigorev and T. D. Shermergor
*The Scattering of Ultrasonic Waves by Polycrystals
(a) Vol. **64**, No. 1, p. 385–394 (1981).
See also 22.8.1, p. 495.

E. L. Albuquerque
Acoustic Wave Scattering by Mixing Modes in Piezoelectric Media
(b) Vol. **104**, No. 2, p. 667–672 (1981).
See also *6*, p. *72*.

A. K. Agyei
*Theory of Second-Harmonic Generation by Soft Ultrasonic Shear Waves in Transforming A-15 Compounds
(b) Vol. **105**, No. 2, p. 469–476 (1981).
See also 6, p. 74; 21.1, p. 359.

V. Z. Zhokhov, A. A. Stepurenko, V. I. Pustovoit, and M. M. Mazur
*Some Features of Acoustoelectronic Interaction in p-Te in Magnetic Field
(a) Vol. **65**, No. 2, p. 491–496 (1981).
See also *14.3.4*, p. *234*; 18, p. 262; 22.1.3, p. 425.

G. K. Chepurnykh and S. V. Tarasenko
Sound Interactions with Nuclear Subsystem in Antiferromagnets
(b) Vol. **106**, No. 1, p. 37–43 (1981).
See also 18.4, p. 283; *19*, p. *290*.

I. M. Silvestrova, Yu. V. Pisarevskii, I. Földvári, Á. Péter, R. Voszka, and J. Janszky
*Refinement of Some Optical and Acoustic Parameters of Paratellurite
(a) Vol. **66**, No. 1, p. K55–K58 (1981).
See also *20.1*, p. *314*; 22.6, p. 478.

A. Planes, J. L. Macqueron, M. Morin, and G. Guenin
*Study of Martensitic Transformation of Cu–Zn–Al Alloy by Coupled Enthalpy and Acoustic Emission Measurements
(a) Vol. **66**, No. 2, p. 717–724 (1981).
See also *1.2*, p. *18*; 21, p. 345; 21.1, p. 362.

S. Ghosh, N. Apte, and V. K. Agarwal
*Excitation of the Hybrid Mode Due to Modulation of an Intense Laser Beam in a Transversely Magnetised Semiconductor Plasma
(b) Vol. **107**, No. 1, p. 75–81 (1981).
See also 18, p. 263; *20*, p. *298*; 22, p. 398.

G. Rowlands
*Damping of Acoustic Modes in One-Dimensional Systems
(b) Vol. **107**, No. 1, p. 157–163 (1981).
See also *6*, p. *78*.

P. Korpiun and R. Tilgner
*The Photoacoustic Effect at Phase Transitions in Thermally Thin and Thick Samples
(a) Vol. **67**, No. 1, p. 201–204 (1981).
See also *1.2*, p. *18*; 21, p. 346.

D. K. Sinha and M. Gupta
*Nonlinear Parametric Interaction in a Piezosemiconducting Medium
(b) Vol. **107**, No. 2, p. 469–478 (1981).
See also 6, p. 79; 22, p. 398.

J. Majszczyk, J. Rązcka, and Z. Czapla
*A New Phase Transition in NH_4HSeO_4 Crystals
(a) Vol. **67**, No. 2, p. K123–K125 (1981).
See also *1.2*, p. *19*; 14.4.1, p. 246; 22.8.1, p. 500.

8. Thermal Properties

Original Papers and Short Notes

M. D. TIWARI
Erratum to: Lattice Specific Heat of α-Phase Alloys Based on Copper and Silver
(b) Vol. **104**, No. 2, p. 751 (1981).
See also *6*, p. *72*; 21.1, p. 356; 21.6, p. 384.

M. STORDEUR and H. T. LANGHAMMER
The Lorenz Number for Many-Valley Semiconductors on Special Transport Conditions
(b) Vol. **104**, No. 2, p. K121–K124 (1981).
See also *14.3*, p. *211*; 22, p. 393.

P. D. PATHAK and R. J. DESAI
*Thermal Properties of Some H.C.P. Metals (II): Debye-Waller Factors of Zinc, Cadmium, and Magnesium
(a) Vol. **64**, No. 2, p. 741–745 (1981).
See also 6, p. 73; 21, p. 341; 21.1, p. 357; 21.3, p. 378.

D. FEKETE
*A Systematic Application of Gyarmati's Wave Theory of Thermodynamics to Thermal Waves in Solids
(b) Vol. **105**, No. 1, p. 161–174 (1981).
See also 6, p. 73.

C. MALINOWSKA-ADAMSKA
Thermodynamic and Transport Properties of Crystals in Anharmonic Approximation
(b) Vol. **105**, No. 1, p. 359–367 (1981).
See also *6*, p. *74*; 14.1, p. 200; 21.2, p. 376.

P. KLEINERT and R. LEIHKAUF
*On the Low-Temperature Specific Heat of Glasses
(b) Vol. **105**, No. 1, p. K67–K70 (1981).
See also *2*, p. *45*; 22.6.1, p. 481.

G. B. TAGGART
Stability Surface for the Blume-Capel Model
(b) Vol. **105**, No. 2, p. 699–705 (1981).
See also *13*, p. *145*; *18*, p. *262*.

V. B. MAGALINSKII and V. I. ZUBOV
*On the Constancy of the Configurational Entropy and Related Properties of Solids along the Melting Curve
(b) Vol. **105**, No. 2, p. K139–K142 (1981).
See also *1.2*, p. *17*; 23, p. 514.

Cz. MARUCHA and J. RAFAŁOWICZ
*The Aluminium Thermal Conductivity and Deviation from Matthiessen's Rule
(a) Vol. **65**, No. 2, p. 523–527 (1981).
See also 21, p. 343.

K. BOHMHAMMEL, P. DEUS, and H. A. SCHNEIDER
*Specific Heat, Debye Temperature, and Related Properties of Compound Semiconductors $A^{II}B^{IV}C_2^V$
(a) Vol. **65**, No. 2, p. 563–569 (1981).
See also *6*, p. *75*; 22.8, p. 487.

C. Bellabarba and S. M. Wasim
*Thermal Conductivity of $AgInTe_2$
(a) Vol. **66**, No. 2, p. K105–K108 (1981).
See also 22.8, p. 489.

E. Fischer, W. Hässler, E. Hegenbarth, and V. I. Fritsberg
*Amorphous Behaviour of PLZT Demonstrated by Thermal Conductivity Measurements
(a) Vol. **66**, No. 2, p. K169–K171 (1981).
See also *14.4.1*, p. *245*; 22.8.1, p. 499.

P. D. Pathak and R. J. Desai
*Thermal Expansion of Magnesium and Temperature Variation of Negative Second Moment of Its Frequency Spectrum
(a) Vol. **66**, No. 2, p. K179–K182 (1981).
See also 21.3, p. 378.

M. Wotcyrz, R. Kubiak, and S. Maciejewski
*X-Ray Investigation of Thermal Expansion and Atomic Thermal Vibrations of Tin, Indium, and Their Alloys
(b) Vol. **107**, No. 1, p. 245–253 (1981).
See also 1.3, p. 25; *6*, p. *78*; 21.7, p. 388.

I. S. Williams, E. S. R. Gopal, and R. Street
*A Note on the Effect of Strain upon the Heat Capacity of Gadolinium near the Curie Temperature $T_C = 293$ K
(a) Vol. **67**, No. 1, p. 83–88 (1981).
See also *10.1*, p. *102*; 21.4, p. 382.

K. J. Bachmann, F. S. L. Hsu, and J. P. Remeika
*The Low Temperature Heat Capacities of SnO_2 and In_2O_3
(a) Vol. **67**, No. 1, p. K39–K42 (1981).
See also 22.6, p. 479.

A. H. Abou El Ela, H. H. A. Labib, and N. Abdelmohsen
Electrical Conductivity, Thermoelectric Power, and Thermal Conductivity of $CuTlSe_2$ in the Solid and Liquid States
(a) Vol. **67**, No. 1, p. K43–K48 (1981).
See also *14.3*, p. *217*; 15, p. 249; 22.8, p. 490.

S. Stokka and V. Samulionis
*Specific Heat near Two Phase Transitions in $CsBi(MoO_4)_2$ Crystals
(a) Vol. **67**, No. 1, p. K89–K92 (1981).
See also 1.2, p. 19; 22.8.1, p. 500.

A. P. Zhernov
*Temperature Dependence of the Righi-Leduc Coefficients for Na and K Metals
(b) Vol. **107**, No. 2, p. 411–421 (1981).
See also 6, p. 79; *14.1*, p. *202*; 21.2, p. 377.

B. D. Indu and R. P. Gairola
*Phonon Heat Conductivity of Solids
(b) Vol. **107**, No. 2, p. K99–K103 (1981).
See also *6*, p. *80*.

9. Diffusion

Original Papers and Short Notes

B. J. Mulder and S. van Heusden
Mechanism of the Darkening of Soda–Lime Glass by a Low-Pressure Mercury Vapour Discharge
(a) Vol. **63**, No. 1, p. 137–141 (1981).
See also *2*, p. *40*; 20.1, p. 302; 22, p. 391.

M. P. Dariel, O. D. McMasters, and K. A. Gschneidner, Jr.
*Diffusion of ^{198}Au in Praseodymium Single Crystals
(a) Vol. **63**, No. 1, p. 329–333 (1981).
See also 21.4, p. 380.

I. Vávra, P. Lobotka, F. Zachar, and J. Osvald
*TEM in-Situ Observation of Electromigration Damage in Al–Cu Strips (I): Constant DC Stressing
(a) Vol. **63**, No. 1, p. 363–370 (1981).
See also 1.5, p. 34; 10.1, p. 96; *14.1*, p. *199*; 21, p. 336; 21.1, p. 352.

V. L. Aksenov and V. Yu. Yushankhai
*On Quantum Diffusion of Positive Muons in Crystals
(b) Vol. **103**, No. 2, p. K147–K150 (1981).

T. D. Dzhafarov and S. V. Khudyakov
*The Influence of Vacancies on Silver Diffusion in Gallium Phosphide
(a) Vol. **63**, No. 2, p. 431–437 (1981).
See also 10.2, p. 107; 22.2.2, p. 432.

P. Lobotka and I. Vávra
*TEM in-Situ Observation of Electromigration Damage in Al–Cu Strips (II): Superimposed AC and DC Stressing
(a) Vol. **63**, No. 2, p. 655–661 (1981).
See also 1.5, p. 34; 10.1, p. 97; *14.1*, p. *199*; 21, p. 337; 21.1, p. 354.

V. D. Krevchik, R. A. Muminov, and A. Ya. Yafasov
*Influence of Ultrasound on Ionic Diffusion Process in Semiconductors
(a) Vol. **63**, No. 2, p. K159–K162 (1981).
See also *7*, p. *82*; 22.1.2, p. 410.

H. Reuther and W. Hinz
Erratum to: Thermotransport in Lithium Silicate Glasses
(a) Vol. **63**, No. 2, p. K217 (1981).
See also *2*, p. *42*; 22.6.1, p. 481.

W. Gust, M. B. Hintz, A. Lodding, H. Odelius, and B. Predel
*Impurity Diffusion of Al in Ni Single Crystals Studied by Secondary Ion Mass Spectrometry (SIMS)
(a) Vol. **64**, No. 1, p. 187–194 (1981).
See also 21.1, p. 355.

R. Fuentes-Samaniego and J. P. Hirth
*Diffusion of Interstitial Solute in a Potential Well Characteristic of a Moving Dislocation
(b) Vol. **106**, No. 1, p. 359–371 (1981).
See also 10, p. 94.

E. M. Trukhanov and L. Zsoldos
Some Features of Generation of Misfit Dislocations in Si during B Diffusion from Doped Oxide
(a) Vol. **66**, No. 1, p. 157–164 (1981).
See also *1.5*, p. *35*; *10.2*, p. *112*; 22.1.2, p. 416.

E. Stracke and Ch. Herzig
*Formation of Slowly Diffusing Gold Defects in Thermomigration Experiments of Gold in Lead
(a) Vol. **66**, No. 1, p. 189–197 (1981).
See also *10.1*, p. *101*; 21, p. 344.

M. T. Kostyshin and Yu. V. Ushenin
*He–Ne Laser Radiation-Stimulated Diffusion of Ag into Vitreous As_2S_3
(a) Vol. **66**, No. 1, p. K47–K50 (1981).
See also 22, p. 397.

R. H. J. Fastenau and P. Penning
On the Point Defect–Dislocation Interaction in Rate Theory
(a) Vol. **66**, No. 2, p. 613–626 (1981).
See also *10*, p. *95*.

L. de Schepper, G. Knuyt, and L. M. Stals
The Influence of Small Amounts of Interstitial Solute Atoms on the Determination of the Activation Energies for Self-Diffusion and Vacancy Formation in α-Iron
(a) Vol. **67**, No. 1, p. 153–162 (1981).
See also *10.1*, p. *102*; 21.1.1, p. 374.

V. Leute, H. M. Schmidtke, W. Stratmann, and W. Winking
*Interdiffusion in Doped and Undoped Semiconductor Alloys $Hg_wCd_{1-w}Te$
(a) Vol. **67**, No. 1, p. 183–192 (1981).
See also *10.2*, p. *116*; 22.4.3, p. 455; 22.4.4, p. 457.

I. J. R. Baumvol
*Radiation-Enhanced Diffusion as a Coating Method for Protection against Corrosion and Wear
(a) Vol. **67**, No. 1, p. 287–298 (1981).
See also *1.6*, p. *39*; 11, p. 128; 21.1.1, p. 374.

H. Takahashi and K. Urban
*A Study of the Point-Defect Mobility in Ag–Cu by High-Voltage Electron Microscopy
(a) Vol. **67**, No. 1, p. 347–359 (1981).
See also *10.1*, p. *102*; 11, p. 128; 21.1, p. 363; 21.6, p. 386.

A. Priemuth
*Electromigration of Cu in V_2VI_3 Compounds
(a) Vol. **67**, No. 2, p. 505–510 (1981).
See also *1*, p. *10*; 14.3, p. 217; 22.7, p. 483.

10. Defect Properties (Irradiation Defects See 11)

Original Papers and Short Notes

A. E. Romanov and V. I. Vladimirov
*Straight Dislinations near a Free Surface (I): Stress Fields
(a) Vol. **63**, No. 1, p. 109–118 (1981).

K. Jagannadham and M. J. Marcinkowski
*Discrete Dislocation Analysis of a Plastic Crack under Repeated Loading
(a) Vol. **63**, No. 1, p. 299–304 (1981).

N. K. Gilra
*Comments: A Model of Creep in Pure Materials
(b) Vol. **103**, No. 2, p. K123–K124 (1981).

A. E. Romanov
*Straight Disclinations near a Free Surface (II): The Interaction between Wedge Disclinations and Surface
(a) Vol. **63**, No. 2, p. 383–388 (1981).

B. Dietrich, J. Härtwig, A. Christoph, and G. Hölzer
*X-Ray Diffraction Contrasts of Section Topographs by Extended Lattice Deformations
(a) Vol. **63**, No. 2, p. 511–517 (1981).
See also *1*, p. *9*.

B. Wielke
*Activation Energy and Activation Volume of Dislocation Movement through Randomly Distributed Point Obstacles
(a) Vol. **64**, No. 1, p. 121–126 (1981).

Yu. A. Tkhorik and L. S. Khazan
On the Possible Mechanism of Misfit Dislocation Generation
(a) Vol. **64**, No. 1, p. K55–K58 (1981).
See also *1.5*, p. *35*; *3*, p. *54*.

G. N. Gaidukov, S. K. Maksimov, and A. P. Filippov
*Effects of a Cloud of Impurity Atoms near a Dislocation on Electron Microscopic Contrast
(a) Vol. **64**, No. 2, p. 519–524 (1981).
See also *1.4*, p. *29*.

B. K. Jap and S. S. Sheinin
*The Validity of the Column Approximation in Image Contrast Calculations of Crystals Containing Stacking Faults
(b) Vol. **105**, No. 1, p. 137–145 (1981).
See also *1*, p. *9*.

P. Rennert
Multiple Scattering Theory in the Small Scattering Centre Approximation. Application to EXAFS, AES, XPS, MEED
(b) Vol. **105**, No. 1, p. 213–221 (1981).
See also *11*, p. *125*; 13.1, p. 155; 17, p. 257.

Yu. N. Gornostirev, B. A. Greenberg, and M. A. Ivanov
*Analysis of the Features of the Plastic Behaviour of Superstructure Ll_2 under Low Temperature Creep
(a) Vol. **66**, No. 2, p. 439–444 (1981).

R. H. J. Fastenau and P. Penning
*On the Point Defect–Dislocation Interaction in Rate Theory
(a) Vol. **66**, No. 2, p. 613–626 (1981).
See also 9, p. 91.

E. Kotomin and V. Kuzovkov
*Kinetics of Defect Accumulation and Recombination (II): Diffusion-Controlled Defect Annihilation
(b) Vol. **108**, No. 1, p. 37–44 (1981).
See also *9*, p. *92*.

M. Wilkens
*A Comment on the Kinematical Theory of Diffraction Contrast of Lattice Defects on Transmission Electron Microscope Images of Crystalline Specimens
(a) Vol. **68**, No. 1, p. 123–128 (1981).
See also *1.4*, p. *32*.

P. V. Petrashen and N. S. Yaroslavskaya
*The Effect of Dislocations on the Pendellösung Fringes in a Bent Crystal
(a) Vol. **68**, No. 1, p. K15–K19 (1981).

R. Sh. Malkovich
*On the Analysis of the Diffusion with Association
(a) Vol. **68**, No. 1, p. K37–K41 (1981).
See also *9*, p. *92*.

K. Z. Botros, R. Perez, and S. S. Sheinin
*On the Contrast Asymmetry Exhibited by Intrinsic and Extrinsic Faults Imaged under Weak-Beam Conditions
(a) Vol. **68**, No. 2, p. 351–356 (1981).

V. M. Chernov
*On the Calculation of Elastic Interaction of Dislocations with Point Defects
(a) Vol. **68**, No. 2, p. 379–384 (1981).

10.1 Metals

Review Article

Original Papers and Short Notes

A. P. Zhernov and T. A. Mamedov
*Theory of the Static Atomic Displacements in Ordinary Metals with Isoelectronic Impurity Atoms
(b) Vol. **103**, No. 2, p. 477–488 (1981).
See also *13*, p. *142*; 21, p. 337.

V. Paidar
*Local Density of Electronic States at the Core of the Screw Dislocation in B.C.C. Lattice
(b) Vol. **103**, No. 2, p. K93–K97 (1981).
See also *13.4*, p. *171*; 21, p. 337.

O. Simmich
*The Influence of the Gradient Free Energy on the Movement of a Flat Interface
(a) Vol. **63**, No. 2, p. 539–545 (1981).
See also *1.3*, p. *22*; 21, p. 337.

E. Bonetti, E. Evangelista, and P. Gondi
*Experimental Observations on Dislocation Link Distribution and Supersaturation during Creep of Al
(a) Vol. **63**, No. 2, p. 645–653 (1981).
See also 12.1, p. 134; 21, p. 337.

P. Lobotka and I. Vávra
TEM in-Situ Observation of Electromigration Damage in Al–Cu Strips (II): Superimposed AC and DC Stressing
(a) Vol. **63**, No. 2, p. 655–661 (1981).
See also 1.5, p. 34; *9*, p. *89*; *14.1*, p. *199*; 21, p. 337; 21.1, p. 354.

Lu. Ganghua and Shen GuangJun
*Identification of a $\frac{1}{6}\langle 310\rangle$ Stair-Rod Dislocation
(a) Vol. **63**, No. 2, p. 705–710 (1981).
See also *1.4*, p. *28*; 21.1, p. 354.

I. Stulíková, V. P. Soldatov, and V. I. Startsev
*Temperature Dependence of Creep of Cd–Ag Single Crystals between 1.55 and 45 K
(a) Vol. **63**, No. 2, p. 717–722 (1981).
See also 21, p. 338; 21.6, p. 384.

A. J. Pedraza and D. Fainstein-Pedraza
*Analytic Approximations for Describing the Irradiation Growth of Zirconium-Base Alloys (I)
(a) Vol. **63**, No. 2, p. 723–735 (1981).
See also *11*, p. *122*; 21, p. 338.

K. Iwasaki
*Determination of the Activation Energies of the Hasiguti Peaks in Pure Copper with the Use of the Differentiated Period of Vibration
(a) Vol. **63**, No. 2, p. K137–K141 (1981).
See also *12.1*, p. *134*; 21.1, p. 354.

F. M. Mansy, N. K. Gobran, and G. Said
*On the Effect of Phase Transformation on the Steady State Creep of Cobalt
(a) Vol. **64**, No. 1, p. K81–K83 (1981).
See also *1.2*, p. *16*; 21.1, p. 356.

P. Mascher, L. Breitenhuber, and W. Puff
*Vacancy Formation Enthalpy in Cadmium by Positron Lifetime Measurements
(b) Vol. **104**, No. 2, p. 601–605 (1981).
See also 21, p. 340.

R. D. Barnard
Some Remarks on Deviations from Matthiessen's Rule in Dilute Alloy Systems
(b) Vol. **104**, No. 2, p. 613–620 (1981).
See also 6, p. 72; 13.2, p. 163; *14.1*, p. *200*; 15, p. 248; 21.6, p. 384.

H. Kronmüller and W. Fernengel
The Role of Internal Stresses in Amorphous Ferromagnetic Alloys
(a) Vol. **64**, No. 2, p. 593–602 (1981).
See also 2, p. 44; 12.1, p. 135; *18.2*, p. *270*; 21.1.1, p. 370.

H.-J. Burmeister and C. Holste
*Change of Activation Area during Cyclic Deformation (II): Quantitative Interpretation with a Model of Heterogeneous Plastic Deformation
(a) Vol. **64**, No. 2, p. 611–624 (1981).
See also 21.1, p. 357.

D. Lepski and P. Burck
Precipitate Structure Analysis by Diffraction Contrast Image Matching for a Metastable Semicoherent Carbide Phase in Molybdenum (I): Generating Inelastic Strain
(a) Vol. **64**, No. 2, p. 625–636 (1981).
See also *1.2*. p. *16*; *1.4*, p. *29*; 12.1, p. 135; 21, p. 340.

V. V. Kirsanov, Yu. S. Pyatiletov, and O. G. Tyupkina
Irradiation Creep Due to Dislocation Glide
(a) Vol. **64**, No. 2, p. 735–740 (1981).
See also *11*, p. *124*; *12.1*, p. *135*; 21, p. 340.

J. H. Basson and H. Booyens
*The Effect of Substrate Relaxation on the Introduction of Misfit Dislocations in Misfitting Epitaxial Layers
(a) Vol. **64**, No. 2, p. 777–786 (1981).
See also 1.5, p. 35; *12.1*, p. *135*; 22.2.1, p. 428; 22.2.2, p. 433; 22.2.4, p. 437.

V. L. Arbuzov, S. M. Klotsman, and I. E. Podchinenov
*Pseudocorrelated Recombination of Interstitials during State II of Annealing of Irradiated Metals
(a) Vol. **64**, No. 2, p. K111–K114 (1981).
See also *11*, p. *124*; 21, p. 341.

A. Vehanen, P. Hautojärvi, and V. S. Mikhalenkov
*Vacancy Clusters in Deformed CuAs Alloy
(a) Vol. **64**, No. 2, p. K145–K148 (1981).
See also 12.1, p. 135; 21.1, p. 357.

H. C. Eaton
*Grain Boundary Dislocation Observations Using the Field Ion Microscope
(a) Vol. **65**, No. 2, p. 497–502 (1981).
See also *17.1*, p. *259*; 21.1, p. 359.

V. D. Natsik and H.-J. Kaufmann
*Specification of the Dislocation Mobility Analysis in the Low-Temperature Range
(a) Vol. **65**, No. 2, p. 571–581 (1981).
See also 21.1, p. 360.

I. Vávra and P. Lobotka
*TEM In-Situ Observation of Electromigration in Al Stripes with Quasi-Bamboo Structure
(a) Vol. **65**, No. 2, p. K107–K108 (1981).
See also *9*, p. *90*; 21, p. 343.

J. Mimault, J. Delafond and J. Grilhé
Comments on the Electric Resistivity and Spinodal Composition of Al–Zn Alloys
(a) Vol. **65**, No. 2, p. K121–K125 (1981).
See also *14.1*, p. *201*; 21.1, p. 360.

H. Löffler and C. Radomský
Some Remarks to the Paper "Comments on the Electric Resistivity and Spinodal Composition of Al–Zn Alloys"
(a) Vol. **65**, No. 2, p. K127–K129 (1981).
See also *14.1*, p. *201*; 21.1, p. 360.

G. Dlubek, O. Brümmer, J. Yli-Kauppila, and J. Johansson
*Recovery of Quenched-in Vacancies in Fe–Al Alloys Studied by Positron Annihilation
(b) Vol. **106**, No. 1, p. K83–K88 (1981).
See also 21.1.1, p. 372.

E. Stracke and Ch. Herzig
*Formation of Slowly Diffusing Gold Defects in Thermomigration Experiments of Gold in Lead
(a) Vol. **66**, No. 1, p. 189–197 (1981).
See also *9*, p. *91*; 21, p. 344.

G. Herbst and H. Kronmüller
*Magnetic After-Effect of Hydrogen in Gadolinium
(a) Vol. **66**, No. 1, p. 255–260 (1981).
See also *18.2*, p. *272*; 21.4, p. 381.

M. A. Semary, S. A. Mahmoud, and F. Mohsen
*Interactions of Domain Walls and Lattice Defects in Ni–5% Mn Alloy
(a) Vol. **66**, No. 1, p. 311–315 (1981).
See also *18.2*, p. *272*; 21.1, p. 360.

Y. Umakoshi and H. Kronmüller
*Observation of Dislocations in Deformed Iron Single Crystals by Means of High-Field Susceptibility Measurements
(a) Vol. **66**, No. 2, p. 509–520 (1981).
See also 18.2, p. 273; 21.1.1, p. 373.

B. Gueffroy and H. Löffler
*On the Reversion Behaviour of an Al–Zn(4.5 at%)–Mg(2.0 at%) Alloy
(a) Vol. **66**, No. 2, p. 585–590 (1981).
See also 21, p. 345; 21.1, p. 362; 21.3, p. 378.

G. N. Gaidukov and J. P. Hirth
*The Influence of Thermodynamic Solution Properties on Solute Atmospheres at Dislocations
(a) Vol. **66**, No. 2, p. 601–606 (1981).
See also *1.3*, p. *25*; 8, p. 86; 21.1.1, p. 373.

T. Ishimasa, Y. Kitano, and Y. Komura
*Faulted Structure in the Fe–Cr σ-Phase
(a) Vol. **66**, No. 2, p. 703–715 (1981).
See also *1.3*, p. *25*; 21.1, p. 362; 21.1.1, p. 374.

R. Riesenberg
*Shear Stress and Adhesion of Film Areas of Various Sizes and Shapes
(a) Vol. **66**, No. 2, p. K93–K96 (1981).
See also 1.5, p. 36; 21.1, p. 362.

S. Ikeno
*Dislocation Behaviour in Niobium Single Crystals with [110] and [100] Tensile Axis at Low Temperatures
(a) Vol. **66**, No. 2, p. K97–K100 (1981).
See also 21, p. 345.

S. P. Mhaskar and P. K. Naik
*Helium Desorption from a Tungsten Single Crystal
(b) Vol. **107**, No. 1, p. 99–103 (1981).
See also *1.6*, p. *39*; 21, p. 345.

I. S. Williams, E. S. R. Gopal, and R. Street
*A Note on the Effect of Strain upon the Heat Capacity of Gadolinium near the Curie Temperature $T_C = 293$ K
(a) Vol. **67**, No. 1, p. 83–88 (1981).
See also *8*, p. *87*; 21.4, p. 382.

L. de Schepper, G. Knuyt, and L. M. Stals
*The Influence of Small Amounts of Interstitial Solute Atoms on the Determination of the Activation Energies for Self-Diffusion and Vacancy Formation in α-Iron
(a) Vol. **67**, No. 1, p. 153–162 (1981).
See also 9, p. 91; 21.1.1, p. 374.

H. Takahashi and K. Urban
*A Study of the Point-Defect Mobility in Ag–Cu by High-Voltage Electron Microscopy
(a) Vol. **67**, No. 1, p. 347–359 (1981).
See also *9*, p. *91*; 11, p. 128; 21.1, p. 363; 21.6, p. 386.

A. M. Eleiche
*Dynamic Strain Ageing in 1100-0 Aluminium Tested Dynamically in Torsion
(a) Vol. **67**, No. 1, p. K63–K66 (1981).
See also 21, p. 346.

E. B. Leiko and A. Luft
*A Combined Study of the Dynamical Properties of Dislocations in Molybdenum Single Crystals by Etch Pitting and Interference Contrast Techniques
(a) Vol. **67**, No. 2, p. 435–442 (1981).
See also 21, p. 347.

R. Danielou, J. N. Daou, E. Ligeon, and P. Vajda
*Lattice Location of Deuterium in Lutetium
(a) Vol. **67**, No. 2, p. 453–460 (1981).
See also 11, p. 130; 21.4, p. 382.

R. Franke, W. Kleinert, and W. Schmidt
*Vacancy Behaviour in the Course of One Loading Cycle in Push–Pull Deformed Polycrystalline Nickel
(a) Vol. **67**, No. 2, p. 469–476 (1981).
See also *14.1*, p. *202*; 21.1, p. 364.

F. de Lima and W. Benoit
*Internal Friction Study of Dislocation Mobility in Deformed Niobium
(a) Vol. **67**, No. 2, p. 565–572 (1981).
See also *12.1*, p. *137*; 21, p. 347.

R. Vetter, R. H. J. Fastenau, and M. I. Baskes
*Atomistic Computer Calculation of the Dilatation Caused by a $\frac{1}{2}\langle 111\rangle\{110\}$ Edge Dislocation in Molybdenum
(a) Vol. **67**, No. 2, p. 585–589 (1981).
See also 21, p. 347.

I. Gorgas, P. Herke, and G. Schoeck
*The Plastic Behaviour of Lithium Single Crystals
(a) Vol. **67**, No. 2, p. 617–623 (1981).
See also 21.2, p. 377.

T. Lepistö, J. Yli-Kauppila, P. Kettunen, and P. Hautojärvi
*Voids in Fatigued Copper Single Crystals
(a) Vol. **67**, No. 2, p. K93–K97 (1981).
See also *1.4*, p. *31*; 21.1, p. 365.

R. Bonnet, G. Marcon, and M. Dupeux
*Comment on "An Approach to the Anisotropic Elastic Fields Related to a Semi-Coherent or Coherent Faceted Rod"
(a) Vol. **67**, No. 2, p. K137–K142 (1981).
See also *12.1*, p. *137*; 21, p. 347.

P. K. Acharyya, A. K. Tewari, S. K. Roy, and D. L. Bhattacharya
Effect of Anharmonicity on the Second-Order Doppler Shift for a Vacancy-Trapped Mössbauer Impurity Atom in Metallic Solid Solution
(b) Vol. **108**, No. 1, p. 211–220 (1981).
See also *5*, p. *65*; 21, p. 347.

A. Korner and H. P. Karnthaler
*Analysis of the Stacking Faults of Dissociated Glide Dislocations in F.C.C. Metals
(a) Vol. **68**, No. 1, p. 19–28 (1981).
See also 21.1, p. 365.

M. Tokuda, J. Kratochvíl, and Y. Ohashi
*Mechanism of Induced Plastic Anisotropy on Polycrystalline Metals
(a) Vol. 68, No. 2, p. 629–636 (1981).
See also *1.4*, p. *32*; 12.1, p. 137; 21, p. 349.

V. Damodara Das and N. Jayaprakash
Initial Lattice Distortion Energy Spectra of Bismuth Thin Films and the Effect of Deposition Rate
(a) Vol. 68, No. 2, p. K159–K164 (1981).
See also *13.4*, p. *185*; *14.1*, p. *204*; 21.7, p. 389.

N. M. Nancheva, T. D. Troev, G. S. Popov, and M. R. Minev
*Investigation of Shock Loaded Aluminium by Positrons
(a) Vol. 68, No. 2, p. K195–K198 (1981).
See also 21, p. 350.

10.2 Non-Metals

Review Article

M. Narayana, V. S. Sivasankar, and S. Radhakrishna
Optical and Magnetic Resonance Studies of Transition Metal Ions and Complexes in Alkali Halides
(b) Vol. **105**, No. 1, p. 11–45 (1981).
See also 13.4, p. 168; *19*, p. *286*; *20.1*, p. *301*; 22.5.2, p. 462.

Original Papers and Short Notes

S. B. S. Sastry and S. Sapru
Thermoluminescence, Fluorescence, and Optical Absorption Studies on Europium-Doped Rubidium Chloride
(b) Vol. **103**, No. 1, p. 185–192 (1981).
See also *20.1*, p. *301*; *20.3*, p. *323*; 22.5.2, p. 462.

H. J. Jodl and F. Bolduan
*Laser Induced Fluorescence and Raman Studies with KI:NO_2^-
(b) Vol. **103**, No. 1, p. 297–310 (1981).
See also *6*, p. *66*; 20.1, p. 302; 20.3, p. 323; 22.5.2, p. 462.

S. R. Bakhchieva, M. G. Kekua, and P. P. Seregin
A Study of Tin Impurity Atoms in Silicon, Germanium, and Silicon–Germanium Solid Solutions by Means of Mössbauer Spectroscopy
(a) Vol. **63**, No. 1, p. 23–30 (1981).
See also *5*, p. *62*; 22.1.1, p. 404; 22.1.2, p. 408.

A. Cavallini, P. Gondi, and A. Castaldini
*Shallow Dislocation States in Ge. Thermal Treatment Effects
(a) Vol. **63**, No. 1, p. 143–149 (1981).
See also *13.4*, p. *169*; 14.3, p. 209; 22.1.1, p. 404.

E. Yu. Gutmanas and P. Haasen
*Thermal Activation Analysis of Dislocation Motion in CdTe
(a) Vol. **63**, No. 1, p. 193–202 (1981).
See also 22.4.3, p. 454.

J. Morawiec and E. Mizera
X-Ray and TEM Investigations of PbTe Crystal
(a) Vol. **63**, No. 1, p. K1–K2 (1981).
See also *1.1*, p. *11*; 22, p. 391.

A. M. Dyachenko, G. V. Berenshtein, and I. E. Maronchuk
*The Internal Stresses in Semiconductors Induced by Microinclusions
(a) Vol. **63**, No. 1, p. K27–K30 (1981).
See also *1.4*, p. *28*; 22.2.1, p. 427; 22.2.4, p. 437.

L. B. Zuev and V. P. Sergeev
*Increase of Dislocation Paths in NaCl:Sr Crystals by Electric Pulses
(a) Vol. **63**, No. 1, p. K119–K121 (1981).
See also 22.5.2, p. 462.

R. S. Lee, J. F. Merklin, C. D. Marrs, and M. H. Richter
Impurity Photoconductivity of γ-Irradiated LiF
(b) Vol. **103**, No. 2, p. 605–611 (1981).
See also *16*, p. *251*; *20.1*, p. *303*; 22.5.2, p. 463.

B. V. R. Chowdari and S. Selvasekarapandian
*Thermoluminescence of Ag^+-Doped CsBr Crystals
(b) Vol. **103**, No. 2, p. 623–628 (1981).
See also *20.3*, p. *324*; 22.5.2, p. 463.

G. Margaritondo, A. D. Katnani, and F. Lévy
*Electronic States of $CdIn_2Se_4$ and $ZnIn_2Se_4$: Role of the Cation Pseudovacancies
(b) Vol. **103**, No. 2, p. 725–731 (1981).
See also *13.1*, p. *153*; 17, p. 256; 22.8, p. 484.

T. Suzuki, W. Skrotzki, and P. Haasen
*Intrinsic Glide Resistance on {100} Planes in Alkali Halides
(b) Vol. **103**, No. 2, p. 763–768 (1981).
See also 22.5.2, p. 463.

A. P. Loghinov, I. D. Mikhailov, and I. V. Kumpanenko
Localization and Mixing of Vibrations in Defect Polymer Solids
(b) Vol. **103**, No. 2, p. K151–K155 (1981).
See also *6*, p. *68*; 22.9, p. 508.

P. L. Kukk, H. A. Aarna, and M. P. Voogne
*High-Temperature Point Defects in CdS:Cu
(a) Vol. **63**, No. 2, p. 389–399 (1981).
See also *14.3*, p. *210*; 22.4.1, p. 447.

T. D. Dzhafarov and S. V. Khudyakov
The Influence of Vacancies on Silver Diffusion in Gallium Phosphide
(a) Vol. **63**, No. 2, p. 431–437 (1981).
See also *9*, p. *89*; 22.2.2, p. 432.

A. I. Kolyubakin and S. A. Shevchenko
*On the Spectrum of Dislocation States in Germanium
(a) Vol. **63**, No. 2, p. 677–685 (1981).
See also *13.4*, p. *171*; 14.3, p. 210; 22.1.1, p. 404.

E. R. Vance, D. K. Agrawal, and J. G. Pepin
*Powder X-Ray Diffraction Study of Halide Sodalites
(a) Vol. **63**, No. 2, p. K189–K193 (1981).
See also *1.1*, p. *11*; 22.8, p. 485.

K. Sierański and J. Szatkowski
Theory of the Free-Carrier Absorption in Zincblende Narrow-Gap Semiconductors
(b) Vol. **104**, No. 1, p. 57–67 (1981).
See also *20.1*, p. *304*; 22.2.4, p. 437; 22.4.4, p. 456.

P. Sridaran, R. C. Bhuniya, and V. V. Ratnam
*Optical and Thermal Stability of Ag Centres in RT X-Irradiated KCl:Ag Crystals
(b) Vol. **104**, No. 2, p. 459–468 (1981).
See also 20.1, p. 306; 20.3, p. 326; 22.5.2, p. 464.

E. Jahne, P. Kleinert, B. Kh. Bairamov, and V. V. Toporov
Impurity Vibrational Properties of Ga in InP Crystals
(b) Vol. **104**, No. 2, p. 531–539 (1981).
See also *6*, p. *71*; *20.1*, p. *307*; 22.2.2, p. 432.

P. A. Kulis, M. J. Springis, I. A. Tale, V. S. Vainer, and J. A. Valbis
*Impurity-Associated Colour Centres in Mg- and Ca-Doped Al_2O_3 Single Crystals
(b) Vol. **104**, No. 2, p. 719–725 (1981).
See also 19, p. 288; 20.3, p. 326; 22.6, p. 476.

G. R. Augst and L. G. Bakhshetsyan
*Calculation of the Heat of Solution in Alkali Halide Crystals by the Lattice Static Method
(b) Vol. **104**, No. 2, p. K85–K88 (1981).
See also 22.5.2, p. 464.

S. Nakamura, K. Inabe, N. Takeuchi, and K.-J. Kos
*Effect of X-Irradiation on the Aggregation of I–V Dipoles in LiF:Mg Crystals
(b) Vol. **104**, No. 2, p. K139–K141 (1981).
See also *11*, p. *124*; 22.5.2, p. 464.

J. Provoost and G. Jacobs
*On the Formation Time of Absorption Centres in KI and KBr with a Stimulated Raman Probe Beam
(b) Vol. **104**, No. 2, p. K149–K154 (1981).
See also 13.5.1, p. 189; 20.1, p. 308; 22.5.2, p. 464.

U. Puri
*Variation of the Schottky Defect Formation Energy with Pressure
(b) Vol. **104**, No. 2, p. K165–K167 (1981).
See also 22.5.2, p. 464.

E. M. Raspopova, M. Závětová, J. Krištofik, and B. T. Kolomiets
The Influence of Defects Generated during Densification on the Properties of the As–Se System
(a) Vol. **64**, No. 2, p. 457–460 (1981).
See also *2*, p. *44*; 14.4, p. 238; *20.1*, p. *308*; 22, p. 393.

A. Turos, Hj. Matzke, and P. Rabette
*The Thermal Behavior of Pt/Al_2O_3 Catalysts Studied with the Channeling Technique
(a) Vol. **64**, No. 2, p. 565–575 (1981).
See also *11*, p. *124*; 22.6, p. 476.

K. Kojima
Erratum to: Slip Systems in Deformed Anthracene Crystals
(a) Vol. **64**, No. 2, p. 797 (1981).
See also 22.9, p. 509.

R. Rauch and E. Liebold
*VUV Absorption Spectra of Oxygenated CaF_2 Crystals
(a) Vol. **64**, No. 2, p. K165–K168 (1981).
See also *20.1*, p. *309*; 22.5.3, p. 471.

L. M. Brescansin and A. Fazzio
Theoretical Study of Single Vacancies in GaSb
(b) Vol. **105**, No. 1, p. 339–345 (1981).
See also *13.4*, p. *175*; 22.2.3, p. 435.

B. V. R. Chowdari and S. Selvasekarapandian
Thermoluminescence of Heavy Metal Ion Doped Cesium Halides X-Irradiated at Low Temperature
(b) Vol. **105**, No. 1, p. 377–383 (1981).
See also *20.3*, p. *327*; 22.5.2, p. 464.

V. Seeman, S. Reifman, and T. Lehto
*A New Type V Centre in SrO
(b) Vol. **105**, No. 1, p. 409–412 (1981).
See also *19*, p. *289*; 22.6, p. 476.

R. Jayaprakash and V. V. Ratnam
*On the Thermoluminescence of Sb-Doped Calcium Fluorophosphate
(a) Vol. **65**, No. 1, p. 147–152 (1981).
See also *20.3*, p. *327*; 22.8.1, p. 496.

B. Jenichen and R. Köhler
Simulated X-Ray Contrast of Striations for Plane Waves in the Laue Case
(a) Vol. **65**, No. 1, p. 245–252 (1981).
See also *1.4*, p. *29*; 22.1.2, p. 413.

V. M. Orera
*The Effect of Lithium Impurity on Flow Stress in MgO
(a) Vol. **65**, No. 1, p. 293–300 (1981).
See also 12.2, p. 140; 22.6, p. 477.

H. Lemke
Properties of Doping Levels of Dislocations in Silicon
(a) Vol. **65**, No. 1, p. 389–401 (1981).
See also *13.4*, p. *176*; 22.1.2, p. 414.

V. D. Akhmetov, I. S. Bzinkovskaya, V. V. Bolotov, and R. N. Lovyagin
*Depth Distribution of Growth Defects in GaP Films on Sapphire Synthesized in Vacuum
(a) Vol. **65**, No. 1, p. K9–K12 (1981).
See also 3, p. 54; *20.1*, p. *310*; 22.2.2, p. 433.

N. Takeuchi, K. Inabe, and S. Nakamura
*Post-Irradiation Annealing Effect on the Growth of the Z_3 Band in LiF:Mg Crystal
(a) Vol. **65**, No. 1, p. K33–K36 (1981).
See also 11, p. 125; 13.4, p. 176; *20.1*, p. *310*; 22.5.2, p. 464.

M. Sharon, R. R. Pradhananga, and C. M. Sunta
*Thermoluminescence Glow Curves of Ca^{2+} and Sr^{2+} Doped LiCl Single Crystals
(a) Vol. **65**, No. 1, p. K67–K70 (1981).
See also *20.3*, p. *327*; 22.5.2, p. 465.

S. P. Kuila, R. K. Gartia, and V. V. Ratnam
*Emission Processes in Thermoluminescence of Li-Doped KCl Single Crystals
(a) Vol. **65**, No. 1, p. K71–K74 (1981).
See also *20.3*, p. *327*; 22.5.2, p. 465.

D. Gwinner and R. Labusch
Photoluminescence in Plastically Twisted Silicon
(a) Vol. **65**, No. 1, p. K99–K101 (1981).
See also *20.3*, p. *327*; 22.1.2, p. 414.

G. Neumann
*On the Defect Structure of Zinc-Doped Zinc Oxide
(b) Vol. **105**, No. 2, p. 605–612 (1981).
See also *19*, p. *290*; 22.4, p. 445.

R. J. Sayer, R. H. Prince, and W. W. Duley
Site Effects on N and O Luminescence in Electron-Irradiated Neon Films
(b) Vol. **105**, No. 2, p. 727–738 (1981).
See also *20.3*, p. *328*; 23, p. 514.

A. A. Predvoditelev, G. I. Nichugovskii, and V. I. Veselov
*Simulation of Dislocation Motion through a Dislocation Forest
(a) Vol. **65**, No. 2, p. 469–478 (1981).
See also *10.1*, p. *100*; 21.1, p. 359; 22.5.2, p. 465.

B. Jenichen and R. Köhler
*Experimental and Simulated X-Ray Contrast of Striations for Nearly Plane Waves in the Laue Case
(a) Vol. **65**, No. 2, p. 535–543 (1981).
See also *1.4*, p. *30*; 22.1.2, p. 415.

G. Oelgart and G. Arnold
*Impurity Profile on GaP Diodes in the Neighbourhood of a p–n Junction
(a) Vol. **65**, No. 2, p. 609–616 (1981).
See also *14.3.3*, p. *227*; 22.2.2, p. 433.

O. E. Aksenov, E. D. Aluker, D. J. Lusis, and I. P. Mezina
*Luminescence of Activated Alkali Halide Crystals at Excitation in the Fundamental Absorption Region
(a) Vol. **65**, No. 2, p. 617–622 (1981).
See also *20.3*, p. *328*; 22.5.2, p. 465.

C. Donolato
*An Analytical Model of SEM and STEM Charge Collection Images of Dislocations in Thin Semiconductor Layers (I): Minority Carrier Generation, Diffusion, and Collection
(a) Vol. **65**, No. 2, p. 649–658 (1981).
See also *14.3.2*, p. *223*; 14.3.4, p. 234; 22.1.2, p. 415.

W. SKROTZKI, G. FROMMEYER, and P. HAASEN
*Plasticity of Polycrystalline Ionic Solids
(a) Vol. **66**, No. 1, p. 219–228 (1981).
See also 22.5.1, p. 461; 22.5.2, p. 465.

H.-J. KAUFMANN, S. V. LUBENETS, and E. I. OSTAPCHUK
*Influence of Temperature on the Initial State of Plastic Deformation of LiF Crystals
(a) Vol. **66**, No. 1, p. 229–238 (1981).
See also 22.5.2, p. 466.

V. S. SHUKLA and D. M. ROWE
Precipitation of Boron from Silicon–Germanium Alloy and Its Effect in the Thermoelectric Transport Properties
(a) Vol. **66**, No. 1, p. 243–248 (1981).
See also *14.3*, p. *215*; *15*, p. *249*; 22.1.1, p. 406; 22.1.2, p. 416.

S. I. STENIN and A. K. GUTAKOVSKII
*The Plastic Deformation Kinetics for Heteroepitaxial Films during the Misfit Dislocation Generation from the Growth Surface
(a) Vol. **66**, No. 1, p. 249–253 (1981).
See also *1.5*, p. *36*; 3, p. 55; 22.1.1, p. 406.

J. REICHEL
*The Distinction of Several Different Kinds of Oxygen Donors in Silicon
(a) Vol. **66**, No. 1, p. 277–282 (1981).
See also 14.3, p. 215; 22.1.2, p. 416.

B. CHANDRA, A. R. LAKSHMANAN, and R. C. BHATT
*Effect of Deep Traps on Sensitization in LiF (TLD-100) Phosphor. A Correlation Study between Peak 7 and 10 Centres
(a) Vol. **66**, No. 1, p. 335–340 (1981).
See also *20.3*, p. *330*; 22.5.2, p. 466.

V. K. JAIN
*Z_3 Centre Thermoluminescence and Sensitization Mechanism in Lithium Fluoride
(a) Vol. **66**, No. 1, p. 341–346 (1981).
See also 20.1, p. 314; *20.3*, p. *330*; 22.5.2, p. 466.

K. ČERMÁK
*Colour Centres in YAG
(a) Vol. **66**, No. 1, p. K37–K39 (1981).
See also *20.1*, p. *314*; 22.8.2, p. 505.

E. YU. BRAILOVSKII, L. A. MATVEEVA, G. N. SEMENOVA, YU. A. TKHORIK, and L. S. KHAZAN
*Radiation-Stimulated Relaxation of Internal Stresses in Heteroepitaxial Systems
(a) Vol. **66**, No. 1, p. K59–K62 (1981).
See also *11*, p. *127*; 22.1.1, p. 406; 22.2.1, p. 430; 22.4.1, p. 449.

V. N. BAPAT and S. P. KATHURIA
Correlation of Optical Absorption and Thermoluminescence in LiF:TLD-100
(a) Vol. **66**, No. 1, p. K67–K70 (1981).
See also *20.1*, p. *314*; *20.3*, p. *330*; 22.5.2, p. 466.

L. N. Aleksandrov and M. I. Zotov
*Dynamic Characteristics of Structural Defects in Semiconductors by Internal Friction Data
(a) Vol. **66**, No. 2, p. 467–475 (1981).
See also *12.2*, p. *140*; 22.1.1, p. 406; 22.1.2, p. 417; 22.2.1, p. 430; 22.4.1, p. 449.

J. Lecomte, J. P. Loup, M. Hervieu, and B. Raveau
*Non-Stoichiometry and Electrical Conductivity of Strontium Niobates with Perovskite Structure (II): Ionic Conductivity of $Sr(Sr_{(1/3)+x}Nb_{(2/3)-x})\,O_{3-(3/2)x}$
(a) Vol. **66**, No. 2, p. 551–558 (1981).
See also *14.4*, p. *240*; 22.8.1, p. 499.

M. Kittler and W. Seifert
*On the Sensitivity of the EBIC Technique as Applied to Defect Investigations in Silicon
(a) Vol. **66**, No. 2, p. 573–583 (1981).
See also 1.4, p. 30; 22.1.2, p. 417.

E. M. Trukhanov and S. I. Stenin
*Equilibrium Dislocation Configurations in Heterostructures
(a) Vol. **66**, No. 2, p. 591–596 (1981).
See also *1.5*, p. *36*; 22.1.1, p. 406; 22.6, p. 478.

J. Tatarkiewicz and K. Wieteska
*IR Spectroscopy and X-Ray Topography Study of Annealing of Proton Bombarded Silicon
(a) Vol. **66**, No. 2, p. K101–K103 (1981).
See also *11*, p. *128*; 20.1, p. 316; 22.1.2, p. 417.

B. R. Sood
Purity of Potassium Iodide Necessary for Its Use in Phasers
(a) Vol. **66**, No. 2, p. K145–K147 (1981).
See also *6*, p. *78*; *20.2*, p. *322*; 22.5.2, p. 467.

K. Vogler
*Two-Photon Absorption of F-Centres in KCl
(b) Vol. **107**, No. 1, p. 195–199 (1981).
See also 13.4, p. 180; *20.3*, p. *332*; 22.5.2, p. 467.

H. J. Hoffmann
*Recombination Kinetics of Electrons with α-Centers in Colored Alkali Halides
(b) Vol. **107**, No. 1, p. 215–224 (1981).
See also 13.4, p. 180; *16*, p. *254*; 22.5.2, p. 467.

B. R. Bułka
*Stability of the Peierls State Due to Impurities
(b) Vol. **107**, No. 1, p. 359–365 (1981).
See also *13.4*, p. *180*; 22.9, p. 511.

K. Toriumi and N. Itoh
*Behaviour of Defects Introduced by a Nanosecond Intense Electron Pulse at High Temperatures in Alkali Halides
(b) Vol. **107**, No. 1, p. 375–386 (1981).
See also 20.1, p. 316; 22.5.2, p. 467.

P. Kukk and A. Erm
*The Structure of Radiative and Non-Radiative Recombination Centres in Activated CdS Phosphors
(a) Vol. 67, No. 2, p. 395–406 (1981).
See also 13.4, p. 182; *20.3*, p. *333*; 22.4.1, p. 450.

D. Wieluńska, L. Wieluński, and A. Turos
*The Effect of Dislocations on the Planar Dechanneling (I): Analytical Model
(a) Vol. 67, No. 2, p. 413–419 (1981).
See also *11*, p. *129*; 22.1.2, p. 420.

V. L. Vinetskii, I. R. Entinzon, and G. A. Kholodar
*Probability of Defect Formation in an Elastic Impact of Atoms in Crystals
(a) Vol. 67, No. 2, p. 477–483 (1981).
See also *11*, p. *130*; 22.1.1, p. 407; 22.1.2, p. 420.

Y. Aoyagi, Y. Segawa, S. Namba, T. Suhara, H. Nishihara, and H. Gamo
*Dynamic Behavior of the Photodarkening Process in As_2S_3 Chalcogenide Glass
(a) Vol. 67, No. 2, p. 669–676 (1981).
See also *2*, p. *50*; 20, p. 299; 22, p. 399.

H. Jain and A. S. Nowick
Electrical Conduction in Paratellurite (TeO_2) Crystals
(a) Vol. 67, No. 2, p. 701–707 (1981).
See also *14.4*, p. *240*; 22.6, p. 480.

G. A. Andreev
Strain-Rate Dependence of Vacancy Cluster Accumulation in KCl and KBr Crystals
(a) Vol. 67, No. 2, p. K127–K130 (1981).
See also 22.5.2, p. 468.

A. H. Harker and J. M. Vail
*Theoretical Analysis of Some Ground State and Optical Properties of Impurity F Centers
(b) Vol. 108, No. 1, p. 87–96 (1981).
See also *13.4*, p. *182*; 20.1, p. 318; 22.5.2, p. 468.

S. Marklund
*Energy Levels of Intrinsic and Extrinsic Stacking Faults in Silicon
(b) Vol. 108, No. 1, p. 97–102 (1981).
See also *13.4*, p. *182*; 22.1.2, p. 421.

G. Nierste and F. Fischer
*Thermally Stimulated Depolarization of NO_3^-, NO_2^-, and PO_2^- Centres
(b) Vol. 108, No. 1, p. 269–274 (1981).
See also *14.4*, p. *241*; 22.5.2, p. 468.

D. N. Popov, N. M. Nancheva, M. R. Minev, and K. I. Terzijsky
*Positron Annihilation in Z_1-Centers in Additively Coloured KCl:Ca
(b) Vol. 108, No. 1, p. K7–K9 (1981).
See also 20.1, p. 319; 22.5.2, p. 468.

E. A. Radyabov and A. I. Nepomnyachikh
*Neutral Magnesium Atoms on Anion Sites in LiF
(b) Vol. **108**, No. 1, p. K75–K78 (1981).
See also 20.3, p. 333; 22.5.2, p. 469.

D. Wieluńska, L. Wieluński, and A. Turos
*The Effect of Dislocations on the Planar Dechanneling (II): Transparency Effect
(a) Vol. **68**, No. 1, p. 45–51 (1981).
See also *11*, p. *130*; 22.1.2, p. 421.

I. E. Bondarenko, V. G. Eremenko, B. Ya. Farber, V. I. Nikitenko, and E. B. Yakimov
*On the Real Structure of Monocrystalline Silicon near Dislocation Slip Planes
(a) Vol. **68**, No. 1, p. 53–60 (1981).
See also 22.1.2, p. 421.

E. A. Radyabov and A. I. Nepomnyachikh
*Magnesium Color Centers at 3.5 and 5.0 eV in Lithium Fluoride
(a) Vol. **68**, No. 1, p. 77–81 (1981).
See also 20.1, p. 319; 20.3, p. 333; 22.5.2, p. 469.

M. Böhm, S. Herrmann, and A. Scharmann
*Dipole Centres in $CaMoO_4$ Crystals
(a) Vol. **68**, No. 1, p. 129–134 (1981).
See also *14.4*, p. *241*; 20.1, p. 319; 22.8.1, p. 500.

G. Döding and R. Labusch
Anisotropic Conductivity of CdS after Plastic Deformation and Conduction along Dislocations (I): Macroscopic Measurements
(a) Vol. **68**, No. 1, p. 143–151 (1981).
See also *14.3*, p. *218*; 22.4.1, p. 450.

H. Peibst and H. Raidt
*Nucleation of Oxygen Precipitations and Efficiency of Internal Gettering Centres in Czochralski Silicon
(a) Vol. **68**, No. 1, p. 253–260 (1981).
See also 14.3.3, p. 230; 22.1.2, p. 422.

L. A. Borisova and M. Pasemann
*The Influence of the Residual Impurities (C, O) on the Crystal Structure of GaAs
(a) Vol. **68**, No. 1, p. 261–267 (1981).
See also *1.4*, p. *32*; 13.4, p. 183; 22.2.1, p. 430.

D. Lapraz and A. Baumer
*Chloroapatite, $Ca_5(PO_4)_3Cl$: Thermoluminescent Properties
(a) Vol. **68**, No. 1, p. 309–319 (1981).
See also *20.3*, p. *334*; 22.8, p. 491.

A. Dauletbekova and A. Elango
*Conversion of F–H_A Pairs into α–I_A Pairs in KBr:Li and KBr:Na X-Rayed at 80 K
(b) Vol. **108**, No. 2, p. 299–305 (1981).
See also *20.1*, p. *320*; 22.5.2, p. 469.

A. L. Shluger and E. A. Kotomin
*Semiempirical Calculations of Defect Properties in LiF Crystals (I): Perfect Crystal
(b) Vol. **108**, No. 2, p. 673–681 (1981).
See also *13.4*, p. *184*; 22.5.2, p. 469.

N. S. Minaev, A. V. Mudrii, and V. D. Tkachev
*Symmetry and Nature of the 1.0186 eV Luminescence Centre in Neutron-Irradiated Silicon
(b) Vol. **108**, No. 2, p. K89–K94 (1981).
See also 11, p. 131; 12.2, p. 141; *20.3*, p. *334*; 22.5.2, p. 469.

S. Shimanuki and N. Nishimaki
*Formation of Interstitial Silver Centers by the Optical Destruction of Ag^- Centers in KCl: Li and KCl: Na Crystals
(b) Vol. **108**, No. 2, p. K95–K100 (1981).
See also *20.1*, p. *320*; 22.5.2, p. 470.

P. H. Leo, K. D. Moore, P. L. Jones, and F. H. Cocks
*Positron Annihilation Studies of Si, a-Si, and a-Si: H Using Doppler Broadening Techniques
(b) Vol. **108**, No. 2, p. K145–K149 (1981).
See also 13.4, p. 184; 22.1.2, p. 422.

D. van Dyck, D. Colaïtis, and S. Amelinckx
*Diffraction from Polysynthetically Twinned Structures: The Geometrical Characteristics of the Diffraction Pattern
(a) Vol. **68**, No. 2, p. 385–398 (1981).
See also 1.3, p. 26; 22.8, p. 492.

L. I. Datsenko, V. I. Khrupa, and E. N. Kislovskii
Borrmann Effect in Elastically Bent Silicon Crystals with Structural Defects
(a) Vol. **68**, No. 2, p. 399–404 (1981).
See also *12.2*, p. *141*; *20*, p. *300*; 22.1.2, p. 423.

G. Döding and R. Labusch
*Anisotropic Conductivity of CdS after Plastic Deformation and Conduction along Dislocations (II): Measurements on a Microscopic Scale
(a) Vol. **68**, No. 2, p. 469–476 (1981).
See also *14.3*, p. *219*; 22.4.1, p. 451.

H. Endert, E. Förster, K. Goetz, M. P. Malashnikov, J. A. Mikhailov, A. V. Rode, G. V. Sklizkov, and W.-D. Zimmer
*Nanosecond X-Ray Topography of Crystalline Solids with a Laser-Produced Plasma Source
(a) Vol. **68**, No. 2, p. 483–488 (1981).
See also *1.6*, p. *39*; 20.2, p. 322; 22.5.3, p. 472.

F. F. Komarov, V. D. Kuryazov, V. S. Solovev, and S. Yu. Shiryaev
*Investigation of Defect Formation in Boron Ion-Implanted Silicon during Annealing
(a) Vol. **68**, No. 2, p. 519–529 (1981).
See also *11*, p. *131*; 22.1.2, p. 423.

11. Irradiation Effects
(X-Ray Diffraction Investigations See 1 and 10)

Review Article

V. V. TITOV
*On the Kinetics of Ion-Implantation-Induced Phase Transitions
(a) Vol. **63**, No. 1, p. 11–22 (1981).
See also *1.2*, p. *14*.

Original Papers and Short Notes

S. V. PLOTNIKOV, D. E. POPOV, E. I. ROZUM, O. G. KOSTAREVA, and S. A. VOROBIEV
*Orientational Population of Quantum States for Fast Electrons in Planar Channeling
(b) Vol. **103**, No. 1, p. 81–87 (1981).
See also 13, p. 142; 22.1.2, p. 408.

F. VAN STEEN, M. VANHAELST, P. MATTHYS, E. BOESMAN, and D. SCHOEMAKER
Isotropic EPR Spectra of Vanadium in Alkali Chlorides
(b) Vol. **103**, No. 1, p. 345–351 (1981).
See also *19*, p. *286*; 22.5.2, p. 462.

J. BUCHAR and Z. BÍLEK
Plastic Flow of Irradiated Steel at High Strain Rates
(a) Vol. **63**, No. 1, p. 259–264 (1981).
See also *10.1*, p. *96*; 21.1.1, p. 368.

E. NOMURA, K. MURATA, and K. NAGAMI
*A Theoretical Estimate of the Voltage Dependence of the Pattern Resolution in Electron-Beam Lithography
(a) Vol. **63**, No. 1, p. 281–288 (1981).
See also *2*, p. *41*; 22.1.2, p. 408; 22.9, p. 508.

M. M. SHUKLA and J. MAHANTY
*Compatibility of Interionic Pair Potentials and Channeling Potentials
(b) Vol. **103**, No. 2, p. K119–K121 (1981).
See also 4, p. 57; 13, p. 143.

J. TATARKIEWICZ
*On Some Proton Radiation Effects in Silicon
(a) Vol. **63**, No. 2, p. 423–429 (1981).
See also *20.1*, p. *304*; 22.1.2, p. 409.

P. H. TSIEN, S. C. TSOU, M. TAKAI, D. RÖSCHENTHALER, M. RAMIN, H. RYSSEL, I. RUGE, and K. WITTMAACK
Annealing of Boron-Implanted Silicon Using a CW CO_2 Laser
(a) Vol. **63**, No. 2, p. 547–555 (1981).
See also *1.5*, p. *34*; *3*, p. *53*; 14.3.1, p. 220; 22.1.2, p. 409.

A. DARLIŃSKI
*Measurements of Angular Distribution of the Backscattered Electrons in the Energy Range of 5 to 30 keV
(a) Vol. **63**, No. 2, p. 663–668 (1981).
See also 21, p. 338; 21.1, p. 354; 21.6, p. 384; 22.1, p. 402.

H. J. Blythe, F. Walz, and H. Kronmüller
*Magnetic Relaxation in Electron-Irradiated Ordered Ni_3Fe
(a) Vol. **64**, No. 1, p. 169–176 (1981).
See also 10.1, p. 98; *18.2*, p. *269*; 21.1, p. 355; 21.1.1, p. 369.

T. Kawakubo
Color Centers in Irradiated Durene Crystals
(a) Vol. **64**, No. 1, p. 303–310 (1981).
See also *10.2*, p. *108*; *20.1*, p. *306*; 22.9, p. 509.

Y. Komem, A. Greenberg, and C. L. Bauer
*Recrystallization near Low-Angle [001] Tilt Boundaries Following Cobalt Implantation in Bicrystalline Thin Films of Gold
(a) Vol. **64**, No. 1, p. 317–324 (1981).
See also *1.5*, p. *35*; 21.6, p. 384.

J. P. Hauck, M. M. Beguwala, C. L. Hayes, and R. F. Palys
*Laser Annealing of Silicon Wafers at 10.6 μm
(a) Vol. **64**, No. 1, p. 407–412 (1981).
See also 14.3, p. 211; 22.1.2, p. 411.

D. P. Yadav, H. N. Acharya, and K. V. Rao
Effect of Quenching and X-Ray Irradiation on the Dielectric Properties of PbI_2 Single Crystals
(a) Vol. **64**, No. 1, p. 413–417 (1981).
See also *14.4*, p. *238*; 22.5, p. 458.

A. Zentko, M. Timko, P. Duhaj, and I. Škorvánek
*The Effect of Neutron Irradiation on the Electrical Resistivity of Amorphous $Fe_{47}Ni_{25}B_{18}Si_{10}$ Alloys
(a) Vol. **64**, No. 1, p. K19–K22 (1981).
See also 2, p. 43; *14.1*, p. *200*; 21.1.1, p. 369.

D. Hildebrandt, R. Manns, and H. Düsterhöft
*The Mean Kinetic Energy of Sputtered Atoms as a Function of the Angle of Emission
(a) Vol. **64**, No. 1, p. K27–K30 (1981).
See also 21, p. 339.

R. Klabes, R. Grötzschel, M. Voelskow, S. Panzer, and H. Bartsch
*High-Speed Electron Beam Annealing of Arsenic and Gallium Implanted Silicon
(a) Vol. **64**, No. 1, p. K73–K75 (1981).
See also 22.1.2, p. 412.

H. Sobotta, V. Riede, C. Ascheron, V. Geist, and D. Oppermann
Infrared Absorption of Hydrogen in Proton-Implanted GaP
(a) Vol. **64**, No. 1, p. K77–K80 (1981).
See also *20.1*, p. *306*; 22.2.2, p. 432.

M. I. Tarasik, V. D. Tkachev, V. U. Yavid, and A. M. Yanchenko
*Capture Mechanism of Holes by Radiation Defects with $E_c - 0.20$ eV Level in Sb-Doped Ge
(b) Vol. **104**, No. 2, p. K117–K120 (1981).
See also *13.4*, p. *173*; 22.1.1, p. 404.

S. Nakamura, K. Inabe, N. Takeuchi, and K.-J. Kos
*Effect of X-Irradiation on the Aggregation of I–V Dipoles in LiF:Mg Crystals
(b) Vol. **104**, No. 2, p. K139 – K141 (1981).
See also *10.2*, p. *109*; 22.5.2, p. 464.

V. K. Kohn, M. V. Kovalchuk, R. M. Imamov, and E. F. Lobanovich
*The Method of Integral Characteristics in X-Ray Diffraction Studies of the Structure of the Surface Layers of Single Crystals
(a) Vol. **64**, No. 2, p. 435 – 442 (1981).
See also *1.6*, p. *38*.

B. Lukasch
*Correction of the Limited Energy Resolution in RBS Spectra
(a) Vol. **64**, No. 2, p. 533 – 538 (1981).
See also 22.1.2, p. 412.

A. Turos, Hj. Matzke, and P. Rabette
*The Thermal Behavior of Pt/Al_2O_3 Catalysts Studied with the Channeling Technique
(a) Vol. **64**, No. 2, p. 565 – 575 (1981).
See also *10.2*, p. *109*; 22.6, p. 476.

T. Reichert, R. Meier-Hirmer, and H. Küpfer
Annealing Behaviour of the Critical Current Density and the Superconducting Parameters T_c, $\varkappa$, and H_{c2} in a Neutron-Irradiated V_3Si Single Crystal
(a) Vol. **64**, No. 2, p. 585 – 592 (1981).
See also *14.2*, p. *206*; 21.1, p. 357.

D. Lukášik and P. Macko
*Creation of Thermal Spikes in the GeS_{11} Glassy System by Fast Neutron Irradiation
(a) Vol. **64**, No. 2, p. 677 – 682 (1981).
See also 2, p. 44; 22, p. 394.

V. V. Kirsanov, Yu. S. Pyatiletov, and O. G. Tyupkina
*Irradiation Creep Due to Dislocation Glide
(a) Vol. **64**, No. 2, p. 735 – 740 (1981).
See also 10.1, p. 99; *12.1*, p. *135*; 21, p. 340.

V. L. Arbuzov, S. M. Klotsman, and I. E. Podchinenov
*Pseudocorrelated Recombination of Interstitials during State II of Annealing of Irradiated Metals
(a) Vol. **64**, No. 2, p. K111 – K114 (1981).
See also *10.1*, p. *99*; 21, p. 341.

A. F. Khokhlov, A. I. Mashin, and A. M. Satanin
The Dependence of Magnetic Ordering Temperature in Amorphous Semiconductors on Paramagnetic Centre Concentration
(b) Vol. **105**, No. 1, p. 129 – 136 (1981).
See also 2, p. 45; *18.1*, p. *266*; *18.4*, p. *282*; 19, p. 289; 22.1.1, p. 405; 22.1.2, p. 413.

A. F. Burenkov, F. F. Komarov, and M. M. Temkin
*Influence of the Interatomic Potential and the Elastic Scattering Cross-Section Approximation on the Spatial Distribution of Implanted Ions
(b) Vol. **105**, No. 1, p. 201–212 (1981).
See also 22.1, p. 402; 22.1.2, p. 413.

P. Rennert
*Multiple Scattering Theory in the Small Scattering Centre Approximation. Application to EXAFS, AES, XPS, MEED
(b) Vol. **105**, No. 1, p. 213–221 (1981).
See also 10, p. 93; 13.1, p. 155; 17, p. 257.

N. I. Zimin, R. Wedell, and O. Greschner
*Theory of Planar Channelling of Relativistic Positrons
(b) Vol. **105**, No. 1, p. 257–264 (1981).
See also 22.1.2, p. 413.

J. Garrido, J. L. Castaño, and J. Piqueras
Photoluminescence in Gallium Arsenide Irradiated with Thermal Neutrons
(a) Vol. **65**, No. 1, p. 103–106 (1981).
See also *14.3*, p. *213*; *20.3*, p. *327*; 22.2.1, p. 428.

M. A. Kumakhov and V. A. Muralev
*Spatial Distribution of Channeled Ions
(a) Vol. **65**, No. 1, p. 107–117 (1981).
See also 22.1.2, p. 413.

W. Wesch, E. Glaser, G. Götz, H. Karge, and R. Prager
*Correlation between Structural Defects and Optical Properties in Ion-Implanted Silicon
(a) Vol. **65**, No. 1, p. 225–232 (1981).
See also *20.1*, p. *310*; 22.1.2, p. 413.

D. Sachelarie, M. Dragan, M. Stoica, and M. Sachelarie
*Radiation Effects of Fast Neutrons in Silicon Diffused Detectors
(a) Vol. **65**, No. 1, p. 379–388 (1981).
See also *13.4*, p. *176*; 14.3.3, p. 227; 22.1.2, p. 414.

I. B. Ermolovich and V. V. Gorbunov
*On Specific Features of the Defect Formation Mechanism in Cadmium Chalcogenides under Thermal Neutron Irradiation
(a) Vol. **65**, No. 1, p. K13–K18 (1981).
See also 22.4.1, p. 448; 22.4.2, p. 452; 22.4.3, p. 454.

N. Takeuchi, K. Inabe, and S. Nakamura
Post-Irradiation Annealing Effect on the Growth of the Z_3 Band in LiF:Mg Crystal
(a) Vol. **65**, No. 1, p. K33–K36 (1981).
See also *10.2*, p. *110*; 13.4, p. 176; *20.1*, p. *310*; 22.5.2, p. 464.

A. F. Akkerman, V. A. Botvin, and G. Ya. Chernoy
*Characteristics of Secondary Electrons Ejected from Solid Targets by Multiply Charged Ions with Energy up to 20 MeV/Nucleon
(b) Vol. **105**, No. 2, p. 461–468 (1981).
See also 22.1.2, p. 414.

L. A. Gevorgian and N. A. Korkhmazian
*On the Theory of Radiation of Non-Relativistic Electrons Grazing into Metal Targets
(b) Vol. **105**, No. 2, p. 623–628 (1981).
See also 13, p. 145; 21, p. 343.

B. Trieman and N. Kristianpoller
*Defects Induced in $RbMgF_3$ Crystals by X- and UV-Irradiation
(b) Vol. **105**, No. 2, p. 739–746 (1981).
See also *20.3*, p. *328*; 22.5, p. 459.

V. Kuzovkov and E. Kotomin
Kinetics of Defect Accumulation and Recombination (I): General Formalism
(b) Vol. **105**, No. 2, p. 789–801 (1981).
See also *10*, p. *94*.

A. B. Danilin, A. V. Dvurechenskii, I. A. Ryazantsev, P. A. Timofeev, and V. D. Verner
*Silicon Amorphization by Ion Beam with Radiation Heating
(a) Vol. **65**, No. 2, p. 453–461 (1981).
See also *2*, p. *46*; 22.1.2, p. 415.

M. A. Kumakhov and V. A. Muralev
*Spatial Distributions of Channeled Ions. Comparison with Experiments
(a) Vol. **65**, No. 2, p. 623–636 (1981).
See also 22.1.2, p. 415.

G. Götz, H. D. Geiler, M. Wagner, K. H. Heinig, and H. Woittenneck
*Laser Annealing of Thin Buried Amorphous Layers in Silicon
(a) Vol. **65**, No. 2, p. 677–682 (1981).
See also *2*, p. *47*; 22.1.2, p. 415.

B. Rauschenbach, E. Richter, and K. Hohmuth
Spinodal Decomposition in Amorphous Boron-Implanted Iron Films
(a) Vol. **65**, No. 2, p. K103–K105 (1981).
See also *2*, p. *47*; 21.1.1, p. 372.

H. Wagner
Influence of Lattice Defects on the Superconducting Behaviour of Implanted PdH_x Thin Film Layers
(a) Vol. **65**, No. 2, p. K169–K174 (1981).
See also *14.2*, p. *206*; 21, p. 343.

Sh. A. Vakhidov, E. M. Ibragimova, I. Nuritdinov, A. F. Rakov, and G. I. Ikramov
*Self-Trapped Particles in Complex Oxide Crystals
(b) Vol. **106**, No. 1, p. 31–35 (1981).
See also 13.5.1, p. 190; 19, p. 290; 20.1, p. 312; 20.3, p. 329; 22.6, p. 477; 22.8, p. 487.

I. B. Ermolovich, V. V. Gorbunov, and G. I. Matvievskaya
*Properties of the R_n and IR-1 Luminescence Bands in Cadmium Sulfide Irradiated with Thermal Neutrons
(b) Vol. **106**, No. 1, p. 45–53 (1981).
See also 13.4, p. 178; 16, p. 254; *20.3*, p. *329*; 22.4.1, p. 448.

A. V. Gorbunov, E. M. Nadgornyi, and S. N. Valkovskii
*Laser Pulse Induced Dislocation Structure in Ionic Crystals (I): Bulk Damage of NaCl
(a) Vol. **66**, No. 1, p. 53–63 (1981).
See also *10.2*, p. *112*; 22.5.2, p. 465.

K. Urban
Ion Damage to Silicon Crystals inside an Electron Microscope
(a) Vol. **66**, No. 1, p. 213–218 (1981).
See also *10.2*, p. *112*; 22.1.2, p. 416.

R. Klabes, J. Matthäi, M. Voelskow, G. A. Kachurin, E. V. Nidaev, and H. Bartsch
*Flash Lamp Annealing of Arsenic Implanted Silicon
(a) Vol. **66**, No. 1, p. 261–266 (1981).
See also *2*, p. *47*; 22.1.2, p. 416.

G. Oelgart, J. Fiddicke, and R. Reulke
Investigation of Minority-Carrier Diffusion Lengths by Means of the Scanning Electron Microprobe (SEM)
(a) Vol. **66**, No. 1, p. 283–292 (1981).
See also *14.3.3*, p. *228*; 22.1.2, p. 416; 22.2.2, p. 433.

E. Yu. Brailovskii, L. A. Matveeva, G. N. Semenova, Yu. A. Tkhorik, and L. S. Khazan
*Radiation-Stimulated Relaxation of Internal Stresses in Heteroepitaxial Systems
(a) Vol. **66**, No. 1, p. K59–K62 (1981).
See also *10.2*, p. *113*; 22.1.1, p. 406; 22.2.1, p. 430; 22.4.1, p. 449.

E. E. Penzina, V. M. Metsik, I. A. Parfinovich, and K. A. Makushev
*Interaction of Power Laser Radiation with Colour Centres in KCl Crystals
(a) Vol. **66**, No. 1, p. K71–K74 (1981).
See also *10.2*, p. *114*; 20.1, p. 314; 22.5.2, p. 466.

B. Cord, A. Hofstaetter, and A. Scharmann
EPR Investigations on the Dynamic Behaviour of Hole Centres in $CaWO_4$
(b) Vol. **106**, No. 2, p. 499–504 (1981).
See also *19*, p. *291*; 22.8.1, p. 498.

A. Polyakov, Yu. Ryabikin, B. Alimbek, and S. Pivovarov
Relaxation Spectra and Their Use in the Structural Studies
(b) Vol. **106**, No. 2, p. K149–K152 (1981).
See also *19*, p. *292*; 22.9, p. 511.

A. V. Gorbunov, E. M. Nadgornyi, and S. N. Valkovskii
*Laser Pulse Induced Dislocation Structure in Ionic Crystals (II): Surface Damage of NaCl and MgO
(a) Vol. **66**, No. 2, p. 455–462 (1981).
See also *10.2*, p. *114*; 22.5.2, p. 467; 22.6, p. 478.

E. F. Krimmel, H. Oppolzer, H. Runge, and W. Wondrak
*Kinetics of Scanned Electron Beam Annealing of High-Energy As Ion Implanted Silicon
(a) Vol. **66**, No. 2, p. 565–571 (1981).
See also *3*, p. *55*; 14.3.3, p. 228; 22.1.2, p. 417.

H. van Swijgenhoven, J. Vanoppen, and L. M. Stals
*Vitrification of Recrystallized $Fe_{40}Ni_{38}Mo_4B_{18}$ during 5 keV Ar^+-Sputtering
(a) Vol. **67**, No. 1, p. K5–K9 (1981).
Erratum
(a) Vol. **68**, No. 2, p. K211 (1981).
See also *2*, p. *49*; 21.1, p. 363; 21.1.1, p. 375.

V. D. Akhmetov, V. V. Bolotov, and L. S. Smirnov
*Optical Properties of Defects Produced at Annealing of Neutron-Induced Clusters in Silicon
(a) Vol. **67**, No. 1, p. K33–K37 (1981).
See also *13.4*, p. *181*; *20.1*, p. *317*; 22.1.2, p. 419.

H. Bahadur and R. Parshad
A New Electronic Method for Detection of Hydrogen Concentration in Quartz Crystals
(a) Vol. **67**, No. 1, p. K53–K57 (1981).
See also *10.2*, p. *116*; 22.6, p. 479.

A. M. Taratin and S. A. Vorobiev
*Planar Channeling of Relativistic Protons in Bent Crystal
(b) Vol. **107**, No. 2, p. 521–528 (1981).
See also 13, p. 148; 22.1.2, p. 419.

E. Kührt and R. Wedell
*Theoretical Description of Planar Channeling Process at Small Depths
(b) Vol. **107**, No. 2, p. 665–675 (1981).
See also 22.1, p. 403; 22.1.2, p. 420.

H. E. Roosendaal, B. Schmiedeskamp, H. H. Hubbes, and H. O. Lutz
*Planar Channeling Trajectories in Diamond and Silicon
(b) Vol. **107**, No. 2, p. 677–687 (1981).
See also 22.1, p. 403; 22.1.2, p. 420.

A. D. Pogrebnyak, V. A. Kuzminikh, V. I. Raszov, and A. S. Rudnev
*Positron Annihilation and Positron Profiles in Si, Irradiated by Super Dense Electron Pulses
(b) Vol. **107**, No. 2, p. K79–K82 (1981).
See also 22.1.2, p. 420.

A. P. Dolgolenko and I. I. Fishchuk
*A-Centres Build-Up Kinetics in the Conductive Matrix of Pulled n-Type Silicon with Calculation of Their Recharges at Defect Clusters
(a) Vol. **67**, No. 2, p. 407–411 (1981).
See also 22.1.2, p. 420.

D. Wieluńska, L. Wieluński, and A. Turos
*The Effect of Dislocations on the Planar Dechanneling (I): Analytical Model
(a) Vol. **67**, No. 2, p. 413–419 (1981).
See also *10.2*, p. *117*; 22.1.2, p. 420.

Yu. G. Chukalkin, V. V. Petrov, and B. N. Goshchitskii
*Radiation Effects in Hexagonal Ferrite $BaFe_{12}O_{19}$
(a) Vol. **67**, No. 2, p. 421–426 (1981).
See also *18.3*, p. *280*; 22.8.2, p. 506.

A. F. Akkerman and S. A. Akkerman
*Range Distribution Characteristics of H^+ and He^{++} Ions with Energies (5 to 2) $\times 10^3$ keV Implanted into Silicon
(a) Vol. **68**, No. 1, p. 69–75 (1981).
See also 22.1.2, p. 421.

K. Wieteska
*X-Ray Diffraction Investigations of High Energy α-Particle Damage in Silicon
(a) Vol. **68**, No. 1, p. 179–185 (1981).
See also *1.1*, p. *13*; 22.1.2, p. 422.

A. A. Bednyakov, Yu. V. Bulgakov, V. S. Nikolaev, and V. L. Chernov
*Energy Straggling of Hydrogen and Helium Ions in Al, C, and Polystyrene at Energies of Tens and Hundreds keV per Nucleon
(a) Vol. **68**, No. 1, p. 187–192 (1981).
See also 21, p. 348; 22.1, p. 403; 22.9, p. 512.

G. Völkel, H.-E. Müller, and W. Flohrer
Critical Anomaly of the Electron Spin–Lattice Relaxation in X-Irradiated Rochelle Salt
(b) Vol. **108**, No. 2, p. 501–506 (1981).
See also 1.2, p. 21; 6, p. 81; *14.4.1*, p. *246*; 22.9, p. 513.

N. S. Minaev, A. V. Mudrii, and V. D. Tkachev
Symmetry and Nature of the 1.0186 eV Luminescence Centre in Neutron-Irradiated Silicon
(b) Vol. **108**, No. 2, p. K89–K93 (1981).
See also *10.2*, p. *119*; 12.2, p. 141; *20.3*, p. *334*; 22.5.2, p. 469.

K. Gärtner and K. Hehl
*Z_1-Dependence of Low Energy Ranges in Si
(b) Vol. **108**, No. 2, p. K109–K111 (1981).
See also 22.1.2, p. 422.

F. F. Komarov, V. D. Kuryazov, V. S. Solovev, and S. Yu. Shiryaev
*Investigation of Defect Formation in Boron Ion-Implanted Silicon during Annealing
(a) Vol. **68**, No. 2, p. 519–529 (1981).
See also *10.2*, p. *119*; 22.1.2, p. 423.

C. A. B. Ball, F. D. Auret, and H. C. Snyman
*Transmission Electron Microscopy and Auger Electron Spectroscopy Studies of Boron Implanted Silicon Single Crystals
(a) Vol. **68**, No. 2, p. 545–553 (1981).
See also 9, p. 92; 10.2, p. 120; 22.1.2, p. 423.

H. G. Bach, W. R. Fahrner, and D. Bräunig
Equivalent Circuit for Ion Implanted Counterdoped Layers as Determined by MOS Admittance and Crosstalk Measurements
(a) Vol. **68**, No. 2, p. 589–601 (1981).
See also *14.3.2*, p. *224*; *14.3.3*, p. *230*; 22.1.2, p. 423.

12. Mechanical Properties
(Plastic Deformations See 10)

Original Paper

I. T. Suprun
The Influence of External Stress on the Interaction of Dislocations with Impurity Atoms
(a) Vol. **66**, No. 1, p. 361–370 (1981).
See also *10*, p. *94*.

12.1 Metals

Original Papers and Short Notes

W. Siebke
*Considerations on the Bulk Modulus of Pure Metals
(a) Vol. **64**, No. 2, p. 577–583 (1981).
See also *4*, p. *58*; 21, p. 340.

H. Kronmüller and W. Fernengel
The Role of Internal Stresses in Amorphous Ferromagnetic Alloys
(a) Vol. **64**, No. 2, p. 593–602 (1981).
See also 2, p. 44; 10.1, p. 99; *18.2*, p. *270*; 21.1.1, p. 370.

D. Lepski and P. Burck
Precipitate Structure Analysis by Diffraction Contrast Image Matching for a Metastable Semicoherent Carbide Phase in Molybdenum (I): Generating Inelastic Strain
(a) Vol. **64**, No. 2, p. 625–636 (1981).
See also *1.2*, p. *16*; *1.4*, p. *29*; 10.1, p. 99; 21, p. 340.

V. V. Kirsanov, Yu. S. Pyatiletov, and O. G. Tyupkina
*Irradiation Creep Due to Dislocation Glide
(a) Vol. **64**, No. 2, p. 735–740 (1981).
See also 10.1, p. 99; *11*, p. *124*; 21, p. 340.

J. H. Basson and H. Booyens
*The Effect of Substrate Relaxation on the Introduction of Misfit Dislocations in Misfitting Epitaxial Layers
(a) Vol. **64**, No. 2, p. 777–786 (1981).
See also 1.5, p. 35; *10.1*, p. *99*; 22.2.1, p. 428; 22.2.2, p. 433; 22.2.4, p. 437.

I. Yu. Bezotosnyi and I. N. Nikolaev
Effect of Pressure on the Hyperfine Magnetic Fields at ^{57}Fe Nuclei in $Pd_{0.95}Fe_{0.05}$ and $Fe_{0.65}Ni_{0.08}Mn_{0.27}$ Alloys
(a) Vol. **64**, No. 2, p. K129–K131 (1981).
See also *18.2*, p. *270*; *18.4*, p. *282*; 21.1, p. 357; 21.1.1, p. 370.

A. Vehanen, P. Hautojärvi, and V. S. Mikhalenkov
Vacancy Clusters in Deformed $\underline{\mathrm{Cu}}$As Alloy
(a) Vol. **64**, No. 2, p. K145–K148 (1981).
See also *10.1*, p. *99*; 21.1, p. 357.

I. I. Papirov, V. V. Vorobev, A. I. Pikalov, and N. I. Moreva
*Anisotropy of Yttrium Microhardness
(a) Vol. **65**, No. 1, p. 141–145 (1981).
See also 21, p. 342.

V. I. Ivanov, B. K. Kardashev, and S. P. Nikanorov
Acoustic Technique Application in Determining Structure and Dislocation–Point Defect Force Interaction Parameters
(a) Vol. **65**, No. 1, p. 335–339 (1981).
See also *10.1*, p. *99*; 21, p. 342.

S. A. Mahmoud, M. A. Semary, and Z. Farid
*Internal Friction Spectra of Some Al–Cu Alloys
(a) Vol. **65**, No. 1, p. K63–K66 (1981).
See also *10.1*, p. *100*; 21, p. 342; 21.1, p. 359.

F. de Lima and W. Benoit
*Internal Friction Study of Dislocation Mobility in Deformed Niobium
(a) Vol. **67**, No. 2, p. 565–572 (1981).
See also *10.1*, p. *103*; 21, p. 347.

H. M. Ledbetter, J. F. LaBrecque, and J. L. Dahnke
*Propagation Errors in Elastic-Constant Inversion
(a) Vol. **67**, No. 2, p. 643–647 (1981).
See also 21, p. 347; 21.1, p. 364; 21.6, p. 386.

E. A. Abou-Saif, M. Abd-Rabo, M. M. Ezzo, A. A. Gamil, and A. A. Mohamed
*Effect of Heat Treatment on the Structure and Mechanical Properties of Aluminium–Tin Alloys
(a) Vol. **67**, No. 2, p. K101–K105 (1981).
See also *1.3*, p. *26*; 21, p. 347.

R. Bonnet, G. Marcon, and M. Dupeux
*Comment on "An Approach to the Anisotropic Elastic Fields Related to a Semi-Coherent or Coherent Faceted Rod"
(a) Vol. **67**, No. 2, p. K137–K142 (1981).
See also *10.1*, p. *103*; 21, p. 347.

A. del Moral
A High-Order Magnetoelastic Hamiltonian for the Conduction Electron Contribution in Cubic Rare-Earth Metals
(b) Vol. **108**, No. 1, p. 45–52 (1981).
See also *18*, p. *263*; 21.4, p. 382.

H. H. Mende and H.-J. Kaiser
*X-Ray Topographic Investigations of Copper Whiskers during the Registration of Stress–Strain Curves
(a) Vol. **68**, No. 1, p. K21–K24 (1981).
See also 10.1, p. 104; 21.1, p. 366.

R. Bonnet
*On Elastic Properties of Dihedral Edge Dislocations
(b) Vol. **108**, No. 2, p. 435–441 (1981).
See also *10.1*, p. *104*; 21.6, p. 386.

M. Tokuda, J. Kratochvíl, and Y. Ohashi
Mechanism of Induced Plastic Anisotropy on Polycrystalline Metals
(a) Vol. **68**, No. 2, p. 629–636 (1981).
See also *1.4*, p. *32*; *10.1*, p. *105*; 21, p. 349.

Y. Waseda and K. T. Jacob
*Refinement of a Correlation between Isothermal Compressibility and Surface Tension of Liquid Metals
(a) Vol. **68**, No. 2, p. K117–K122 (1981).
See also *2*, p. *51*; 21, p. 349.

12.2 Non-Metals

Original Papers and Short Notes

J. Shanker and K. Singh
*The Anderson Parameter and Temperature Dependence of Bulk Modulus for Alkali Halide Crystals
(b) Vol. **103**, No. 1, p. 151–157 (1981).
See also 8, p. 84; 22.5.2, p. 462.

R. K. Singh and P. Khare
Study of Cohesion and Allied Properties of Silver Halide Crystals
(b) Vol. **103**, No. 1, p. 337–344 (1981).
See also *4*, p. *57*; 22.5.1, p. 461.

D. Houde and D. Cheeke
*Acoustic Mode Softening in $K_2Pt(CN)_4Br_{0.3} \cdot 3\ H_2O$ (KCP)
(b) Vol. **103**, No. 1, p. K21–K25 (1981).
See also *6*, p. *67*; 22.8, p. 484.

A. Krell
*Alumina Structure with Improved Fracture Properties
(a) Vol. **63**, No. 1, p. 183–192 (1981).
See also 22.6, p. 474.

C. Ecolivet and H. Poignant
Berlinite and Quartz $\alpha \leftrightarrow \beta$ Phase Transition Analogy as Seen by Brillouin Scattering
(a) Vol. **63**, No. 2, p. K107–K109 (1981).
See also *1.2*, p. *14*; 22.8.1, p. 494.

A. Basu, S. Sengupta, and A. N. Basu
*Three Body Interaction in the Fluorite Structure Ionic Crystals
(b) Vol. **103**, No. 2, p. 535–546 (1981).
See also *4*, p. *57*; 22.5.3, p. 471.

J. Shanker, J. P. Singh, and V. P. Gupta
*Evaluation of the Fourth-Order Elastic Constants of NaCl-Structure Alkali Halide Crystals
(b) Vol. **103**, No. 2, p. 573–580 (1981).
See also 22.5.2, p. 462.

S. Ghosh, A. N. Basu, and S. Sengupta
*Extended Deformable Shell Model Calculation of the Lattice Statics and Dynamics of LiH and LiD Crystals
(b) Vol. **103**, No. 2, p. 595–603 (1981).
See also *6*, p. *67*; 22, p. 391.

A. V. Singh and J. Shanker
Effect of Three-Body Potential on the Dielectric Behaviour and Optic Mode Grüneisen Parameters of Alkali Halide Crystals
(b) Vol. **103**, No. 2, p. 733–738 (1981).
See also *4*, p. *57*; 6, p. 68; 14.4, p. 238; 22.5.2, p. 463.

A. Kulczycki
Lattice Dynamics of $[Ni(NH_3)_6](ClO_4)_2$, $[Ni(NH_3)_6](BF_4)_2$, and $[Ni(NH_3)_6](PF_6)_2$
(b) Vol. **103**, No. 2, p. 753–756 (1981).
See also *6*, p. *68*; 22.8, p. 484.

V. I. Khrupa, E. N. Kislovskii, and L. I. Datsenko
*Energy Absorption at the X-Ray Interference Transmission in an Elastically Bent Thin Crystal
(a) Vol. **63**, No. 2, p. 439–443 (1981).
See also 22.1.2, p. 409.

J. Shanker, G. D. Jain, and S. K. Sharma
*Evaluation of Third-Order Elastic Constants and Pressure Derivatives of Second-Order Elastic Constants for AgCl–AgBr Mixed Crystals
(b) Vol. **104**, No. 1, p. 75–80 (1981).
See also 22.5.4, p. 473.

T. Soma
*Pressure Derivatives of Elastic Stiffness Constants under Pressure of Si and Ge
(b) Vol. **104**, No. 1, p. 293–298 (1981).
See also 22.1.1, p. 404; 22.1.2, p. 410.

R. W. Munn
*External and Internal Strains in Piezoelectric Crystals
(b) Vol. **104**, No. 1, p. 325–330 (1981).
See also *14.4*, p. *238*.

L. B. Harris and N. G. Cutmore
*Factors Affecting Microhardness Measurements on NaCl Crystals
(a) Vol. **64**, No. 1, p. 105–112 (1981).
See also 22.5.2, p. 463.

B. P. Chandra
*Mechanoluminescence and Piezoelectric Behaviour of Molecular Crystals
(a) Vol. **64**, No. 1, p. 395–405 (1981).
See also 14.3, p. 211; *20.3*, p. *325*; 22.9, p. 509.

E. Anastassakis and M. Cardona
*Internal Strains Raman-Active Optical Phonons
(b) Vol. **104**, No. 2, p. 589–600 (1981).
See also *6*, p. *71*; 22.1, p. 402; 22.5.3, p. 471; 22.6, p. 475.

A. Selmi, P. Gibart, and G. Weill
Pressure Dependence of the Galvanomagnetic Properties of n- and p-Type Ferromagnetic Semiconductors $HgCr_2Se_4$
(a) Vol. **64**, No. 2, p. 665–668 (1981).
See also *14.3*, p. *212*; *18.2*, p. *270*; 22.8.2, p. 503.

V. Balynas, Z. Dobrovolskis, W. Hoerstel, and A. Krotkus
The Effect of Hydrostatic Pressure on the Anomalous Sign Reversal of the Hall Coefficient in Tellurium
(a) Vol. **64**, No. 2, p. K115–K117 (1981).
See also 13.1, p. 155; *14.3*, p. *212*; 22.1.3, p. 425.

V. Ramachandran, M. M. Ibrahim, V. C. Padaki, and E. S. R. Gopal
*Elastic Constants of Sodium Nitrate
(a) Vol. **67**, No. 1, p. K49–K52 (1981).
See also 22.8.1, p. 499.

H. Bahadur and R. Parshad
*Some New Results on the Frequency Characteristics of Quartz Crystals Irradiated by Ionizing and Particle Radiations
(a) Vol. **67**, No. 2, p. 683–692 (1981).
See also 11, p. 130; 22.6, p. 479.

G. Sorge, A. Almeida, L. A. Shuvalov, and R. M. Fedosyuk
Influence of the Mechanical Stress Component T_5 on the Elastic Behaviour of $KH_3(SeO_3)_2$/$KD_3(SeO_3)_2$ Crystals in the Vicinity of Their Ferroelastic Phase Transition
(a) Vol. **68**, No. 1, p. 245–251 (1981).
See also *1.2*, p. *20*; *14.4.1*, p. *246*; 22.8, p. 491.

W. Gehlhoff and K. H. Segsa
*Changes of EPR Spectra by Inhomogeneous Plane Stress Distributions in a Two-Layer System
(b) Vol. **108**, No. 2, p. 421–433 (1981).
See also *19*, p. *294*; 22, p. 400.

N. S. Minaev, A. V. Mudrii, and V. D. Tkachev
Symmetry and Nature of the 1.0186 eV Luminescence Centre in Neutron-Irradiated Silicon
(b) Vol. **108**, No. 2, p. K89–K94 (1981).
See also *10.2*, p. *119*; 11, p. 131; *20.3*, p. *334*; 22.5.2, p. 469.

A. M. Ghodgaonkar and K. Ramani
*Evaluation of Bulk Modulus, Pressure Derivative of Bulk Modulus, and Cohesive Energy by Modified Morse-Kratzer Potential
(b) Vol. **108**, No. 2, p. K133–K138 (1981).
See also *4*, p. *61*; 22, p. 400; 22.5.1, p. 461; 22.5.3, p. 472.

L. I. Datsenko, V. I. Khrupa, and E. N. Kislovskii
*Borrmann Effect in Elastically Bent Silicon Crystals with Structural Defects
(a) Vol. **68**, No. 2, p. 399–404 (1981).
See also 10.2, p. 119; *20*, p. *300*; 22.1.2, p. 423.

H. Koizumi and T. Suzuki
*The Influence of Yield Stress on Fracture of NaCl Crystals
(a) Vol. **68**, No. 2, p. 579–588 (1981).
See also 22.5.2, p. 470.

13. Electron States

Original Papers and Short Notes

E. KOLLEY, W. KOLLEY, and A. BECKMANN
Transition Temperature of Superconducting Alloys with Random Transfer Integrals
(b) Vol. **103**, No. 2, p. 775–781 (1981).
See also *14.2*, p. *205*; 21.1, p. 353.

M. M. SHUKLA and J. MAHANTY
Compatibility of Interionic Pair Potentials and Channeling Potentials
(b) Vol. **103**, No. 2, p. K119–K121 (1981).
See also 4, p. 57; *11*, p. *121*.

S. GUHA and P. SEN
Parametric Decay of Alfén Waves in Piezoelectric Semiconductors
(b) Vol. **103**, No. 2, p. K131–K136 (1981).
See also *20*, p. *296*; 22.2.3, p. 435.

M. BOHL
Electronic Structure, UV Absorption, and Dispersion of Optical Glasses
(a) Vol. **63**, No. 2, p. 529–538 (1981).
See also *2*, p. *41*; *20.1*, p. *304*; 22.6, p. 474.

W. EBELING, W. RICHERT, W. D. KRAEFT, and W. STOLZMANN
*Padé Approximations for the Thermodynamic Functions of Weakly Interacting Coulombic Quantum Systems
(b) Vol. **104**, No. 1, p. 193–202 (1981).
See also 13.5, p. 186; 22.1.1, p. 404.

U. PIETSCH and K. UNGER
*The Temperature Dependence and Dynamical Properties of the Bond Charge in GaAs and InSb
(b) Vol. **104**, No. 1, p. 253–260 (1981).
See also *4*, p. *58*; 6, p. 69; 22.2.1, p. 427; 22.2.3, p. 435.

J. M. LÓPEZ and J. A. ALONSO
*Volume Dependence of the Bulk Modulus in Alkali Metals
(b) Vol. **104**, No. 1, p. 307–312 (1981).
See also *12.1*, p. *134*; 21.2, p. 376.

V. I. ZUBOV and S. SH. SOULAYMAN
*Unsymmetrized Self-Consistent Field Approximation for Strongly Anharmonic Crystals with Taking into Account the Polarizability of the Ions
(b) Vol. **104**, No. 1, p. 383–392 (1981).
See also *6*, p. *70*.

F. CLARO
*Spectrum of Tight Binding Electrons in a Square Lattice with Magnetic Field
(b) Vol. **104**, No. 1, p. K31–K34 (1981).

N. G. BEBENIN
*The Critical Dynamics of Hot Electrons in a Ferromagnetic Semiconductor at $T = T_c$
(b) Vol. **104**, No. 1, p. K65–K70 (1981).
See also *14.3.4*, p. *233*; 18.2, p. 269; 22, p. 393.

B. Lorenz
*Charge Order–Disorder Transition in the Extended Hubbard Model. Finite Bandwidth Effect
(b) Vol. **105**, No. 1, p. 233–241 (1981).
See also 18, p. 261; 22.6, p. 476.

V. G. Baru and E. V. Grekov
*Van der Waals Theory of the Phase Transition in Quasi-Two-Dimensional Electron Gas in Magnetic Semiconductors
(b) Vol. **105**, No. 1, p. 369–376 (1981).
See also 14.3.3, p. 227; 18.2, p. 270; 22, p. 394.

V. I. Kruglov
*Non-Markovian Evolution of Crystal Electrons in a Strong Electromagnetic Wave Field
(b) Vol. **105**, No. 1, p. 425–435 (1981).
See also *6*, p. *74*.

R. Engelke, Th. Roy, H.-G. Neumann, and K. Hübner
Chemical Bond and Related Properties of SiO_2 (VIII): Experimental and Theoretical Investigation of the Structure of SiO_x
(a) Vol. **65**, No. 1, p. 271–280 (1981).
See also *2*, p. *46*; *4*, p. *58*; 22.6, p. 477.

K. N. Khanna
*Wigner-Seitz Radius Dependent Model Potential
(b) Vol. **105**, No. 2, p. 485–490 (1981).
See also 1, p. 9; 4, p. 59; 14.1, p. 200; 21, p. 342; 21.1, p. 359; 21.2, p. 376; 21.3, p. 378.

K. Elk
*On the Electrical Conductivity of the Periodic Anderson Model
(b) Vol. **105**, No. 2, p. 507–513 (1981).
Erratum
(b) Vol. **107**, No. 1, p. 387 (1981).
See also *14.1*, p. *201*; 21, p. 343.

D. Ihle and B. Lorenz
*Spectral Density Method for the Spinless Fermion Model
(b) Vol. **105**, No. 2, p. 557–566 (1981).

L. A. Gevorgian and N. A. Korkhmazian
On the Theory of Radiation of Non-Relativistic Electrons Grazing into Metal Targets
(b) Vol. **105**, No. 2, p. 623–628 (1981).
See also *11*, p. *126*; 21, p. 343.

G. B. Taggart
*Stability Surface for the Blume-Capel Model
(b) Vol. **105**, No. 2, p. 699–705 (1981).
See also 8, p. 85; *18*, p. *262*.

P. F. Tua and G. D. Mahan
*Variational Hartree-Fock Calculations of the Ground State Energy of Metallic Hydrogen
(b) Vol. **105**, No. 2, p. 769–775 (1981).
See also 21, p. 343.

K. Eapen
*Antiferromagnetism in the Random Hubbard Model
(b) Vol. **106**, No. 2, p. K123–K125 (1981).
See also *18.4*, p. *284*.

J. Filipecki, J. Dworakowski, and B. Całusiński
Investigation of the Metallic Glass $Fe_{40}Ni_{40}B_{20}$ with the Use of Positron Annihilation Methods
(a) Vol. **66**, No. 2, p. 463–466 (1981).
See also *2*, p. *48*; 21.1.1, p. 373.

J. Sivardière
*Cyclic and Dihedral Point Groups as Subgroups of the Limiting Groups
(b) Vol. **107**, No. 1, p. 117–125 (1981).
See also *1*, p. *10*.

P. Schlottmann and L. M. Falicov
Local Susceptibility Calculation for Intermediate-Valence Solids
(b) Vol. **107**, No. 1, p. 165–176 (1981).
See also *18.1*, p. *266*.

L. E. Oliveira
*Local Field Effects and a Model Dielectric Response Matrix $\varepsilon_{G,G'}(\boldsymbol{q})$ for a Covalent Semiconductor
(b) Vol. **107**, No. 1, p. 255–265 (1981).
See also *14.4*, p. *240*; 22.1, p. 402.

H. G. Brühl
*Die Bestimmung der Valenzladungsdichte von Elementhalbleitern und III–V-Verbindungen mit Röntgenbeugungsmethoden (I): Grundlagen
(Determination of the Bulk Charge Density of Elemental Semiconductors and III–V Compounds by X-Ray Diffraction Methods (I): Fundamentals)
(b) Vol. **107**, No. 1, p. 289–295 (1981).
See also *1*, p. *10*; 22.1, p. 402; 22.2, p. 426.

L. Skála
*The CNDO/BW Study of Lithium Clusters (I): Two-, Three-, and Four-Atomic Clusters
(b) Vol. **107**, No. 1, p. 351–357 (1981).
See also *4*, p. *60*; 21.2, p. 377.

R. L. Jacobs
*Model Electronic Structure and Saturation Magnetisation of Some Substituted Fe–Si–B Metallic Glasses
(b) Vol. **107**, No. 1, p. K13–K17 (1981).
See also 2, p. 49; *18.2*, p. *273*; 21.1.1, p. 374.

U. Pietsch and K. Unger
Asphericity and Temperature Dependence of the Bond Charge
(b) Vol. **107**, No. 1, p. K19–K23 (1981).
See also *4*, p. *60*; 22.1.2, p. 418.

E. F. Kustov, L. I. Uskova, and S. A. Klinchikov
*Estimation of Covalency Effects in the Optical Spectra of Crystals Containing Iron Group Ions (II)
(b) Vol. 108, No. 1, p. 53–58 (1981).
See also *20.1*, p. *318*; 22.5, p. 460; 22.6, p. 480.

A. I. Landa, V. E. Panin, and M. F. Zhorovkov
The Pseudopotential Method and Thermodynamics of Eutectic Systems
(b) Vol. 108, No. 1, p. 113–122 (1981).
See also *1.2*, p. *20*; *2*, p. *50*, 8, p. 88; 21, p. 347; 21.1, p. 365.

G. S. Dubey and D. K. Chaturvedi
*Static and Dynamic Structure Factors of Classical Plasma in Weak and Intermediate Couplings
(b) Vol. 108, No. 1, p. 123–129 (1981).
See also *1*, p. *10*.

V. V. Rozhkov and P. B. Rutkevich
The Theory of Thermal Dechanneling from Axial Channels
(b) Vol. 108, No. 1, p. 145–151 (1981).
See also *11*, p. *130*.

I. Gorczyca
*Calculations of Deformation-Potential Constants in InSb
(b) Vol. 108, No. 1, p. 205–209 (1981).
See also *6*, p. *80*; 22.2.3, p. 436.

K. Saermark, J. Lebech, W. Braune, and N. Kubicki
*Non-Local Calculations of the Dispersion of Electromagnetic Waves in Bismuth. Dispersion Relation for the Voight Configuration: $\boldsymbol{q} \parallel \boldsymbol{n} \parallel \boldsymbol{C}_1, \boldsymbol{q} \perp \boldsymbol{B}$
(b) Vol. 108, No. 1, p. 275–280 (1981).
See also *20*, p. *299*; 21.7, p. 389.

A. V. Tulupov
Some Characteristic Features of the Kumakhov Radiation in an X-Ray Range
(b) Vol. 108, No. 1, p. K11–K14 (1981).
See also *11*, p. *130*; *20*, p. *299*; 22.1.2, p. 421.

V. Ilkovič and L. Mucha
*Bogolyubov Method for the Randomly Located Spins with Oscillatory Interaction
(b) Vol. 108, No. 1, p. K37–K39 (1981).
See also *18.2*, p. *274*.

R. Piasecki
*Generalized Model of an Anisotropic Ferromagnet with Two Spins per Site. Magnetic Excitations
(b) Vol. 108, No. 1, p. K41–K45 (1981).
See also *18.2*, p. *274*.

J. Spałek, A. M. Oleś, and K. A. Chao
*Magnetic Phases of Strongly Correlated Electron in a Nearly Half-Filled Narrow Band
(b) Vol. 108, No. 2, p. 329–340 (1981).
See also *18*, p. *264*; 21, p. 348.

S. NARA, A. FREUDENHAMMER, and G. HEBER
*Dynamics of Two-Spin Cluster in Spin Glass (I): Simple Model Calculation for Remanent Magnetization
(b) Vol. **108**, No. 2, p. 389–398 (1981).
See also *18*, p. *264*; 22.8, p. 491.

B. LORENZ
*Cluster Approximation and Perturbation Theory for Band Models with Coulomb Interaction
(b) Vol. **108**, No. 2, p. 411–419 (1981).
See also *18*, p. *264*.

W. NOLTING and A. M. OLÉS
*Simple Spectral Density Approach to the Problem of Mixed Valency
(b) Vol. **108**, No. 2, p. 459–466 (1981).
See also 22, p. 400.

V. V. VALKOV and S. G. OVCHINNIKOV
Mixed Valence Phase Transition in Ferromagnets
(b) Vol. **108**, No. 2, p. 601–608 (1981).
See also *18.2*, p. *275*.

M. SZUSZKIEWICZ, B. ROZENFELD, and J. CHOJCAN
*Localization of 3d-Like Electrons in Cr–Fe and V–Fe Alloys
(b) Vol. **108**, No. 2, p. 623–630 (1981).
See also 21.1, p. 366; 21.1.1, p. 375.

M. BACCI
*A Semi-Clasical Approach to the Interpretation of Spin Equilibria
(b) Vol. **108**, No. 2, p. K123–K128 (1981).
See also *18.1*, p. *266*.

H.-G. BRÜHL and U. PIETSCH
*Investigation of the Bond Charge in GaP with X-Ray Data
(a) Vol. **68**, No. 2, p. 689–695 (1981).
See also *4*, p. *61*; 22.2.2, p. 434.

A. MATULIONIS, J. POŽELA, and E. STARIKOV
*Hot Electron Velocity Correlation and Diffusion in a Many Layered Heterostructure
(a) Vol. **68**, No. 2, p. K149–K152 (1981).
See also *14.3.4*, p. *236*; 22.2.1, p. 431; 22.2.4, p. 439.

13.1 Band Structure

Review Article

M. Asche and O. G. Sarbei
*Electron–Phonon Interaction in n-Si
(b) Vol. **103**, No. 1, p. 11–50 (1981).
See also *6*, p. *66*; 14.3.4, p. 232; 22.1.2, p. 408.

Original Papers and Short Notes

V. Čápek
*Role of Magnetization in the Electron Quasiparticle Energy in the s–d Model of Ferromagnetic Semiconductors
(b) Vol. **103**, No. 1, p. 107–113 (1981).
See also *18.2*, p. *267*; 22, p. 390.

G. B. Dubrovskii and V. I. Belyavskii
*Selection Rules and Absorption Lineshapes for Electron Transitions between Minizones in a Superlattice
(b) Vol. **103**, No. 1, p. 131–137 (1981).
See also *20.1*, p. *301*; 22.3, p. 440.

V. Szöcs, V. Čápek, and B. Velický
*Simple Two-Band s–d Model for Ferromagnetic Semiconductors with Hybridization (II): Further Development and Optical Properties
(b) Vol. **103**, No. 1, p. 139–149 (1981).
See also 18.2, p. 267; 22, p. 390.

Nguyen Ba An, Nguyen Van Hieu, Nguyen Toan Thang, and Nguyen Ai Viet
*Interband Electronic Raman Scattering of Polarized Light in CuCl
(b) Vol. **103**, No. 1, p. 205–210 (1981).
See also 13.5.1, p. 188; *20*, p. *296*; 22.5, p. 458.

O. Ziep
*Band-to-Band Recombination via Spontaneous Emission of Phonon–Plasmon Coupled Modes in Lead Chalcogenides
(b) Vol. **103**, No. 1, p. 239–245 (1981).
See also 6, p. 66; 13.5.2, p. 194; 22.8, p. 484.

V. Dose and H. Scheidt
*Isochromat and SXAPS Measurements on Hafnium, Tantalum, and Tungsten
(b) Vol. **103**, No. 1, p. 247–252 (1981).
See also 20, p. 296; 21, p. 336.

Z. W. Gortel, J. Szymański, and L. Świerkowski
Acceptor States in Zero-Gap Semiconductors in Magnetic Field. Theory for Short-Range Defect Potential
(b) Vol. **103**, No. 1, p. 429–439 (1981).
See also *13.4*, p. *169*; 18, p. 260; 22, p. 390.

C. Ance et A. Amamou
*Spectre de photoémission de l'alliage ternaire $Ga_{1-x}Al_xSb$
(Photoemission Spectra of the Ternary $Ga_{1-x}Al_xSb$ Alloy)
(b) Vol. **103**, No. 1, p. 441 – 449 (1981).
See also 17, p. 256; 22.2.4, p. 437.

A. N. Chakravarti, K. P. Ghatak, K. K. Ghosh, S. Ghosh, and A. Dhar
Effect of Carrier Degeneracy on the Screening Length in n-Cd_3As_2
(b) Vol. **103**, No. 1, p. K55 – K60 (1981).
See also *14.3*, p. *209*; 22.3, p. 440.

T. P. Humphreys and G. P. Srivastava
*Electronic Charge Density for Two Isoelectronic Series: Ge–GaAs–ZnSe and α-Sn–InSb–CdTe
(b) Vol. **103**, No. 1, p. K85 – K90 (1981).
See also 4, p. 57; 22, p. 391.

G. L. Belenkii, E. Yu. Salaev, R. A. Suleimanov, and E. I. Mirzoev
*The Influence of Pressure on the Electronic Spectrum of Layered Crystals
(a) Vol. **63**, No. 1, p. 97 – 102 (1981).
See also 20.1, p. 302; 20.3, p. 324; 22.3, p. 440.

V. V. Sobolev, O. G. Maksimova, and S. G. Kroitoru
*Reflectivity Spectra and Band Structure of the ZnTe–CdTe System
(b) Vol. **103**, No. 2, p. 499 – 509 (1981).
See also *20.1*, p. *303*; 22.4.4, p. 456.

L. M. Sandratskii, R. F. Egorov, and A. A. Berdyshev
*Energy Band Structure and Electronic Properties of NiAs Type Compounds (I): Ferromagnetic Manganese Arsenide
(b) Vol. **103**, No. 2, p. 511 – 518 (1981).
See also 18.2, p. 267; 21.1, p. 353.

A. N. Ermoshkin, E. A. Kotomin, and R. A. Evarestov
*Semiempirical Calculations of the Impurity Level Positions with Respect to the Perfect Crystal Bands
(b) Vol. **103**, No. 2, p. 581 – 587 (1981).
See also *13.4*, p. *170*; 22.5.2, p. 463.

H. V. Sharma and I. Singh
*Phase Transition in the Extended Falicov-Kimball Model with Finite f-Bandwidth
(b) Vol. **103**, No. 2, p. 629 – 635 (1981).
See also 21, p. 337.

S. E. Kulkova, V. E. Egorushkin, I. N. Anokhina, and V. P. Fadin
*The Change of Electronic Structure and Optical Properties in FeCo at the Order–Disorder Transition
(b) Vol. **103**, No. 2, p. 653 – 658 (1981).
See also *20.1*, p. *303*; 21.1, p. 353; 21.1.1, p. 368.

N. V. Starostin, M. P. Shepilov, and A. B. Alekseev
*Energy Structure of the Alkaline-Earth Fluorides
(b) Vol. **103**, No. 2, p. 717 – 723 (1981).
See also 22.5.3, p. 471.

G. MARGARITONDO, A. D. KATNANI, and F. LÉVY
*Electronic States of $CdIn_2Se_4$ and $ZnIn_2Se_4$: Role of the Cation Pseudovacancies
(b) Vol. **103**, No. 2, p. 725–731 (1981).
See also *10.2*, p. *107*; 17, p. 256; 22.8, p. 484.

S. TEWARI and R. M. SINGRU
*Electronic States and Fermi Surface Properties of NiH
(b) Vol. **103**, No. 2, p. 745–751 (1981).
See also *13.2*, p. *163*; 22, p. 392.

J. MERTSCHING and H. J. FISCHBECK
*The Incommensurate Peierls Phase of the Quasi-One-Dimensional Fröhlich Model with a Nearly Half-Filled Band
(b) Vol. **103**, No. 2, p. 783–795 (1981).
See also *1.2*, p. *15*.

J. HORÁK, L. TICHÝ, and P. LOŠŤÁK
Electrical Conductivity of BiTeI Crystals
(a) Vol. **63**, No. 2, p. 407–416 (1981).
See also *14.3*, p. *210*; 15, p. 247; 22.8, p. 485.

A. A. TUKHFATULLIN, V. E. PANIN, V. I. SIMAKOV, V. S. DEMIDENKO, and A. P. KALYANOV
*Electron Structure and the Nature of Co_3Ti Phase Formation
(b) Vol. **104**, No. 1, p. 69–74 (1981).
See also 1.2, p. 15; 21.1, p. 355.

P. BOGUSŁAWSKI
*The Theory of the Electron Spin–Lattice Relaxation in Zincblende Semiconductors
(b) Vol. **104**, No. 1, p. 89–95 (1981).
See also *6*, p. *69*; 22.2, p. 426; 22.4, p. 445.

L. M. SANDRATSKII, R. F. EGOROV, and A. A. BERDYSHEV
*Energy Band Structure and Electronic Properties of NiAs-Type Compounds (II): Antiferromagnetic Manganese Telluride
(b) Vol. **104**, No. 1, p. 103–107 (1981).
See also 18.4, p. 281; 22, p. 392.

TH. HARBICH and G. MAHLER
*The Electron–Hole Plasma in a Strong Electromagnetic Field. Optical Properties in a Simplified Model
(b) Vol. **104**, No. 1, p. 185–191 (1981).
See also *13.5*, p. *186*; 20.1, p. 305; 22.2.1, p. 427.

U. BEHN
*Nearest-Neighbour Poisson Approximation to a Functional Approach for the Hubbard Model
(b) Vol. **104**, No. 1, p. 203–208 (1981).
See also 14.1, p. 199.

B. FELLMUTH, H. KRÜGER, R. RUDOLPH, and R. HERRMANN
*Composition Dependence of the Maximum Cyclotron Mass m_1^c in n-Type Semiconducting $Bi_{1-x}Sb_x$
(b) Vol. **104**, No. 1, p. 209–217 (1981).
See also *19*, p. *288*; 22.7, p. 483.

A. R. Hassan and R. Raouf
Polarization Dependence of Three-Photon Absorption in Solids
(b) Vol. **104**, No. 2, p. 703–709 (1981).
See also *20.1*, p. *307*; 22.5, p. 458.

M. Baleva and S. D. Stoyanov
*On the Optical Properties and Electronic Structure of Solid Solutions of $(Ag_{x/2}Pb_{1-x}Sb_{x/2})$ Te Type
(b) Vol. **104**, No. 2, p. 737–741 (1981).
See also *20.1*, p. *307*; 22.8, p. 486.

V. M. Koleshko and N. Yu. Trifonov
*Fine Impurity Band Structure of One-Dimensional Disordered Crystal
(b) Vol. **104**, No. 2, p. K105–K108 (1981).
See also *13.4*, p. *173*; 22, p. 393.

V. N. Borzunov, A. E. Kadyshevich, and A. P. Rusakov
*Self-Consistent Band Structure Calculation of CuCl
(b) Vol. **104**, No. 2, p. K113–K116 (1981).
See also 22.5, p. 458.

U. Dębska, M. Dietl, G. Grabecki, E. Janik, E. Kierzek-Pecold, and M. Klimkiewicz
Energy Band Gaps of $Hg_xCd_yMn_zTe$ $(x + y + z = 1)$ Epitaxial Layers
(a) Vol. **64**, No. 2, p. 707–713 (1981).
See also 3, p. 54; *20.1*, p. *308*; 22.4.4, p. 456.

V. Dolocan and F. Iova
Optical Properties of Bi_2O_3 Thin Films
(a) Vol. **64**, No. 2, p. 755–759 (1981).
See also *20.1*, p. *309*; 22.6, p. 476.

V. Balynas, Z. Dobrovolskis, W. Hoerstel, and A. Krotkus
The Effect of Hydrostatic Pressure on the Anomalous Sign Reversal of the Hall Coefficient in Tellurium
(a) Vol. **64**, No. 2, p. K115–K117 (1981).
See also 12.2, p. 139; *14.3*, p. *212*; 22.1.3, p. 425.

E. P. Domashevskaya and V. A. Terekhov
*d–s,p Resonance and Electronic Structure of Compounds, Alloys and Solid Solutions
(b) Vol. **105**, No. 1, p. 121–127 (1981).
See also 17, p. 257; 20, p. 297; 21, p. 341; 21.1, p. 358.

P. Rennert
Multiple Scattering Theory in the Small Scattering Centre Approximation. Application to EXAFS, AES, XPS, MEED
(b) Vol. **105**, No. 1, p. 213–221 (1981).
See also 10, p. 93; *11*, p. *125*; 17, p. 257.

Sh. Sh. Bashkirov and V. V. Parfenov
*Conduction Band Width of Mn–Zn Ferrites
(b) Vol. **105**, No. 1, p. K27–K29 (1981).
See also 16, p. 252; 20.1, p. 310; 22.8.2, p. 503.

T. G. KERIMOVA, SH. S. MAMEDOV, and R. KH. NANI
*On the Band Structure of $CdGa_2S_4$
(b) Vol. **105**, No. 1, p. K39–K44 (1981).
See also 20.1, p. 310; 22.8, p. 487.

N. M. RAVINDRA, V. K. SRIVASTAVA, and B. MISHRA
Temperature Dependent Electrical Conductivity of Natural Galena Crystals
(a) Vol. **65**, No. 1, p. K59–K61 (1981).
See also *14.3*, p. *213*; 22, p. 395.

M. DOBROWOLSKA, W. DOBROWOLSKI, R. R. GAŁĄZKA, and A. MYCIELSKI
Interband Magnetooptical Studies on the Semimagnetic Semiconductor $Hg_{1-x}Mn_xSe$
(b) Vol. **105**, No. 2, p. 477–484 (1981).
See also *18*, p. *262*; *20*, p. *297*; 22.4.2, p. 456.

J. SZCZYRBOWSKI
*The Exponential Shape of the Optical Absorption Edge Tail
(b) Vol. **105**, No. 2, p. 515–524 (1981).
See also *20.1*, p. *310*; 22, p. 395.

G. JUNGK
*On the Experimental Determination of Many-Particle Effects at Interband Transitions
(b) Vol. **105**, No. 2, p. 551–556 (1981).
See also 13.5.1, p. 190; *20.1*, p. *311*; 22.2.1, p. 429; 22.2.2, p. 433.

P. RENNERT
*Calculation of the Compton Profile of Beryllium
(b) Vol. **105**, No. 2, p. 567–575 (1981).
See also 13.2, p. 163; *20*, p. *297*; 21.3, p. 378.

H. NEUMANN, YU. I. TSIPIVKA, and K. UNGER
Influence of Dielectric Screening and Band Nonparabolicity on Ionized Impurity Scattering in Doped Semiconductors
(b) Vol. **105**, No. 2, p. 577–583 (1981).
See also *14.3*, p. *214*; 14.4, p. 239; 22.2.1, p. 429; 22.2.3, p. 436.

D. G. KANHERE, R. M. SINGRU, and R. HARTHOORN
*Electron Momentum Distribution and Compton Profiles of Ag
(b) Vol. **105**, No. 2, p. 715–722 (1981).
See also 13.2, p. 163; *20*, p. *298*; 21.6, p. 385.

O. V. FARBEROVICH and S. V. VLASOV
*APW Research of Europium Monochalcogenides (I): Band Structures and Electron-Electron Interaction in Ferromagnetic Semiconductors
(b) Vol. **105**, No. 2, p. 755–768 (1981).
See also 18.2, p. 271; 20.1, p. 311; 22, p. 396.

A. K. SAXENA
*Non-Γ Deep Levels and the Conduction Band Structure of $Ga_{1-x}Al_xAs$ Alloys
(b) Vol. **105**, No. 2, p. 777–787 (1981).
See also *13.4*, p. *177*; 14.3, p. 214; 22.2.4, p. 437.

I. V. Bodnar, B. V. Korzun, and A. I. Lukomskii
*Composition Dependence of the Band Gap of $CuInS_{2x}Se_{2(1-x)}$
(b) Vol. **105**, No. 2, p. K143–K145 (1981).
See also *20.1*, p. *311*; 22.8, p. 487.

V. F. Kovalenko, G. P. Peka, and L. G. Shepel
Photoluminescent Methods for Determination of Graded Band-Gap Semiconductor Parameters (I): Analysis of Methods
(a) Vol. **65**, No. 2, p. 529–534 (1981).
See also 13.3, p. 165; *20.3*, p. *328*; 22.2.4, p. 437.

H. Van Cong
Heavy Doping Effects in Bipolar Silicon Transistors and p–n Junction Silicon Solar Cells
(a) Vol. **65**, No. 2, p. 683–693 (1981).
See also *14.3.3*, p. *227*; 22.1.2, p. 415.

S. Acharya and O. N. Srivastava
*Electronic Behaviour of SnS_2 Crystals
(a) Vol. **65**, No. 2, p. 717–723 (1981).
See also 1.2, p. 17; *14.3*, p. *214*; 22, p. 396.

N. M. Ravindra and V. K. Srivastava
*Temperature Dependence of the Energy Gap in Pyrite (FeS_2)
(a) Vol. **65**, No. 2, p. 737–742 (1981).
See also 22, p. 396.

M. Matlak and J. Zieliński
*On the Electrical Conductivity and the Density of States for the Narrow-Band Periodic Anderson Model in the Symmetric Case
(b) Vol. **106**, No. 1, p. 73–77 (1981).
See also *14.1*, p. *201*; 21, p. 344.

G. Ambrazevičius, G. Babonas, Yu. V. Rud, and A. Šileika
*The Fundamental Absorption Edge of $MgSiP_2$
(b) Vol. **106**, No. 1, p. 85–89 (1981).
See also *20.1*, p. *312*; 22.8, p. 488.

V. I. Chizhikov, V. L. Panyutin, B. E. Ponedelnikov, et A. E. Rozenson
*Structure de bande électronique de $CdAl_2Te_4$ dans les modifications avec les groupes spatiaux D_{2d}^{11}, D_{2d}^{9}, D_{2d}^{1} et S_4^2
(Electronic Energy Band Structure of $CdAl_2Te_4$ Modifications with Space Group D_{2d}^{11}, D_{2d}^{9}, D_{2d}^{1}, and S_4^2)
(b) Vol. **106**, No. 1, p. 91–98 (1981).
See also 22.8, p. 488.

S. W. Koch, S. Schmitt-Rink, and H. Haug
Theory of Optical Nonlinearities in InSb
(b) Vol. **106**, No. 1, p. 135–140 (1981).
See also *20.1*, p. *312*; 22.2.3, p. 436.

J. S. Brown and A. H. M. Lopez-Escobar
*A Muffin-Tin Calculation of the Electronic d-Bandwidth of F.C.C. Palladium
(b) Vol. **106**, No. 1, p. K33–K38 (1981).
See also 21, p. 344.

F. M. Gashimzade and G. S. Orudzhev
*Effect of Pressure and Temperature on the Band Structure of TlSe
(b) Vol. **106**, No. 1, p. K67–K71 (1981).
See also 22.3, p. 442.

V. F. Kovalenko, G. P. Peka, and L. G. Shepel
*Photoluminescent Methods for Determination of Graded Band-Gap Semiconductor Parameters (II): Experimental Investigation of $Al_xGa_{1-x}As$
(a) Vol. **66**, No. 1, p. 407–414 (1981).
See also 13.3, p. 166; *20.3*, p. *330*; 22.2.4, p. 438.

E. P. Domashevskaya, L. N. Marshakova, V. A. Terekhov, A. N. Lukin, Ya. A. Ugai, V. I. Nefedov, and Ya. V. Salyn
*Role of Noble Metal d-States in the Formation of the Electron Structure of Ternary Sulphides
(b) Vol. **106**, No. 2, p. 429–435 (1981).
See also 17, p. 257; 20, p. 298; 22.8, p. 488.

S. I. Kurganskii and O. V. Farberovich
*Problems of the OPW Method (III): Rare Earths
(b) Vol. **106**, No. 2, p. 437–445 (1981).
See also 21.4, p. 382; 22.6, p. 478.

A. M. Berezhkovskii, A. A. Ovchinnikov, and R. A. Suris
A New Method for Detecting the Wannier-Stark Ladder in a Semiconductor in a Strong Electric Field
(b) Vol. **106**, No. 2, p. 461--466 (1981).
See also *14.3.4*, p. *235*; *20.1*, p. *314*; 22, p. 397.

D. M. Hatch
*Order Parameter Symmetry for the β–γ Transition in $Pb_3(VO_4)_2$
(b) Vol. **106**, No. 2, p. 473–479 (1981).
See also *1.2*, p. *18*; 22.8.1, p. 498.

A. Rauh and H. J. Schellnhuber
*Diamagnetic Band Structure. Comparison of Second-Order Perturbation with a First-Principles Calculation
(b) Vol. **106**, No. 2, p. 537–544 (1981).
See also 18, p. 263.

B. Fellmuth, H. Krüger, R. Rudolph, and R. Herrmann
*Composition Dependence of the Anisotropy of the Cyclotron Masses in n-Type Semiconducting $Bi_{1-x}Sb_x$.
(b) Vol. **106**, No. 2, p. 561–567 (1981).
See also *13.2*, p. *163*; 21.7, p. 388.

M. Wöhlecke and G. Borstel
*Electron-Spin Polarization in Hexagonal Crystals Induced by Circularly Polarized Light
(b) Vol. **106**, No. 2, p. 593–599 (1981).
See also 17, p. 257.

S. B. Trickey, A. K. Ray, and J. P. Worth
*Adequacy of Local-Exchange Excitation Hamiltonians in Insulators
(b) Vol. **106**, No. 2, p. 613–620 (1981).
See also 23, p. 514.

V. D. Iskra
*Phonon Assisted Interband Optical Transitions in Semiconductors Substaining a Coulomb Random Field
(b) Vol. **106**, No. 2, p. 627–633 (1981).
See also 6, p. 77; *20.1*, p. *315*; 22, p. 397.

V. I. Belinicher and V. N. Novikov
Photogalvanic Effect in Piezoelectrics. Quantitative Theory for Interband Transitions in Gallium Arsenide
(b) Vol. **107**, No. 1, p. 61–68 (1981).
See also *16*, p. *254*; 22.2.1, p. 430.

G. Simon and W. Eichler
*Investigations on a Two-Valence Band Model for Sb_2Te_3
(b) Vol. **107**, No. 1, p. 201–206 (1981).
See also 14.3, p. 216; 22.7, p. 483.

A. Heinrich, W. Cordts, and J. Monecke
*Electronic Band Structure of Pseudodirect Chalcopyrite Semiconductors (II): $ZnGeP_2$ and $ZnSiAs_2$
(b) Vol. **107**, No. 1, p. 319–333 (1981).
See also 22.8, p. 489.

V. I. Chizhikov, V. L. Panyutin, B. E. Ponedelnikov et A. E. Rozenson
*Masse effective des électrons et $E_g(x)$ dans la solution solide $CdIn_2(Se_{1-x}Te_x)_4$
(Effective Masses of Electrons and $E_g(x)$ in $CdIn_2(Se_{1-x}Te_x)_4$ Solid Solutions)
(b) Vol. **107**, No. 1, p. 347–350 (1981).
See also 22.8, p. 489.

W. Braune, N. Kubicki, U. Straubel, and R. Herrmann
*Electronic Peculiarities in $Bi_{1-x}Sb_x$ Alloys Close to the Semiconductor–Semimetal Transition at $x \approx 0.22$
(b) Vol. **107**, No. 1, p. K67–K70 (1981).
See also *19*, p. *292*; 21.7, p. 388.

Y. Yamada
*Band Calculation of Copper by an Interpolation Scheme Based on Recent Pseudopotential Methods
(b) Vol. **107**, No. 2, p. 503–512 (1981).
See also 21.1, p. 364.

I. Bartoš and F. Máca
On Self-Deconvolutions in Auger Spectroscopy
(b) Vol. **107**, No. 2, p. 513–519 (1981).
See also *17*, p. *257*.

H. Puff
On the Influence of Fluctuations on the Peierls Transition of the Quasi-One-Dimensional Fröhlich Model with a Nearly Half-Filled Band
(b) Vol. **107**, No. 2, p. 543–552 (1981).
See also *1.2*, p. *19*.

S. Manninen, B. K. Sharma, T. Paakkari, S. Rundqvist, and M. W. Richardson
*Electron Momentum Distribution and Charge Transfer Study of FeAl, CoAl, and NiAl
(b) Vol. **107**, No. 2, p. 749–755 (1981).
See also 13.2, p. 164; 21.1, p. 364; 21.1.1, p. 375.

G. P. Betteridge
*Comparison of the s Band Structure of Face Centred Cubic and Hexagonal Close Packed Solids (I): The Common Zone Approach
(b) Vol. **107**, No. 2, p. 791–799 (1981).

J. Humlíček and E. Schmidt
*Precise Determination of Gap Energies from Modulation Spectra
(b) Vol. **107**, No. 2, p. K105–K108 (1981).
See also 20.1, p. 318; 22.3, p. 442.

Yu. A. Pospelov and G. S. Grachev
*About the Character of Temperature Variations of Bi-Group Semimetal Electron Spectra
(b) Vol. **107**, No. 2, p. K155–K158 (1981).
See also 21.7, p. 389.

G. P. Betteridge
*Comparison of the s Band Structure of Face Centred Cubic and Hexagonal Close Packed Solids (II): Energy Correspondence
(b) Vol. **108**, No. 1, p. 245–253 (1981).

W. Grevendonk, L. van den Abeele, P. van den Keybus, B. Vanhuyse, and G. Ruymbeek
Faraday Rotation in Quartz and Oxide Glasses
(b) Vol. **108**, No. 1, p. K53–K56 (1981).
See also 2, p. 50; *18*, p. *264*; *20*, p. *299*; 22.6, p. 480; 22.6.1, p. 482.

M. Lähdeniemi, E. Ojala, and M. Okochi
*XPS Band Emission and Density of States of CoAl
(b) Vol. **108**, No. 1, p. K61–K64 (1981).
See also *17*, p. *258*; 21, p. 348; 21.1, p. 365.

H. van Cong, S. Brunet, and S. Charar
Effective Electric Field in the n-Type Heavily Doped Emitter Region of Silicon p–n Junction Solar Cells
(a) Vol. **68**, No. 1, p. K49–K53 (1981).
See also *14.3.3*, p. *230*; 22.1.2, p. 422.

B. R. NAG, A. N. CHAKRAVARTI, and P. K. BASU
Generalized Einstein Relation for the Diffusivity–Mobility Ratio in Multi-Band Degenerate Semiconductors
(a) Vol. **68**, No. 1, p. K75–K80 (1981).
See also *14.3*, p. *218*; 22, p. 400.

E. LAZO, J. RÖESSLER, and L. A. MORAGA
*Phase Transition and Spatial Correlation of Localized States in the Falicov-Ramírez-Kimball Model
(b) Vol. **108**, No. 2, p. 341–350 (1981).
See also *1.2*, p. *21*; 22.8, p. 491.

A. N. GUSATINSKII, M. A. BLOKHIN, P. SHUMICHEN, M. A. BUNIN, and S. S. MIKHAILOVA
*X-Ray L_1 Spectra and XPS Investigation of the Electronic Structure of Indium Compounds
(b) Vol. **108**, No. 2, p. 405–410 (1981).
See also *20*, p. *299*; 21, p. 349; 22, p. 400; 22.2, p. 426; 22.3, p. 442.

I. SINGH and G. P. SRIVASTAVA
*Semiempirical Pseudopotential Calculations for the Electronic Structure of Zincblende Semiconductors
(b) Vol. **108**, No. 2, p. 467–473 (1981).
See also 22.2.1, p. 431; 22.2.2, p. 434; 22.4.3, p. 455.

M. SURMA, S. MACIEJEWSKI, R. KUBIAK, and M. WOŁCYRZ
*Electronic Structure of the $Cd_{65}Hg_{35}$ Alloy and Its Temperature Dependence near the Phase Transition
(b) Vol. **108**, No. 2, p. 475–482 (1981).
See also 1.2, p. 21; 21, p. 349.

H. NEUMANN, H. SOBOTTA, W. KISSINGER, V. RIEDE, and G. KÜHN
*Hole Effective Masses in $CuInSe_2$
(b) Vol. **108**, No. 2, p. 483–487 (1981).
See also *20.1*, p. *320*; 22.8, p. 492.

U. BERG, T. CHASSÉ, and O. BRÜMMER
*Investigation of the XPS Valence Band Structure from Sn Chalcogenides
(b) Vol. **108**, No. 2, p. 507–510 (1981).
See also 17, p. 258; 22, p. 400.

A. BALDERESCHI, K. MASCHKE, A. MILCHEV, R. PICKENHAIN, and K. UNGER
*Thomas-Fermi Approximation for the Valence Electron Densities in Cubic Semiconductors and Insulators
(b) Vol. **108**, No. 2, p. 511–520 (1981).
See also 14.3, p. 218; 21.7, p. 389; 22.1.1, p. 407; 22.1.2, p. 422; 22.2.1, p. 431; 22.4.2, p. 453; 22.5.2, p. 469.

A. C. SHARMA and S. AULUCK
*Small Wave Vector $\boldsymbol{q}$ and Frequency ω Dependent Longitudinal Dielectric Function for a Model Semiconductor
(b) Vol. **108**, No. 2, p. 587–600 (1981).
See also *14.4*, p. *241*; 22.1, p. 403; 22.2, p. 426; 22.4, p. 446.

H. Boudriot, R. Gründler, B. Kubier, K. Deus, and H. A. Schneider
*Optical Properties of Single-Crystalline $ZnSiP_2$ from Electron Energy Loss Measurements between 3 and 40 eV
(b) Vol. **108**, No. 2, p. K129–K132 (1981).
See also *20.1*, p. *320*; 22.8, p. 492.

13.2 Fermi Surfaces

Original Papers and Short Notes

S. TEWARI and R. M. SINGRU
*Electronic States and Fermi Surface Properties of NiH
(b) Vol. **103**, No. 2, p. 745–751 (1981).
See also *13.1*, p. *153*; 22, p. 392.

S. B. SHRIVASTAVA, N. SHRIVASTAVA, and V. K. OJHA
*Positron Annihilation Rates and the Enhancement Factor in Metals
(b) Vol. **104**, No. 1, p. K61 K64 (1981).
Erratum
(b) Vol. **105**, No. 1, p. K71 (1981).
See also 21, p. 339.

R. D. BARNARD
Some Remarks on Deviations from Matthiessen's Rule in Dilute Alloy Systems
(b) Vol. **104**, No. 2, p. 613–620 (1981).
See also 6, p. 72; 10.1, p. 99; *14.1*, p. *200*; 15, p. 248; 21.6, p. 384.

R. HERRMANN, B. FELLMUTH, R. WEISSFLOG, and V. G. GOLUBEV
*Erratum to: Effect of the Anisotropic of the Fermi Surface on the Propagation of Helicon Waves Parallel to the Trigonal Axis
(b) Vol. **104**, No. 2, p. K169 (1981).
See also 14.3, p. 212; *19*, p. *288*; 21.7, p. 387.

P. RENNERT
Calculation of the Compton Profile of Beryllium
(b) Vol. **105**, No. 2, p. 567–575 (1981).
See also *13.1*, p. *156*; *20*, p. *297*; 21.3, p. 378.

D. G. KANHERE, R. M. SINGRU, and R. HARTHOORN
Electron Momentum Distribution and Compton Profiles of Ag
(b) Vol. **105**, No. 2, p. 715–722 (1981).
See also *13.1*, p. *156*; *20*, p. *298*; 21.6, p. 385.

B. FELLMUTH, H. KRÜGER, R. RUDOLPH, and R. HERRMANN
*Composition Dependence of the Anisotropy of the Cyclotron Masses in n-Type Semiconducting $Bi_{1-x}Sb_x$
(b) Vol. **106**, No. 2, p. 561–567 (1981).
See also *13.1*, p. *158*; 21.7, p. 388.

V. V. NEMOSHKALENKO, A. V. ZHALKO-TITARENKO, V. N. ANTONOV, VL. N. ANTONOV, and E. MROSAN
*Anisotropy of Cyclotron Masses in Gold
(b) Vol. **106**, No. 2, p. K117–K121 (1981).
See also *19*, p. *292*; 21.6, p. 385.

H. OVERHOF and W. BEYER
*Density of States in Amorphous Si:H
(b) Vol. **107**, No. 1, p. 207–213 (1981).
See also 14.3, p. 216; 15, p. 249; 22.1.2, p. 418.

13.3 Surface and Interface States

Original Papers and Short Notes

R. J. Jerrard, H. Ueba, and S. G. Davison
*Virtual Surface States of Thin Films
(b) Vol. **103**, No. 1, p. 353–359 (1981).
See also 22, p. 390.

K. I. Kirov and K. B. Radev
*A Simple Charge-Based DLTS Technique
(a) Vol. **63**, No. 2, p. 711–716 (1981).
See also *14.3.2*, p. *223*; 22.1.2, p. 410.

G. F. Golovanova, V. F. Kiselev, S. N. Kozlov, A. S. Petrov, and E. A. Silaev
*On the Effect of Halogens on Proton Processes in the Si–SiO_2 Structures
(a) Vol. **64**, No. 1, p. 177–186 (1981).
See also 19, p. 288; 22.1.2, p. 411; 22.6, p. 475.

J. Marczewski
*Low-Temperature Annealing of Surface States at the Si–SiO_2 Interfaces
(a) Vol. **64**, No. 2, p. K107–K110 (1981).
See also *14.3.2*, p. *223*; 22.1.2, p. 413.

S. N. Kozlov and N. L. Levshin
*On the Distribution Function of Activation Barriers Surmounted by Charge Carriers when Captured to Slow Surface States of a Semiconductor
(a) Vol. **64**, No. 2, p. K169–K172 (1981).
See also *14.3.2*, p. *223*; 22, p. 394.

A. Kiejna
*Simple Non-Local Calculation of Jellium Metal Surface Properties
(b) Vol. **105**, No. 1, p. 147–153 (1981).
See also 21, p. 341.

A. I. Bobrysheva, V. T. Zyukov, and S. A. Moskalenko
*The Excited States of the Surface Biexciton
(b) Vol. **105**, No. 1, p. K45–K49 (1981).
See also *13.5.1*, p. *190*; 22, p. 395.

H. M. Przewłocki, S. Krawczyk, and A. Jakubowski
A Simple Technique of Work Function Difference Determination in MOS Structure
(a) Vol. **65**, No. 1, p. 253–257 (1981).
See also *14.3.3*, p. *227*; 22.1.2, p. 414.

V. F. Kovalenko, G. P. Peka, and L. G. Shepel
Photoluminescent Methods for Determination of Graded Band-Gap Semiconductor Parameters (I): Analysis of Methods
(a) Vol. **65**, No. 2, p. 529–534 (1981).
See also 13.1, p. 157; *20.3*, p. *328*; 22.2.4, p. 437.

G. D. Bagratishvili, R. B. Dzhanelidze, D. A. Jishiashvili, L. V. Piskanovskii, A. N. Zyuganov, V. N. Mikhelashvili, and P. S. Smertenko
Mechanism of Charge Flow through the M–Ge_3N_4–GaAs Structure
(a) Vol. **65**, No. 2, p. 701–707 (1981).
See also *14.3.4*, p. *234*; 22.2.1, p. 429.

V. F. Kisilev, S. N. Kozlov, and N. L. Levshin
*On the Mechanism of Dissipation of Energy Released in the Capture of Charge Carriers to Adsorptive Slow States of a Semiconductor
(a) Vol. **66**, No. 1, p. 93–101 (1981).
See also 14.3.2, p. 224; 22.1.1, p. 405.

V. F. Kovalenko, G. P. Peka, and L. G. Shepel
Photoluminescent Methods for Determination of Graded Band-Gap Semiconductor Parameters (II): Experimental Investigation of $Al_xGa_{1-x}As$
(a) Vol. **66**, No. 1, p. 407–414 (1981).
See also 13.1, p. 158; *20.3*, p. *330*; 22.2.4, p. 438.

V. I. Grinev and V. F. Kiselev
*On the Nature and Energy Spectrum of Surface Recombination Centres
(a) Vol. **66**, No. 2, p. 493–502 (1981).
See also 22.1.1, p. 406.

J. Szuber
*On the Correlation of Paramagnetic Properties and Electronic Structure of (111) Si Surfaces Baked at Low Temperature in High Vacuum
(a) Vol. **66**, No. 2, p. K141–K144 (1981).
See also 17, p. 257; 22.1.2, p. 418.

K. Heilig, H. Flietner, J. Reineke, and Lê T. T. Tuyên
*Surface State Distributions at Silicon Surfaces, Influenced by Hydrogen Fluoride
(a) Vol. **66**, No. 2, p. K165–K168 (1981).
See also 22.1.2, p. 418.

L. N. Safronov, L. N. Aleksandrov, and R. N. Lovyagin
*Luminescent Analysis of an Impurity in the Near-Surface Layer of Thermally Evaporated Silicon
(b) Vol. **107**, No. 2, p. 461–468 (1981).
See also 13.4, p. 181; *20.3*, p. *332*; 22.1.2, p. 419.

K. Masuda
*Calculation of Atomic Relaxation near the (111) 1 × 1 Surface of Covalent Semiconductors: Tight-Binding Green's Function Approach
(b) Vol. **107**, No. 2, p. 529–533 (1981).
See also *1.6*, p. *39*; 22.1, p. 402; 22.1.1, p. 407; 22.1.2, p. 419.

M. Taut and M. Schubert
*Anisotropy of the Dipole Barrier of Metals
(b) Vol. **107**, No. 2, p. K139–K144 (1981).
See also 1.6, p. 39; 21, p. 346.

V. J. Rao, W. A. Anderson, G. Rajeswaran, and M. Thayer
Grain Boundary Effects in Poly-Si as Determined by Low Frequency Capacitance and Conductance Data
(a) Vol. **67**, No. 2, p. 709–715 (1981).
See also *14.3.3*, p. *230*; 22.1.2, p. 421.

Shen Yuyi
*Relativistic Effect on a Deformed Semi-Infinite Crystalline Surface
(b) Vol. **108**, No. 1, p. 255–260 (1981).

13.4 Impurity and Defect States

Review Articles

A. B. Roitsin
Generalized Spin-Hamiltonian and Low-Symmetry Effects in Paramagnetic Resonance
(b) Vol. **104**, No. 1, p. 11–35 (1981).
See also *19*, p. *286*; 22, p. 390.

E. Spenke and W. Heywang
Twenty Five Years of Semiconductor-Grade Silicon
(a) Vol. **64**, No. 1, p. 11–44 (1981).
See also *3*, p. *53*; *14.3*, p. *209*; 14.3.3, p. 225; 22.1.2, p. 408.

M. Narayana, V. S. Sivasankar, and S. Radhakrishna
Optical and Magnetic Resonance Studies of Transition Metal Ions and Complexes in Alkali Halides
(b) Vol. **105**, No. 1, p. 11–45 (1981).
See also 10.2, p. 106; *19*, p. *286*; *20.1*, p. *301*; 22.5.2, p. 462.

C. A. Hogarth and T. Iqbal
The Electrical and Structural Properties of Thin Films of Polymers Deposited from the Vapour Phase, with Special Reference to Polypropylene
(a) Vol. **65**, No. 1, p. 11–37 (1981).
See also *1.5*, p. *34*; *14.3.1*, p. *220*; 14.3.4, p. 232; 22.9, p. 507.

K. W. Böer
The CdS/Cu_2S Solar Cell (II): The Junction and Junction-Emitter Relation
(a) Vol. **66**, No. 1, p. 11–43 (1981).
See also *14.3.3*, p. *225*; 14.3.4, p. 232; *16*, p. *251*; 22.4.1, p. 447.

H. Bässler
*Localized States and Electronic Transport in Single Component Organic Solids with Diagonal Disorder
(b) Vol. **107**, No. 1, p. 9–54 (1981).
See also 2, p. 40; 13.5.1, p. 188; *14.3*, p. *209*; 14.3.4, p. 232; 20.1, p. 301; 20.3, p. 323; 22.9, p. 507.

Original Papers and Short Notes

Y. Maruyama and H. Bässler
Mobility and Recombination Coefficient of Charage Carriers in Disordered Pentacene
(b) Vol. **103**, No. 1, p. 263–268 (1981).
See also *16*, p. *251*; 22.9, p. 507.

J. Monecke, W. Siegel, E. Ziegler, and G. Kühnel
*On the Concentration Dependence of the Thermal Impurity-to-Band Activation Energies in Semiconductors
(b) Vol. **103**, No. 1, p. 269–279 (1981).
See also *14.3*, p. *209*; 22.2.2, p. 432; 22.8, p. 484.

R. C. du Varney, J. R. Niklas, and J. M. Spaeth
*ENDOR Study of Cubic Fe^{3+} Centres in $KMgF_3$
(b) Vol. **103**, No. 1, p. 329–336 (1981).
See also *19*, p. *286*; 22.8, p. 484.

Z. W. Gortel, J. Szymański, and L. Swierkowski
*Acceptor States in Zero-Gap Semiconductors in Magnetic Field. Theory for Short-Range Defect Potential
(b) Vol. **103**, No. 1, p. 429–439 (1981).
See also 13.1, p. 151; 18, p. 260; 22, p. 390.

M. Moreno
A Study about the Optical Absorption Spectra of KCl:Hg^+
(b) Vol. **103**, No. 1, p. K1–K3 (1981).
See also *20.1*, p. *302*; 22.5.2, p. 462.

E. V. Goncharova, V. V. Kaminskii, V. A. Kapustin, and I. A. Smirnov
Defects and Isomorphic Semiconductor–Metal Phase Transition in Samarium Monosulphide
(b) Vol. **103**, No. 1, p. K9–K11 (1981).
See also *1.2*, p. *14*; 22, p. 391.

N. T. Bagraev, L. S. Vlasenko, A. A. Lebedev, I. A. Merkulov, and P. Yusupov
*Exchange Impurity Levels and Photoconductivity of Semiconductors
(b) Vol. **103**, No. 1, p. K51–K54 (1981).
See also *16*, p. *251*; 22.1.2, p. 408.

J. Zieliński and M. Matlak
*On the Magnitude of the Local Spin in the Symmetric Anderson Model
(b) Vol. **103**, No. 1, p. K71–K74 (1981).
See also 21, p. 336.

H. Lemke
*Zum Nachweis von Rekombinationszentren in den Basiszonen von Si-Gleichrichtern
(On the Evidence of Recombination Centres in the Base Regions of Si Rectifiers)
(a) Vol. **63**, No. 1, p. 127–136 (1981).
See also 14.3.3, p. 225; 22.1.2, p. 408.

A. Cavallini, P. Gondi, and A. Castaldini
*Shallow Dislocation States in Ge. Thermal Treatment Effects
(a) Vol. **63**, No. 1, p. 143–149 (1981).
See also *10.2*, p. *106*; 14.3, p. 209; 22.1.1, p. 404.

M. J. J. Lammers and G. Blasse
*Luminescence of Sodium Molybdate ($Na_2Mo_2O_7$) and Sodium Tungstate ($Na_2W_2O_7$)
(a) Vol. **63**, No. 1, p. 157–161 (1981).
See also *20.3*, p. *324*; 22.8.1, p. 493.

I. L. Baginskii and E. G. Kostsov
*Transient Single-Injection Currents in Dielectrics at Fast Trapping
(a) Vol. **63**, No. 1, p. 271–279 (1981).
See also *14.3.4*, p. *232*; 14.4, p. 237.

V. Paidar
*Local Density of Electronic States at the Core of the Screw Dislocation in B.C.C. Lattice
(b) Vol. **103**, No. 2, p. K93–K97 (1981).
See also *10.1*, p. *97*; 21, p. 337.

S. Yu. Shashkin and A. E. Nikiforov
*Calculation of Splitting of 3d Transition-Metal Optical-Absorption Bands in Low-Symmetry Crystals
(b) Vol. **103**, No. 2, p. K105–K108 (1981).
See also *20.1*, p. *304*; 22, p. 392.

J. Monecke
*On the Insulator–Metal Transition in Doped Semiconductors
(b) Vol. **103**, No. 2, p. K115–K117 (1981).
See also *14.3*, p. *210*; 22, p. 392.

S. W. Biernacki and H.-J. Schulz
*Determination of the Crystal-Field Splitting Parameter Dq: Example of Ti^{2+} in CdTe
(b) Vol. **103**, No. 2, p. K163–K167 (1981).
See also *20.1*, p. *304*; 22.4.3, p. 454.

M. Krystek
*Determination of Model Parameters from Complex Spectra
(a) Vol. **63**, No. 2, p. 519–528 (1981).
See also 20.3, p. 324; 22.8.1, p. 494.

V. M. Lomako and P. Ya. Starostin
*Simulation of Electrical Characteristics and Properties of Defect Clusters in Irradiated Semiconductors (I): Approximation of Electric Charge Density
(a) Vol. **63**, No. 2, p. 585–593 (1981).
See also 14.3.4, p. 233; 22.2.1, p. 427.

A. I. Kolyubakin and S. A. Shevchenko
*On the Spectrum of Dislocation States in Germanium
(a) Vol. **63**, No. 2, p. 677–685 (1981).
See also *10.2*, p. *107*; 14.3, p. 210; 22.1.1, p. 404.

K. Asami, T. Naka, and M. Ishiguro
*Assignment of the Optical Absorption Bands in CsI:Tl and CsBr:Tl by Application of High Pressure
(b) Vol. **104**, No. 1, p. 145–149 (1981).
See also 1.2, p. 15; *20.1*, p. *305*; 22.5.2, p. 463.

A. C. van der Steen and L. T. F. Dijcks
The Luminescence Properties of Alkaline-Earth Oxides Activated with $6s^2$ Ions
(b) Vol. **104**, No. 1, p. 283–292 (1981).
See also 6, p. 70; *20.3*, p. *325*; 22.6, p. 475.

J. Pörsch and N. B. Manson
*Magnetic Circular Dichroism of the $^4A_{2g} \rightarrow {}^4T_{2g}$ Zero-Phonon Line in $MgO:Cr^{3+}$
(b) Vol. **104**, No. 1, p. K7–K10 (1981).
See also *20.1*, p. *305*; 22.6, p. 475.

G. Blasse
*Some Remarks on the Ce^{3+} and Bi^{3+} Luminescence in Alkaline-Earth Chalcogenides
(b) Vol. **104**, No. 1, p. K57–K59 (1981).
See also *20.3*, p. *325*; 22.4.1, p. 447; 22.6, p. 475.

M. Höhne and A. A. Lebedev
Changes in the EPR of Gold in Silicon Induced by Light
(b) Vol. **104**, No. 1, p. K79–K83 (1981).
See also *19*, p. *288*; 22.1.2, p. 410.

D. G. Galimov, K. D. Tarzimanov, and N. R. Yafaev
On the Nature of Colour Centres in Ultramarine and Blue Alkali Borate Glasses Containing Sulphur
(a) Vol. **64**, No. 1, p. 65–71 (1981).
See also *10.2*, p. *108*; 19, p. 288; 20.1, p. 306; 22.8.1, p. 495.

A. le Bloa, P. N. Favennec, and Y. Colin
*DLTS Method Using a Single Temperature Scanning
(a) Vol. **64**, No. 1, p. 85–93 (1981).
Erratum
(a) Vol. **66**, No. 2, p. 807 (1981).
See also 14.3, p. 210; 22.1.2, p. 411; 22.2.1, p. 428.

K. Schmalz and P. Gaworzewski
*On the Donor Activity of Oxygen in Silicon at Temperatures from 500 to 800 °C
(a) Vol. **64**, No. 1, p. 151–158 (1981).
See also 14.3, p. 211; 22.1.2, p. 411.

D. Schalch, A. Scharmann, and H. Weber
Electrical Conduction in Reactively Sputtered Silicon Nitride Films at Medium Field Strengths
(a) Vol. **64**, No. 1, p. 159–167 (1981).
See also *2*, p. *42*; *14.3.3*, p. *226*; 20.1, p. 306; 22, p. 393.

H. Lemke
*Dotierungseigenschaften von Eisen in Silizium
(Doping Levels of Iron in Silicon)
(a) Vol. **64**, No. 1, p. 215–224 (1981).
See also 14.3, p. 211; 22.1.2, p. 411.

S. V. Joshi and M. C. Joshi
*Investigation of the E_c − 0.55 eV Level of Iron in Silicon
(a) Vol. **64**, No. 1, p. K7–K11 (1981).
See also 22.1.2, p. 411.

P. M. Oliveira and B. Maffeo
*V_k Center in Cesium Halides by the MSXα Method
(b) Vol. **104**, No. 2, p. 453–458 (1981).
See also 22.5.2, p. 464.

B. Esser and I. P. Zvyagin
*On the Time Dependence of the Capture of Non-Equilibrium Hopping Charge Carriers
(b) Vol. **104**, No. 2, p. 523–529 (1981).
See also *2*, p. *43*; 22, p. 393.

W. Nolting and A. M. Oleś
*Local Quasiparticles in the Anderson Model
(b) Vol. **104**, No. 2, p. 563–573 (1981).

I. Yaeger, R. Shuker, and B. M. Wanklyn
*Optical Studies of Er^{3+} Ions in $Er_2Ge_2O_7$ Single Crystals
(b) Vol. **104**, No. 2, p. 621–625 (1981).
See also *20.1*, p. *307*; 22.8, p. 486.

B. Gaillard, C. Blanchard, A. Deville, and J. Livage
*Evidence of Two Paramagnetic Defects in WO_3-Doped V_2O_5
(b) Vol. **104**, No. 2, p. 627–630 (1981).
See also *19*, p. *288*; 22.6, p. 476.

Yu. G. Semenov and B. D. Shanina
*Exchange Interaction between Paramagnetic Centers and Valence Band Electrons in Semiconductors with Cubic Lattices
(b) Vol. **104**, No. 2, p. 631–639 (1981).
See also *13.1*, p. *154*; 18.1, p. 265; 22.4.2, p. 452; 22.4.3, p. 454.

G. Gevers, J. Pichon, and A. S. Barrière
Optical Absorption Study of Nickel Fluoride Thin Films
(b) Vol. **104**, No. 2, p. 641–647 (1981).
See also *20.1*, p. *307*; 22.5, p. 458.

S. Porowski, L. Konczewcz, J. Kowalski, R. L. Aulombard, and J. L. Robert
*High Magnetic Field Observation of the Resonance Donor States of S in InSb
(b) Vol. **104**, No. 2, p. 657–666 (1981).
See also *14.3*, p. *211*; 22.2.3, p. 435.

T. Tsuboi
*The Parity-Forbidden $4f^5 \rightarrow 4f^5$ Transition of the Sm^{3+} Ion in CaF_2
(b) Vol. **104**, No. 2, p. 697–701 (1981).
See also *20.1*, p. *307*; 22.5.3, p. 471.

V. M. Koleshko and N. Yu. Trifonov
*Fine Impurity Band Structure of One-Dimensional Disordered Crystal
(b) Vol. **104**, No. 2, p. K105–K108 (1981).
See also *13.1*, p. *155*; 22, p. 393.

M. I. Tarasik, V. D. Tkachev, V. U. Yavid, and A. M. Yanchenko
*Capture Mechanism of Holes by Radiation Defects with $E_c - 0.20$ eV Level in Sb-Doped Ge
(b) Vol. **104**, No. 2, p. K117–K120 (1981).
See also *11*, p. *123*; 22.1.1, p. 404.

D. A. Aronov and V. Zaitova
The Influence of Surface Recombination on the Photoelectromagnetic (PEM) Effect in Semiconductors with Deep Traps
(b) Vol. **104**, No. 2, p. K159–K162 (1981).
See also *16*, p. *252*; 22.2.1, p. 428.

F. Stöckmann
*An Exact Evaluation of Steady-State Space-Charge-Limited Currents for Arbitrary Trap Distributions
(a) Vol. **64**, No. 2, p. 475–483 (1981).
See also *14.3.4*, p. *233*; 22, p. 393.

M. Kitao, H. Akao, T. Ishikawa, and S. Yamada
Influence of Copper Addition on Electrical and Optical Properties of Amorphous As_2Se_3
(a) Vol. **64**, No. 2, p. 493–498 (1981).
See also 2, p. 44; *14.3*, p. *212*; *20.1*, p. *308*; 22, p. 394.

H. Lemke
*Eigenschaften der Dotierungsniveaus von Mangan und Vanadium in Silizium
(Properties of Manganese and Vanadium Doping Levels in Silicon)
(a) Vol. **64**, No. 2, p. 549–556 (1981).
See also 22.1.2, p. 412.

K. D. Glinchuk, K. Lukat, and A. V. Prokhorovich
*Non-Trivial Excitation Dependence of the Kinetics of the Extrinsic Luminescence Decay in GaAs
(a) Vol. **64**, No. 2, p. K139–K143 (1981).
See also *20.3*, p. *326*; 22.2.1, p. 428.

I. N. Timchenko and D. D. Nedeoglo
*Hopping Conduction and Energy Spectrum of Localized States in Low-Doped Intermediate-Compensated n-ZnSe
(b) Vol. **105**, No. 1, p. 55–61 (1981).
See also *14.3*, p. *212*; 22.4.2, p. 452.

R. K. Sinha and M. L. Mukherjee
*On the Kinetics of Phosphorescence and High-Temperature Glow Peak in CaF_2:Pr Single Crystals
(b) Vol. **105**, No. 1, p. 69–75 (1981).
See also *20.3*, p. *326*; 22.5.3, p. 471.

M. Kibler and G. Grenet
*The Configuration f^N in Cubic Symmetry: A Symmetry Adapted Weak-Field Approach
(b) Vol. **105**, No. 1, p. 185–191 (1981).

N. J. Trappeniers, F. S. Stibbe, and J. L. Rao
*Optical Absorption Spectrum of Cr^{3+} Ions Doped in CsCl
(b) Vol. **105**, No. 1, p. 243–247 (1981).
See also *20.1*, p. *309*; 22.5.2, p. 464.

H. Dersch, J. Stuke, and J. Beichler
*Electron Spin Resonance of Doped Glow-Discharge Amorphous Silicon
(b) Vol. **105**, No. 1, p. 265–274 (1981).
See also 2, p. 45; *19*, p. *289*; 22.1.2, p. 413.

Y. Ishihara and I. Nakada
*Electrical Conduction of GeSe at Low Temperatures
(b) Vol. **105**, No. 1, p. 285–290 (1981).
See also *14.3*, p. *212*; 22, p. 394.

S. Asano and N. Yamashita
Luminescence and Electron-Phonon Interaction in the MgS:Bi^{3+} Phosphor (I): Experiments
(b) Vol. **105**, No. 1, p. 305–310 (1981).
See also *6*, p. *74*; *20.3*, p. *327*; 22.4.1, p. 448; 22.6, p. 476.

V. S. Mironov, G. V. Ionova, V. S. Lubimov, I. N. Groznov, and S. P. Ionov
*Cooperative Jahn-Teller Effect and Spin-Orbit Coupling in d-Systems
(b) Vol. **105**, No. 1, p. 333–338 (1981).
See also 18.2, p. 270; 18.4, p. 282; 22.5, p. 459.

L. M. Brescansin and A. Fazzio
*Theoretical Study of Single Vacancies in GaSb
(b) Vol. **105**, No. 1, p. 339–345 (1981).
See also 10.2, p. 110; 22.2.3, p. 435.

E. Tataru and I. Boskovitz
*The Use of Magnetic Circular Dichroism Measurements in Determining the Reduced Matrix Elements of the Electric-Dipole Momentum Operator
(b) Vol. **105**, No. 1, p. K1–K4 (1981).
See also 18, p. 261; *20*, p. *297*; 22.8, p. 486.

S. J. Pearton
*Magnetic Field Dependence of Defect State in GaAs
(b) Vol. **105**, No. 1, p. K19–K22 (1981).
See also 14.3, p. 213; 18, p. 261; 22.2.1, p. 428.

L. V. Chernysh, V. A. Kovarskii, and E. P. Sinyavskii
*Laser Induced Auger Recombination in Doped Semiconductors
(b) Vol. **105**, No. 1, p. K51–K53 (1981).
See also 6, p. 74; 16, p. 253; 22, p. 395.

V. M. Lomako and P. Ya. Starostin
*Simulation of Electrical Characteristics and Properties of Defect Clusters in Irradiated Semiconductors (II): Self-Consistent Solution
(a) Vol. **65**, No. 1, p. 39–43 (1981).
See also 14.3.4, p. 234; 22.2.1, p. 428.

D. Shaw and H. R. Moghaddam
*A New Method for the Analysis of Thermally Stimulated Current Data
(a) Vol. **65**, No. 1, p. 167–176 (1981).
See also *14.3*, p. *213*.

S. Asano
Luminescence and Electron–Phonon Interaction in the MgS:Bi^{3+} Phosphor (II): Theory of $^3T_{1u}$ and $^3A_{1u}$ Mixing
(b) Vol. **105**, No. 2, p. 613–621 (1981).
See also *6*, p. *74*; *20.3*, p. *328*; 22.4.1, p. 448.

A. K. Saxena
*Non-Γ Deep Levels and the Conduction Band Structure of $Ga_{1-x}Al_xAs$ Alloys
(b) Vol. **105**, No. 2, p. 777–787 (1981).
See also *13.1*, p. *156*; 14.3, p. 214; 22.2.4, p. 437.

J. R. Meyer and F. J. Bartoli
*Ionized Impurity Scattering in n-Type ZnSe from Partial Wave Phase Shifts
(b) Vol. **105**, No. 2, p. K77–K80 (1981).
See also *14.3*, p. *214*; 22.4.2, p. 452.

P. Bussemer and A. Taubert
*To the Asymmetry of the Absorption Spectrum at Localized Centres
(b) Vol. **105**, No. 2, p. K107–K110 (1981).
See also 6, p. 75; *20.1*, p. *311*.

Yu. A. Gorokhovatskii, A. N. Gubkin, and A. I. Margulev
Thermally Stimulated Depolarization for Quasi-Continuous Distribution of Relaxators According to Activation Energy and Frequency Factor
(a) Vol. **65**, No. 2, p. 485–490 (1981).
See also *14.4*, p. *239*; 22.6, p. 477.

T. C. Leslie and J. W. Allen
Thermal Quenching of Photoluminescence in ZnS:Mn and ZnSe:Mn
(a) Vol. **65**, No. 2, p. 545–551 (1981).
See also *20.3*, p. *328*; 22.4.1, p. 448; 22.4.2, p. 452.

H. Nickel and E. Kuphal
*Deep Level Spectroscopy and Schottky Barrier Characteristics of LPE n- and p-InP
(a) Vol. **65**, No. 2, p. 583–588 (1981).
See also *14.3.4*, p. *234*; 22.2.2, p. 433.

V. S. Ballandovich, Yu. M. Tairov, and G. N. Violina
*Acceptor Boron in α-SiC (6H): Investigation by the Photocapacitance Method
(a) Vol. **65**, No. 2, p. 709–716 (1981).
See also *16*, p. *253*; 20.3, p. 329; 22.3, p. 441.

R. P. Khare and M. Khare
*A New Method of Analysis of Traps Taking Part in the Electroluminescence Process
(a) Vol. **65**, No. 2, p. 725–729 (1981).
See also *20.3*, p. *329*; 22.6, p. 477.

Z. Januškevičius, N. Korech, A. Sakalas, and I. Tychina
Electrical Properties of CdP_2 Single Crystals
(a) Vol. **65**, No 2, p. K149–K151 (1981).
See also *14.3*, p. *215*; 22.3, p. 441.

G. C. Barker and J. W. Tucker
*Orbit–Lattice Interaction Hamiltonian for Iron-Group Ions in Non-Cubic Sites
(b) Vol. **106**, No. 2, p. 529–536 (1981).
See also *6*, p. *77*; 22.6, p. 478.

V. Hizhnyakov, T. Soovik and S. Zazubovich
*Origin of Radiative Transitions from Metastable Minima of the Excited State of $Sn^{2+}v_c^-$ Centres in Alkali Halides
(b) Vol. **106**. No. 2, p. 581–586 (1981).
See also *20.3*, p. *331*; 22.5.2, p. 466.

T. Bhaskar Rao and M. Narayana
*A Quantitative Estimation of the Admixture in the Ground State Wave Function of Cu^{2+} in Low-Symmetry Crystal Fields
(b) Vol. **106**, No. 2, p. 601–606 (1981).

Yu. P. Gnatenko, Z. D. Kovalyuk, and P. A. Skubenko
*Influence of Ni Impurity on the Absorption Spectrum of Layered GaSe Crystals
(b) Vol. **106**, No. 2, p. 621–626 (1981).
See also *20.1*, p. *315*; 22.3, p. 442.

J. F. Pouradier and G. Grimouille
*Growth of Semiconducting CdF_2:Ho^{3+} Crystals and Study of Their Optical and Transport Properties
(b) Vol. **106**, No. 2, p. 657–665 (1981).
See also *3*, p. *55*; *14.3*, p. *215*; 20.1, p. 315; 22.5.3, p. 471.

O. Sild and I. Sildos
*Zeeman Effect in Low-Temperature Luminescence Spectra of Molecular Ions O_2^- and S_2^- in Alkali Halide Crystals
(b) Vol. **106**, No. 2, p. 675–682 (1981).
See also *20.3*, p. *331*; 22.5.2, p. 466.

H. Lemke
*Eigenschaften repulsiver Zentren in Silizium
(Properties of Repulsive Centres in Silicon)
(a) Vol. **66**, No. 2, p. 541–549 (1981).
See also 22.1.2, p. 417.

T. Takebe, H. Ono, J. Saraie, and T. Tanaka
*Photocapacitance Study of Deep Levels in n-CdTe Single Crystals
(a) Vol. **66**, No. 2, p. 725–736 (1981).
See also 16, p. 254; 22.4.3, p. 455.

N. Singh, G. L. Marwaha, and V. K. Mathur
Luminescence Centres and Charge Compensation in CaS Phosphors
(a) Vol. **66**, No. 2, p. 761–765 (1981).
See also *20.3*, p. *331*; 22, p. 397.

S. V. Moharil
*A New Method for Obtaining Trapping Parameters from Glow Curves
(a) Vol. **66**, No. 2, p. 767–772 (1981).
See also *20.3*, p. *331*; 22.5.2, p. 467.

V. D. Akhmetov, V. V. Bolotov, and L. S. Smirnov
*Optical Properties of Defects Produced at Annealing of Neutron-Induced Clusters in Silicon
(a) Vol. **67**, No. 1, p. K33–K37 (1981).
See also 11, p. 129; *20.1*, p. *317*; 22.1.2, p. 419.

B. B. Baizakov, R. A. Moominov, S. H. Shamirzaiev, and A. Ya. Yafasov
*Bias Current in Semiconductors during Overcharging of Impurity Centres
(a) Vol. **67**, No. 1, p. K59–K62 (1981).
See also *14.3*, p. *217*; 22, p. 398.

B. Olejníková, Ľ. Hrivnák, and M. Kedro
*The Model Potential for Charged Centres in Semiconductors
(b) Vol. **107**, No. 2, p. 451–460 (1981).
See also 22, p. 398.

L. N. Safronov, L. N. Aleksandrov, and R. N. Lovyagin
Luminescent Analysis of an Impurity in the Near-Surface Layer of Thermally Evaporated Silicon
(b) Vol. **107**, No. 2, p. 461–468 (1981).
See also *13.3*, p. *166*; *20.3*, p. *332*; 22.1.2, p. 419.

G. R. Augst and L. G. Bakhshetsyan
*Substitutional and Interstitial Hydrogen Impurities in Alkali Halide Crystals
(b) Vol. **107**, No. 2, p. 497–501 (1981).
See also *10.2*, p. *116*; 22.5.2, p. 268.

J. Dziesiaty, D. Backs, and D. Mantwitz
*Photosensitive EPR Spectra of Pb-Doped ZnS Single Crystals
(b) Vol. **107**, No. 2, p. 553–557 (1981).
See also *19*, p. *293*; 22.4.1, p. 449.

A. C. Mezger and C. Jaccard
*Electron Transfer within F Center Pairs in KCl from Optical Emission and Absorption
(b) Vol. **107**, No. 2, p. 689–697 (1981).
See also 20.1, p. 317; *20.3*, p. *332*; 22.5.2, p. 468.

K. Sierański and J. Szatkowski
Disorder Scattering in Mixed $Pb_{1-x}Sn_xTe$-Type Semiconductors
(b) Vol. **107**, No. 2, p. K87–K90 (1981).
See also *14.3*, p. *217*; 22.8, p. 490.

M. Baran, A. A. Mirzakhanyan, E. G. Sharoyan, and H. Szymczak
*Zero Field Splitting of the Mo^{3+} Ground State in Corundum
(b) Vol. **107**, No. 2, p. K129–K132 (1981).
See also *19*, p. *293*; 22.6, p. 479.

A. V. Voitsekhovskii and Yu. V. Lilenko
Recombination Effects in Narrow-Gap Semiconductors p-$Hg_{1-x}Cd_xTe$
(a) Vol. **67**, No. 2, p. 381–386 (1981).
See also *16*, p. *255*; 22.4.4, p. 457.

P. Kukk and A. Erm
The Structure of Radiative and Non-Radiative Recombination Centres in Activated CdS Phosphors
(a) Vol. **67**, No. 2, p. 395–406 (1981).
See also *10.2*, p. *117*; *20.3*, p. *333*; 22.4.1, p. 450.

J. Sinkkonen
*Photoconductivity of an Inhomogeneous Semiconductor
(a) Vol. **67**, No. 2, p. 555–564 (1981).
See also *16*, p. *255*; 22.1.3, p. 425; 22.4.1, p. 450.

X. Aymerich-Humet, F. Serra-Mestres, and J. Millán
On the Resonant Broadening of the Trap Assisted Tunneling in MIM Structures
(a) Vol. **67**, No. 2, p. 597–602 (1981).
See also *14.3.3*, p. *230*; *14.3.4*, p. *235*; 22.6, p. 479.

E. Klier
*A Simple Method of Evaluation of TSC Measurements
(a) Vol. **67**, No. 2, p. K151–K154 (1981).
See also *14.3*, p. *217*; 22, p. 399.

B. Olejníková
*The Influence of the Electric Field on the Electron Capture Coefficient of Screened Coulomb Centres
(b) Vol. **108**, No. 1, p. 79–85 (1981).
See also *14.3.4*, p. *235*; 22, p. 399.

A. H. Harker and J. M. Vail
*Theoretical Analysis of Some Ground State and Optical Properties of Impurity F Centers
(b) Vol. **108**, No. 1, p. 87–96 (1981).
See also *10.2*, p. *117*; 20.1, p. 318; 22.5.2, p. 408.

S. Marklund
*Energy Levels of Intrinsic and Extrinsic Stacking Faults in Silicon
(b) Vol. **108**, No. 1, p. 97–102 (1981).
See also *10.2*, p. *117*; 22.1.2, p. 421.

B. Esser
*On the Lifetime of Excited States in One-Dimensional Disordered Systems
(b) Vol. **108**, No. 1, p. 195–204 (1981).
See also *2*, p. *50*; 20.3, p. 333; 22, p. 399.

S. Asano, N. Yamashita, Y. Nakao et Y. Matsushima
*Luminescence et résonance paramagnétique de l'ion Mn^{2+} dans le luminophore MgS
(Luminescence and Paramagnetic Resonance of the Mn^{2+} Ion in MgS Phosphor)
(b) Vol. **108**, No. 1, p. 229–239 (1981).
See also 19, p. 294; 20.1, p. 319; 20.3, p. 333; 22.4.1, p. 450.

A. Andreeff, E. A. Goremychkin, H. Griessmann, L. P. Kaun, B. Lippold, W. Matz, O. D. Chistyakov, E. M. Savitskii, and P. G. Ivanitskii
*The Crystals Field in the Hexagonal Compound $PrCu_5$
(b) Vol. **108**, No. 1, p. 261–267 (1981).
See also 8, p. 88; 18, p. 264; 21.1, p. 365; 21.4, p. 383.

C. T. Ho and F. V. Wald
*Some Aspects of the Effect of Heat Treatment on the Minority Carrier Diffusion Length in Low Resistivity p-Type Silicon
(a) Vol. 68, No. 1, p. 103–107 (1981).
See also 14.3, p. 218; 22.1.2, p. 422.

H. L. Oczkowski
*Thermoluminescence in Conductive n-Type ZnSe Crystals
(a) Vol. 68, No. 1, p. 199–207 (1981).
See also *20.3*, p. *333*; 22.4.2, p. 453.

L. A. Borisova and M. Pasemann
The Influence of the Residual Impurities (C, O) on the Crystals Structure of GaAs
(a) Vol. 68, No. 1, p. 261–267 (1981).
See also *1.4*, p. *32*; *10.2*, p. *118*; 22.2.1, p. 430.

G. Bacquet, Vo Quang Truong, M. Vignoles, and G. Bonel
*ESR of the F^+ Centre in B-Type Carbonated Hydroxyapatite
(a) Vol. 68, No. 1, p. K71–K74 (1981).
See also *19*, p. *294*; 22.8.1, p. 501.

R. Kesavamoorthy and R. Krishnaswamy
*Effect of Solid Dilution and Temperature on the Crystal Field Spectra of Chromium Ions in Single Crystals of Chrome Alum
(b) Vol. 108, No. 2, p. 307–313 (1981).
See also *20.1*, p. *320*; 22.8, p. 491.

S. Öberg
*Electronic State Calculations of Dislocations in Cadmium Telluride
(b) Vol. 108, No. 2, p. 357–362 (1981).
See also 22.4.3, p. 455.

R. L. Kleinhenz, Y. H. Lee, J. W. Corbett, E. G. Sieverts, S. H. Muller, and C. A. J. Ammerlaan
*EPR Observation of an Au–Fe Complex in Silicon (I): Experimental Data
(b) Vol. 108, No. 2, p. 363–371 (1981).
See also *19*, p. *294*; 22.1.2, p. 422.

I. D. Mikhailov, S. V. Novikov, and A. V. Vannikov
*The Coulomb Shift of the Charge Transfer Band in Donor–Acceptor Compounds
(b) Vol. 108, No. 2, p. 399–404 (1981).

A. Haug
*Auger Recombination with Deep Impurities in Indirect Band Gap Semiconductors
(b) Vol. 108, No. 2, p. 443–448 (1981).
See also 20.3, p. 334; 22.1.2, p. 422.

F. Kellendonk and G. Blasse
*On the Luminescence of Sm^{3+}, Gd^{3+}, and Dy^{3+} in Yttrium Aluminium Borate
(b) Vol. 108, No. 2, p. 541–548 (1981).
See also *20.3*, p. *334*; 22.8.1, p. 501.

V. Damodara Das and N. Jayaprakash

*Initial Lattice Distortion Energy Spectra of Bismuth Thin Films and the Effect of Deposition Rate

(a) Vol. **68**, No. 2, p. K159–K164 (1981).

See also 10.1, p. 105; *14.1*, p. *204*; 21.7, p. 389.

13.5 Elementary Excitations (Phonons See 6)

Original Papers and Short Notes

U. Lindner
*On Solitary Solutions in a Modified Hubbard Chain
(b) Vol. **106**, No. 1, p. K77–K81 (1981).
See also 18.4, p. 283.

Ch. Högfors, J. Kubát, and M. Rigdahl
A Cooperative Model of Stress Relaxation in Solids
(b) Vol. **107**, No. 1, p. 147–155 (1981).
See also *12.1*, p. *136*; *12.2*, p. *140*; 21, p. 345; 22.5.2, p. 467; 22.9, p. 511.

F. A. Oliveira, M. G. Cottam, and D. R. Tilley
*Green Functions, Power, Spectra, and Raman Scattering of Thin Film Polaritons
(b) Vol. **107**, No. 2, p. 737–747 (1981).
See also *6*, p. *79*; *20*, p. *299*; 22.5.2, p. 468.

S. Modesti, A. Frova, J. L. Staehli, M. Guzzi, and M. Capizzi
*Electron–Hole Plasma Expansion in Direct-Gap $GaAs_{1-x}P_x$
(b) Vol. **108**, No. 1, p. 281–288 (1981).
See also 20.3, p. 333; 22.2.4, p. 438.

A. Czachor and A. Holas
*RPA Collective Excitations in the Two-Dimensional Fermion Systems
(b) Vol. **108**, No. 1, p. K33–K36 (1981).

K. Handrich and J. Resch
Nuclear Quadrupole Spin–Lattice Relaxation Time T_1 of Amorphous Solids
(b) Vol. **108**, No. 1, p. K57–K60 (1981).
See also *2*, p. *50*; 6, p. 81; *18*, p. *264*.

13.5.1 Excitons

Review Article

Original Papers and Short Notes

J. Biellmann, V. Gubanov, B. Meyer, and C. Schwab
*Anisotropy of Exciton Polaritons and Polariton–Phonon Interaction in 2H–PbI_2 Crystals (II): Anisotropy of the Photoluminescence of 2H–PbI_2
(b) Vol. **103**, No. 2, p. 697 – 701 (1981).
See also 6, p. 68; *20.3*, p. *324*; 22.5, p. 458.

P. Dub
*Reflection of Light on a Spatially Dispersive Crystal
(b) Vol. **104**, No. 1, p. 109 – 117 (1981).
See also *20.1*, p. *305*; 22.4.1, p. 447.

P. Petelenz
*Charge Carrier Detrapping by Triplet Excitons
(b) Vol. **104**, No. 1, p. 133 – 144 (1981).
See also 22.9, p. 509.

K. Hamada, T. Taguchi and S. Fujita
*Annealing Behaviour of Oxygen-Bound Exciton in Electron-Irradiated ZnTe
(b) Vol. **104**, No. 1, p. K27 – K30 (1981).
See also 11, p. 122; *20.3*, p. *325*; 22.4.3, p. 454.

M. Ya. Valakh, A. P. Litvinchuk, G. S. Pekar, and G. N. Polysskii
Multiple-Phonon Resonant Raman Scattering in $Zn_{1-x}Cd_xSe$ Crystals
(b) Vol. **104**, No. 2, p. 743 – 750 (1981).
See also *6*, p. *72*; *20.1*, p. *307*; 22.4.4, p. 456.

F. Thuselt
Erratum to: On the Phase Separation of Carriers in Nitrogen-Doped GaP
(b) Vol. **104**, No. 2, p. 752 (1981).
See also 20.3, p. 326; 22.2.2, p. 432.

J. Provoost and G. Jacobs
On the Formation Time of Absorption Centres in KI and KBr with a Stimulated Raman Probe Beam
(b) Vol. **104**, No. 2, p. K149 – K154 (1981).
See also *10.2*, p. *109*; 20.1, p. 308; 22.5.2, p. 464.

V. Angelini, M. Casalboni, M. Grandolfo, and P. Vecchia
*Exciton Thermoabsorption of $GaSe_{1-x}S_x$ Solid Solutions
(b) Vol. **105**, No. 1, p. 63 – 68 (1981).
See also *20.1*, p. *309*; 22.3.1, p. 444.

J. M. Hvam and A. Bivas
*Erratum to: Optical Gain and Induced Absorption in High-Density Exciton System in CuCl
(b) Vol. **105**, No. 1, p. 439 (1981).
See also *20.1*, p. *309*; 22.5, p. 459.

Yu. P. Gnatenko and P. A. Skubenko
*Exciton Absorption of GaSe Crystals in the Indirect Transition Region
(b) Vol. **105**, No. 1, p. K9 – K12 (1981).
See also *20.1*, p. *309*; 22.3, p. 441.

V. A. Benderskii, V. Kh. Brikenshtein, P. G. Filippov, and I. A. Shamovskii
*Non-Linear Fluorescence Quenching in Molecular Crystals (III): Recombination of Static Excitations
(b) Vol. **106**, No. 1, p. 407–418 (1981).
See also *20.3*, p. *330*; 22.9, p. 510.

S. Permogorov, A. Reznitzskii, V. Travnikov, S. Verbin, G. O. Müller, P. Flögel, and M. Nikiforova
*Emission of Localized Excitons in Mixed $CdS_{1-x}Se_x$ Crystals
(b) Vol. **106**, No. 1, p. K57–K60 (1981).
See also *20.3*, p. *330*; 22.4.4, p. 457.

A. Stahl
*Model of the Excitonic Surface Layer in Normal Incidence Reflection
(b) Vol. **106**, No. 2, p. 575–580 (1981).
See also *20.1*, p. *314*; 22.2.1, p. 430; 22.4.1, p. 449.

C. W. M. Timmermans and G. Blasse
On the Luminescence of $Cs_3Bi_2Br_9$ Single Crystals
(b) Vol. **106**, No. 2, p. 647–655 (1981).
See also *20.3*, p. *331*; 22.8, p. 488.

Yu. M. Fedorov, A. A. Leksikov, A. E. Aksyonov, and I. S. Edelman
Magnetic Linear Dichroism and Birefringence of $FeBO_3$ in the $^6A_{1g} \rightarrow {}^6T_{1g}$ Transition Region
(b) Vol. **106**, No. 2, p. K127–K130 (1981).
See also 6, p. 78; *20.1*, p. *315*; 22.8.1, p. 498.

Nguyen Van Hieu
*Resonant Photon–Photon Scattering and Approximate Spherical Symmetry in CuCl
(b) Vol. **107**, No. 1, p. 69–73 (1981).
See also 22.5, p. 459.

R. Kuhnert, R. Helbig, and K. Hümmer
*Photoluminescence of ZnO Single Crystals. Influence of k-Linear Interaction and Non-Analytic Long-Range Exchange on the Emission Spectrum of the Free A-Exciton
(b) Vol. **107**, No. 1, p. 83–92 (1981).
See also *20.3*, p. *331*; 22.4, p. 445.

R. J. Bartlett, C. G. Olson, and D. W. Lynch
Absorption Spectra of MgF_2 and AlF_3 in the Region of the Mg and Al $L_{2,3}$ Excitations
(b) Vol. **107**, No. 1, p. 93–97 (1981).
See also *20.1*, p. *316*; 22.5.3, p. 472.

G. Blattner, C. Klingshirn, R. Helbig, and R. Meinl
*The Influence of a Magnetic Field on the Ground and Excited States of Bound Exciton Complexes in ZnO
(b) Vol. **107**, No. 1, p. 105–115 (1981).
See also 20.1, p. 316; 20.3, p. 332; 22.4, p. 445.

V. V. Hizhnyakov, V. G. Plekhanov, V. V. Shepelev, and G. S. Zavt
*Hot Luminescence of Self-Trapped Excitons in Alkali Halide Crystals
(b) Vol. **108**, No. 2, p. 531–540 (1981).
See also *20.3*, p. *334*; 22.5.2, p. 469.

M. Paić and V. Paić
Diffuse Reflectance Spectra and Polymorphism of Mercury(II) Iodide
(a) Vol. **68**, No. 2, p. 447–455 (1981).
See also 1, p. 10; *20.1*, p. *321*; *20.3*, p. *334*; 22.5, p. 460.

13.5.2 Plasmons

Original Papers and Short Notes

O. Ziep
Band-to-Band Recombination via Spontaneous Emission of Phonon–Plasmon Coupled Modes in Lead Chalcogenides
(b) Vol. **103**, No. 1, p. 239–245 (1981).
See also 6, p. 66; *13.1*, p. *151*; 22.8, p. 484.

M. Rösler and W. Brauer
Theory of Secondary Electron Emission (I): General Theory for Nearly-Free-Electron Metals
(b) Vol. **104**, No. 1, p. 161–175 (1981).
See also *17*, p. *256*; 21, p. 338.

D. N. Tripathy and R. K. Pal
*Plasmon Dispersion in Lithium for Large Wave Vectors
(b) Vol. **104**, No. 1, p. 345–356 (1981).
See also 21, p. 338.

K. S. Srivastava and O. K. Harsh
*Double Plasmon High Energy Satellites in the X-Ray Emission Spectra of Metals
(b) Vol. **104**, No. 1, p. K35–K38 (1981).
See also *20*, p. *296*; 21, p. 339; 21.1, p. 355; 21.2, p. 376.

K. Berndt, C. Wenzel, T. Dörre, T. Herzel, and A. Neumann
*Electron-Energy Loss Spectra of Chromium in the Reflection Mode
(b) Vol. **104**, No. 1, p. K49–K52 (1981).
See also 21.1, p. 355.

B. Kh. Ishmukhametov, M. I. Katsnelson, V. N. Larionov, and A. M. Ustyuzhanin
*Spectrum of Oscillations of the Inhomogeneous Electronic Plasma
(b) Vol. **104**, No. 1, p. K75–K78 (1981).

M. Rösler and W. Brauer
Theory of Secondary Electron Emission (II): Application to Aluminum
(b) Vol. **104**, No. 2, p. 575–587 (1981).
See also *17*, p. *256*; 21, p. 340.

G. D. Priftis and J. Boviatsis
*Negative Volume Plasmon Dispersion in Be for $k > k_F$
(b) Vol. **104**, No. 2, p. 673–677 (1981).
See also 20, p. 297; 21.3, p. 378.

W. Braune, U. Straubel, N. Kubicki, J. Lebech, and K. Saermark
*Relaxation Time and Effective Masses for Charge Carriers in Semimetallic BiSb Alloys versus Alloy Composition
(b) Vol. **106**, No. 2, p. 587–592 (1981).
See also *13.5.4*, p. *197*; 19, p. 291; 21.7, p. 388.

M. J. Gelten and L. A. Bosch
*A New Method to Determine the Coupled Mode Parameters of a Plasmon–Multiphonon System
(b) Vol. **106**, No. 2, p. 635–645 (1981).
See also *6*, p. *77*; 20.1, p. 315; 22, p. 397.

B. Kubier, H. Boudriot, K. Deus, R. Gründler, H. Zscheile, and E. Buhrig
*Plasmon Data, Dielectric Function, and Optical Constants of $CdGeP_2$ from Electron Energy Loss Measurements between 4.5 and 35 eV
(b) Vol. **106**, No. 2, p. K103–K106 (1981).
See also *20.1*, p. *315*; 22.8, p. 488.

P. F. Robusto and R. Braunstein
*Optical Properties of the Surface Plasmon of n^+ Silicon
(b) Vol. **107**, No. 1, p. 127–138 (1981).
See also *20.1*, p. *316*; 22.1.2, p. 418.

P. F. Robusto and R. Braunstein
*Optical Measurements of the Surface Plasmon of Copper
(b) Vol. **107**, No. 2, p. 443–449 (1981).
See also *20.1*, p. *317*; 21.1, p. 364.

R. Zimmermann, M. Rösler, and V. M. Asnin
Direct Absorption inside the Electron–Hole Liquid in Ge: Warping Effects and Final State Interactions
(b) Vol. **107**, No. 2, p. 579–590 (1981).
See also *13.5.1*, p. *192*; *20.1*, p. *317*; 22.1.1, p. 407.

U. Nowak, W. Richter, and G. Sachs
*LO-Phonon–Plasmon Dispersion in GaAs. Hydrodynamical Theory and Experimental Results
(b) Vol. **108**, No. 1, p. 131–143 (1981).
See also *6*, p. *80*; 20.1, p. 318; 22.2.1, p. 430.

M. Kobayashi and I. Yokota
*Collective Motions in a Metal-Doped Molten Salt
(b) Vol. **108**, No. 2, p. 351–355 (1981).
See also *6*, p. *81*; 22.5.1, p. 461.

A. M. Brodskii, L. I. Daikhin, and M. I. Urbakh
On the Polarization Dependence of Photoemission from Metals
(b) Vol. **108**, No. 2, p. 653–662 (1981).
See also *17*, p. *258*; 21, p. 349.

13.5.3 Polarons

Original Papers and Short Notes

G. Bader and Y. Lépine
*A Two-Particle Model for the Ground State Energy of the Polaron
(b) Vol. **103**, No. 1, p. 177–184 (1981).
See also 22, p. 390.

A. A. Gogolin
*Mobility of a Small Polaron at Low Temperatures
(b) Vol. **103**, No. 1, p. 397–409 (1981).
See also *14.3*, p. *209*; 22, p. 390.

P. Kuivalainen, J. Sinkkonen, and T. Stubb
*Bound Magnetic Polaron in Europium Chalcogenides. Magnetic Properties
(b) Vol. **104**, No. 1, p. 299–306 (1981).
See also *18.2*, p. *268*; 22, p. 392.

W. Elflein, W. Käppler, and G. Arlt
Drift Mobility in Insulating Barium Titanate Ceramics
(a) Vol. **64**, No. 1, p. K1–K4 (1981).
See also *14.4.1*, p. *242*; 22.8.1, p. 495.

J. van Royen, F. Beleznay, and J. T. Devreese
On the Equivalence of Different Approximations in Linear Electron Transport Theory
(b) Vol. **107**, No. 1, p. 335–345 (1981).
See also *14.3*, p. *216*; 22, p. 398.

A. V. Sherman
*Enhancement of Polaron Effect in Thin Dielectric Films
(b) Vol. **108**, No. 1, p. 71–78 (1981).

F. M. Peeters and J. T. Devreese
*Analytic Expression for the Zero Temperature Solution of the Polaron Boltzmann Equation
(b) Vol. **108**, No. 1, p. K23–K27 (1981).

J. T. Devreese and F. Brosens
*Momentum Distribution Function of Polarons in the Ohmic Regime
(b) Vol. **108**, No. 1, p. K29–K32 (1981).
See also 14.3, p.218.

13.5.4 Magnons

Original Papers and Short Notes

DIEP-THE-HUNG
Temperature-Dependent Surface Magnetization and Critical Temperature of Ferromagnetic Thin Films
(b) Vol. **103**, No. 2, p. 809–815 (1981).
See also *18.2.1*, p. *276*.

C. G. PATIL and B. S. KRISHNAMURTHY
Magnetic Field Dependent Carrier Mobility in Ferromagnetic Semiconductors
(b) Vol. **105**, No. 1, p. 391–397 (1981).
See also *14.3*, p. *212*; *18.2*, p. *270*; 22.6, p. 476.

K. M. HÄUSLER
*Infrared Spectra of Magnetic Excitations in CoF_2
(b) Vol. **105**, No. 2, p. K81–K83 (1981).
See also *20.1*, p. *311*; 22.5, p. 459.

E. KOLLEY and W. KOLLEY
*Functional Integral Approach to the Electron–Magnon Interaction in Itinerant Ferromagnetism
(b) Vol. **105**, No. 2, p. K85–K89 (1981).
See also 18.2, p. 271; 21, p. 343.

W. BRAUNE, U. STRAUBEL, N. KUBICKI, J. LEBECH, and K. SAERMARK
*Relaxation Time and Effective Masses for Charge Carriers in Semimetallic BiSb Alloys versus Alloy Composition
(b) Vol. **106**, No. 2, p. 587–592 (1981).
See also *13.5.2*, p. *194*; 19, p. 291; 21.7, p. 388.

14. Electrical Properties. Transport Phenomena

Original Papers and Short Notes

V. Madhusudan, S. G. Sathyanarayan, and G. Sivarama Sastry
*Electrical Conductivity and Stimulated Thermocurrent Studies in Dicalcium Barium Propionate Single Crystals
(a) Vol. **66**, No. 1, p. 371–376 (1981).
See also 15, p. 249; 22.9, p. 511.

R. S. Sorbello
*Boltzmann Equation and Force–Force Correlation Function for Electrical Resistivity
(b) Vol. **107**, No. 1, p. 233–243 (1981).

M. K. Kostov, N. P. Ivanchev, and E. M. Kostova
*Experimental Study of Electrogyration in the Centrosymmetric Crystals $Pb_5GeO_4(VO_4)_2$ and $Pb_5SiO_4(VO_4)_2$
(a) Vol. **67**, No. 1, p. K71–K74 (1981).
See also *20*, p. *298*; 22.8, p. 490.

G. Röpke and F. E. Höhne
*Quantum Transport Theory of Impurity and Electron–Electron Scattering Resistivity in n-Type Semiconductors Using the Correlation Function Method
(b) Vol. **107**, No. 2, p. 603–615 (1981).
See also *13*, p. *148*.

M. Wagner
*Aspects of Electrical Conductivity in a Moving Base Approach
(b) Vol. **107**, No. 2, p. 617–636 (1981).
See also *13*, p. *148*; 14.2, p. 207.

L. A. Manakova, B. G. Zemskov, and S. P. Ionov
Some Peculiarities of the Temperature Dependence of the Electronic Conductivity in Charge-Ordered Crystals
(b) Vol. **107**, No. 2, p. 801–812 (1981).
See also *13*, p. *148*; 21, p. 346.

V. A. Kazakov and A. M. Satanin
*A Renormalization Group Approach for Randomly Distributed Conductance
(b) Vol. **108**, No. 1, p. 19–28 (1981).

14.1 Metals. Semi-Metals

Original Papers and Short Notes

I. Vávra, P. Lobotka, F. Zachar, and J. Osvald
*TEM in-Situ Observation of Electromigration Damage in Al–Cu Strips (I): Constant DC Stressing
(a) Vol. **63**, No. 1, p. 363 – 370 (1981).
See also 1.5, p. 34; *9*, p. *89*; 10.1, p. 96; 21, p. 336; 21.1, p. 352.

P. V. Ashrit and M. A. Angadi
*Electrical Properties of Thin Ytterbium Films
(a) Vol. **63**, No. 1, p. K77 – K82 (1981).
See also 21.4, p. 380.

P. Středa and P. Vašek
Thermopower in Strong Magnetic Fields and the Mott Rule
(b) Vol. **103**, No. 2, p. K137 – K140 (1981).
See also *15*, p. *247*; 21.7, p. 387.

R. Vetter and J. Vuik
*Resistivity and Curie Point of NiRh Alloys
(a) Vol. **63**, No. 2, p. 637 – 643 (1981).
See also 18.2, p. 268; 21.1, p. 354.

P. Lobotka and I. Vávra
*TEM in-Situ Observation of Electromigration Damage in Al–Cu Strips (II): Superimposed AC and DC Stressing
(a) Vol. **63**, No. 2, p. 655 – 661 (1981).
See also 1.5, p. 34; *9*, p. *89*; 10.1, p. 97; 21, p. 337; 21.1, p. 354.

U. Behn
Nearest-Neighbour Poisson Approximation to a Functional Approach for the Hubbard Model
(b) Vol. **104**, No. 1, p. 203 – 208 (1981).
See also *13.1*, p. *153*.

J. M. Roland, X. Quillard, and A. Moreau
Study of the α-Phase of Cu–Al Alloys. Investigations of the Near-Equilibrium Disordered State
(a) Vol. **64**, No. 1, p. 45 – 51 (1981).
See also *1.3*, p. *22*; 21.1, p. 355.

Y. D. Yao and S. Arajs
*Electrical Resistivity of Fe-Rich Fe–Pd Alloys between 77 and 1250 K
(a) Vol. **64**, No. 1, p. 95 – 104 (1981).
See also 1.2, p. 15; 21.1.1, p. 369.

L. G. Khvostantsev and V. A. Sidorov
*High-Pressure Polymorphism of Antimony. Thermoelectric Properties and Electrical Resistance Studies
(a) Vol. **64**, No. 1, p. 379 – 384 (1981).
See also 1.2, p. 16; *15*, p. *247*; 21.7, p. 387.

K. Elk
*On the Electrical Conductivity of the Periodic Anderson Model
(b) Vol. **105**, No. 2, p. 507–513 (1981).
Erratum
(b) Vol. **107**, No. 1, p. 387 (1981).
See also *13*, p. *145*; 21, p. 343.

W. Kraak, R. Kuhl, F. Thom, H. Haefner, and P. Schmidt
*Size Effects and Temperature Dependence of the Electrical Resistance in Bismuth Single Crystals of High Perfection
(b) Vol. **105**, No. 2, p. 597–604 (1981).
See also 21.7, p. 387.

D. V. Gitsu, F. M. Muntyanu, and M. I. Onu
*Quantum Oscillations of Thermomagnetic Power and Magnetoresistance in Antimony-Bismuth Alloys
(b) Vol. **105**, No. 2, p. K95–K98 (1981).
See also *15*, p. *248*; 21.7, p. 387.

H. Gruber, E. Krautz, and H. P. Fritzer
Magnetic Susceptibility and Electrical Properties of the $\varkappa$-Phase $Mo_{17}O_{47}$ with Substitutions of Molybdenum
(a) Vol. **65**, No. 2, p. 589–594 (1981).
See also *14.3*, p. *214*; *18*, p. *262*; 18.4, p. 283; 22.6.1, p. 481.

J. Mimault, J. Delafond et J. Grilhé
*Commentaires sur la résistivité électrique des alliages Al–Zn et la décomposition spinodale
(Comments on the Electric Resistivity and Spinodal Composition of Al–Zn Alloys)
(a) Vol. **65**, No. 2, p. K121–K125 (1981).
See also 10.1, p. 101; 21.1, p. 360.

H. Löffler und C. Radomský
*Einige Bemerkungen zur Arbeit "Commentaires sur la résistivité électrique des alliages Al–Zn et la décomposition spinodale"
(Some Remarks to the Paper "Comments on the Electric Resistivity and Spinodal Composition of Al–Zn Alloys")
(a) Vol. **65**, No. 2, p. K127–K129 (1981).
See also 10.1, p. 101; 21.1, p. 360.

W. Heimke and V. I. Vovikov
*On the Transport Properties of Chromium–Alluminium Alloys at Low Temperatures
(a) Vol. **65**, No. 2, p. K139–K142 (1981).
See also 8, p. 86; *15*, p. *248*; 21.1, p. 360.

M. Matlak and J. Zieliński
*On the Electrical Conductivity and the Density of States for the Narrow-Band Periodic Anderson Model in the Symmetric Case
(b) Vol. **106**, No. 1, p. 73–77 (1981).
See also *13.1*, p. *157*; 21, p. 344.

L. B. Dubovskii
Density of States in Multilayer Sandwiches
(b) Vol. **106**, No. 1, p. 315–325 (1981).
See also *13*, p. *146*; 14.2, p. 206; 21, p. 344; 21.7, p. 388.

V. F. Degtyareva, V. E. Antonov, I. T. Belash, and E. G. Ponyatovskii
Distortions of F.C.C. Crystal Lattice of $Pd_{60}Cu_{40}$ Alloy Hydrogenated under High Pressure
(a) Vol. **66**, No. 1, p. 77–86 (1981).
See also *1.2*, p. *17*; 12.1, p. 136; 21.1, p. 360.

J. M. Roland, X. Quillard, and S. K. Marya
On the Kinetics of α-Phase Ordering in Cooper–Aluminium Alloys
(a) Vol. **66**, No. 1, p. 347–360 (1981).
See also *1.3*, p. *24*; 21.1, p. 360.

F. Ogiba, W. H. Schreiner, J. V. Kunzler, and D. E. Brandão
*Electrical Resistivity of the Ni_2MnSn Heusler Alloy
(a) Vol. **66**, No. 1, p. K83–K86 (1981).
See also 21.1, p. 361.

G. Frommeyer and H. G. Brion
*Hall Effect in Directionally Solidified and Deformed Eutectic Silver–Copper Alloys with Fibre Structure
(a) Vol. **66**, No. 2, p. 559–564 (1981).
See also 21.1, p. 362; 21.6, p. 385.

A. Handstein, J. Kadlecová, and V. Gregor
*Magnetoresistance of Amorphous $(Fe_xNi_{100-x})_{80}B_{10}P_{10}$ Alloys in Magnetic Fields up to 14.5 T
(a) Vol. **66**, No. 2, p. 657–663 (1981).
See also *2*, p. *48*; 18.2, p. 273; 21.1, p. 362; 21.1.1, p. 373.

A. P. Zhernov
*Temperature Dependence of the Righi-Leduc Coefficients for Na and K Metals
(b) Vol. **107**, No. 2, p. 411–421 (1981).
See also 6, p. 79; *8*, p. *87*; 21.2, p. 377.

D. Sen and S. K. Sarkar
*Unified Study of Metals on Pseudopotential Theory: Lead
(b) Vol. **107**, No. 2, p. 559–567 (1981).
See also 2, p. 49; *6*, p. *79*; 12.1, p. 136; 15, p. 249; 21, p. 346.

M. Idrees, Farid A. Khwaja, and M. S. K. Razmi
*A Comparative Study of the Pseudopotentials for Cu, Ag, and Au
(b) Vol. **107**, No. 2, p. 769–778 (1981).
See also *13*, p. *148*; 15, p. 249; 21.1, p. 364; 21.6, p. 386.

M. Matlak and J. Zieliński
*Electrical Resistivity and Insulator-to-Metal Transition for the Periodic Anderson Model
(b) Vol. **107**, No. 2, p. K133–K137 (1981).
See also *14.3*, p. *217*.

R. Franke, W. Kleinert, and W. Schmidt
*Vacancy Behaviour in the Course of One Loading Cycle in Push–Pull Deformed Polycrystalline Nickel
(a) Vol. **67**, No. 2, p. 469–476 (1981).
See also *10.1*, p. *103*; 21.1, p. 364.

L. Vatskitchev
Planar Hall Effect in Double Layer Magnetic Films with Unidirectional Anisotropy
(a) Vol. **67**, No. 2, p. 591–596 (1981).
See also *18.2.1*, p. *277*; *18.4*, p. *284*; 21.1, p. 364; 21.1.1, p. 375.

M. A. Angadi and P. V. Ashrit
*The Effect of Substrate Temperature and Grain Boundary Scattering on the Electrical Resistivity of Thin Samarium Films
(a) Vol. **67**, No. 2, p. K119–K122 (1981).
See also 21.4, p. 382.

H. K. Lachowicz
*Remarks on the Extraordinary Hall Effect in Amorphous $Cd_{1-x}Co_x$ Sputtered Films
(a) Vol. **67**, No. 2, p. K131–K135 (1981).
See also *18.2.1*, p. *277*; 21.1, p. 365; 21.4, p. 382.

A. H. Abou El Ela, S. Mahmoud, and M. A. Mahmoud
*Electrical Conduction of Thin Bismuth Films
(a) Vol. **67**, No. 2, p. K155–K160 (1981).
See also 21.7, p. 389.

J. Pietrzak, J. Kudyńska, and Z. Kruczyński
Phase Transition Study in VO_2 Using the EPR Method
(a) Vol. **67**, No. 2, p. K175–K178 (1981).
See also 14.3, p. 218; *19*, p. *293*; 22.6, p. 480.

J. S. Brown
*Electrical Resistivity of Liquid Ti at Its Melting Point
(b) Vol. **108**, No. 1, p. K65–K68 (1981).
See also *2*, p. *50*; 21, p. 348.

H. Haller, D. Korn, and G. Zibold
*Electrical Resistivity of CuCo Solid Solutions at Low Temperature
(a) Vol. **68**, No. 1, p. 135–141 (1981).
See also 1.5, p. 37; 21.1, p. 366; 21.6, p. 386.

I. A. Tsoukalas
*On the Atomic Ordering of Fe-Rich, $Fe_{1-x}Ni_x$ Binary Alloys
(a) Vol. **68**, No. 1, p. K67–K70 (1981).
See also *18.2*, p. *274*; 21.1, p. 366; 21.1.1, p. 375.

C. R. Pichard, C. R. Tellier, L. Ouarbya, and A. J. Tosser
*Perturbations in Transport Properties of Thin Metal Films Due to Energy-Dependent Relaxation Time
(a) Vol. **68**, No. 2, p. 477–482 (1981).
See also *15*, p. *250*; 21, p. 349.

L. Wojtczak, A. Urbaniak-Kucharczyk, J. Mielnicki, and B. Mrygon
*The Magnetic Contribution to Electrical Resistivity in Some Thin Ferromagnetic Films
(a) Vol. **68**, No. 2, p. 537–544 (1981).
See also 18.2.1, p. 277.

14.2 Superconductivity. Superconducting Materials and Devices

Review Article

H. R. KHAN and K. LÜDERS
Nuclear Magnetic Resonance on Amorphous Alloys
(b) Vol. **108**, No. 1, p. 9–18 (1981).
See also *2*, p. *40*; *19*, p. *286*; 21, p. 336; 21.1, p. 351; 21.3, p. 378.

Original Papers and Short Notes

H.-G. MEYER and P. SEIDEL
*Theoretical Investigation of Current Amplitudes of Niobium–Lead Tunnel Junctions
(b) Vol. **103**, No. 1, p. 281–287 (1981).

E. M. SAVITSKII, V. B. GRIBULYA, and N. N. KISELEVA
*Forecasting of Superconducting Compounds
(a) Vol. **63**, No. 1, p. K67–K72 (1981).

N. F. FEDOROV and C. HEIDEN
*Some Observations on the Influence of Oxygen on the Resistive Behavior of Superconducting Vanadium Foils
(a) Vol. **63**, No. 1, p. K83–K86 (1981).
See also 21.1, p. 353.

E. KOLLEY, W. KOLLEY, A. BECKMANN
*Transition Temperature of Superconducting Alloys with Random Transfer Integrals
(b) Vol. **103**, No. 2, p. 775–781 (1981).
See also 13, p. 143; 21.1, p. 353.

E. K. NAIMI
*Acoustoelectric and Acoustomagnetic Effects in Superconductors
(b) Vol. **103**, No. 2, p. K109–K113 (1981).
See also *7*, p. *82*.

J. SOSNOWSKI, M. DRYŚ, and T. MYDLARZ
*The Influence of Cr and Ni Impurities on Superconducting Properties of $Nb_{80}Ga_{20}$ Alloys with A15 type Structure
(a) Vol. **63**, No. 2, p. 693–698 (1981).
See also 21, p. 338.

H.-U. HABERMEIER
*Structure and Superconducting Properties of Nb_3Ge Prepared in a UHV System
(a) Vol. **64**, No. 1, p 195–206 (1981).
See also *1.3*, p. *23*; 10.1, p. 98; 21, p. 339.

M. CISZEK, G. KOZŁOWSKI, and P. TEKIEL
*Occurrence of AC Loss Minimum in Type II Superconductors
(a) Vol. **64**, No. 1, p. K35–K38 (1981).

M. Glugla, H. G. Severin, and G. Sicking
Low-Temperature Heat Capacities of Pd–Al Alloys at Low Al Contents
(b) Vol. **106**, No. 2, p. 505–509 (1981).
See also *8*, p. *86*; 21, p. 344.

M. V. Medvedev and Yu. N. Skryabin
*Influence of Short-Range Chemical Order on the Superconducting Transition Temperature in Magnetic Alloys
(b) Vol. **106**, No. 2, p. K111–K116 (1981).
See also 18.1, p. 266.

W. Zahn
Critical Current versus Magnetic Flux Characteristic of DC Tunnel SQUIDs
(a) Vol. **66**, No. 2, p. 649–656 (1981).

S. C. Jain and C. M. Kachhava
*Pressure Dependence of Superconductivity in Non-Transition Metals
(b) Vol. **107**, No. 1, p. 139–145 (1981).
See also 21, p. 345.

W. Klein and H. Kronmüller
*Pinning Forces in Neutron-Irradiated Nb_3Sn Diffusion Layers
(a) Vol. **67**, No. 1, p. 109–118 (1981).
See also 11, p. 128; 21, p. 346.

M. Wagner
Aspects of Electrical Conductivitiy in a Moving Base Approach
(b) Vol. **107**, No. 2, p. 617–636 (1981).
See also *13*, p. *148*; *14*, p. *198*.

H. K. Phan
*Dynamic Measurement of Thermal Relaxation Times of Superconducting Films at the Transition Temperature
(a) Vol. **67**, No. 2, p. 371–375 (1981).

H.-J. Köhler, P. Seidel, P. Weber, K. Blüthner, S. Linke, and K.-H. Berthel
*Influence of the Niobium Base Electrode on the Temperature Dependence of the Effective Niobium Energy Gap and the Critical Josephson Current in Nb–NbO_x–Pb Tunnel Junctions
(a) Vol. **67**, No. 2, p. 497–503 (1981).

G. Dharmadurai and V. Chopra
*An Axiomatic Heating Model for Thin Film Superconductors
(a) Vol. **67**, No. 2, p. 677–681 (1981).

V. P. Iordatii and L. Z. Kon
*Supercurrent Effect on the Electric Field Penetration Depth in Superconductors
(b) Vol. **108**, No. 1, p. K1–K5 (1981).

E. Krüger
*Magnetic Order in the Forced Normal State of Superconductors
(b) Vol. **108**, No. 1, p. K69–K70 (1981).
See also *18.4*, p. *284*; 21, p. 348.

14.3 Semiconductors

Review Articles

E. Spenke and W. Heywang
*Twenty Five Years of Semiconductor-Grade Silicon
(a) Vol. **64**, No. 1, p. 11–44 (1981).
See also *3*, p. *53*; 13.4, p. 168; 14.3.3, p. 225; 22.1.2, p. 408.

K. M. van Vliet and H. Mehta
*Theory of Transport Noise in Semiconductors
(b) Vol. **106**, No. 1, p. 11–30 (1981).
See also 22, p. 390.

H. Bässler
*Localized States and Electronic Transport in Single Component Organic Solids with Diagonal Disorder
(b) Vol. **107**, No. 1, p. 9–54 (1981).
See also 2, p. 40; *13.4*, p. *168*; 13.5.1, p. 188; 14.3.4, p. 232; 20.1, p. 301; 20.3, p. 323; 22.9, p. 507.

Original Papers and Short Notes

J. Monecke, W. Siegel, E. Ziegler, and G. Kühnel
*On the Concentration Dependence of the Thermal Impurity-to-Band Activation Energies in Semiconductors
(b) Vol. **103**, No. 1, p. 269–279 (1981).
See also *13.4*, p. *168*; 22.2.2, p. 432; 22.8, p. 484.

G. Simon and W. Eichler
*Galvanomagnetische und thermoelektrische Transportuntersuchungen an Sb_2Te_3
(Galvanomagnetic and Thermoelectric Transport Investigations on Sb_2Te_3)
(b) Vol. **103**, No. 1, p. 289–295 (1981).
See also *15*, p. *247*; 22.7, p. 483.

A. A. Gogolin
*Mobility of a Small Polaron at Low Temperatures
(b) Vol. **103**, No. 1, p. 397–409 (1981).
See also *13.5.3*, p. *196*; 22, p. 390.

A. N. Chakravarti, K. P. Ghatak, K. K. Ghosh, S. Ghosh, and A. Dhar
*Effect of Carrier Degeneracy on the Screening Length in n-Cd_3As_2
(b) Vol. **103**, No. 1, p. K55–K60 (1981).
See als 13.1, p. 152; 22.3, p. 440.

A. Cavallini, P. Gondi, and A. Castaldini
Shallow Dislocation States in Ge. Thermal Treatment Effects
(a) Vol. **63**, No. 1, p. 143–149 (1981).
See also *10.2*, p. *106*; *13.4*, p. *169*; 22.1.1, p. 404.

I. Gorczyca
Scattering on Short-Range Potentials in InSb. A Pseudopotential Calculation
(b) Vol. **103**, No. 2, p. 529–533 (1981).
See also *13.4*, p. *170*; 15, p. 247; 22.2.3, p. 435.

Sh. A. Gulamov, Yu. A. Katulevskii, and R. A. Muminov
*Peculiarities of an Electromotive Force Arising in a Semiconductor in a Non-Homogeneous Magnetic Field (I)
(b) Vol. **103**, No. 2, p. 567–572 (1981).
See also 18, p. 260; 22, p. 391.

G. N. Ivanova, D. D. Nedeoglo, A. V. Simashkevich, and I. N. Timchenko
*Negative Magnetoresistance in Zinc Selenide
(b) Vol. **103**, No. 2, p. 643–652 (1981).
See also 13.4, p. 170; 18, p. 260; 22.4.2, p. 452.

G. Srinivasan and C. M. Srivastava
*Electrical Conductivity Mechanism in Zinc and Copper Substituted Magnetite
(b) Vol. **103**, No. 2, p. 665–671 (1981).
See also 15, p. 247; 22.8.2, p. 502.

J. Monecke
*On the Insulator–Metal Transition in Doped Semiconductors
(b) Vol. **103**, No. 2, p. K115–K117 (1981).
See also *13.4*, p. *171*; 22, p. 392.

P. L. Kukk, H. A. Aarna, and M. P. Voogne
*High-Temperature Point Defects in CdS:Cu
(a) Vol. **63**, No. 2, p. 389–399 (1981).
See also *10.2*, p. *107*; 22.4.1, p. 447.

J. Horák, L. Tichý, and P. Lošťák
*Electrical Conductivity of BiTeI Crystals
(a) Vol. **63**, No. 2, p. 407–416 (1981).
See also 13.1, p. 153; 15, p. 247; 22.8, p. 485.

A. I. Kolyubakin and A. S. Shevchenko
On the Spectrum of Dislocation States in Germanium
(a) Vol. **63**, No. 2, p. 677–685 (1981).
See also *10.2*, p. *107*; *13.4*, p. *171*; 22.1.1, p. 404.

V. K. Arora
*Effect of Spin-Splitting on the Magnetoresistance in Parabolic Semiconductors
(b) Vol. **104**, No. 1, p. K71–K73 (1981).
See also *13.1*, p. *154*; 18, p. 260; 22, p. 393.

A. Le Bloa, P. N. Favennec, and Y. Colin
DLTS Method Using a Single Temperature Scanning
(a) Vol. **64**, No. 1, p. 85–93 (1981).
Erratum
(a) Vol. **66**, No. 2, p. 807 (1981).
See also *13.4*, p. *172*; 22.1.2, p. 411; 22.2.1, p. 428.

C. Wagner, A. El-Sadek, and H.-J. Mechelke
*The Annealing Characteristics of Arsenic-Implanted Silicon Investigated at Low Temperatures
(a) Vol. **64**, No. 1, p. 143–150 (1981).
See also *11*, p. *122*; 22.1.2, p. 411.

K. Schmalz and P. Gaworzewski
On the Donor Activity of Oxygen in Silicon at Temperatures from 500 to 800 °C
(a) Vol. **64**, No. 1, p. 151–158 (1981).
See also *13.4*, p. *172*; 22.1.2, p. 411.

H. Lemke
Doping Levels of Iron in Silicon
(a) Vol. **64**, No. 1, p. 215–224 (1981).
See also *13.4*, p. *172*; 22.1.2, p. 411.

F. Menil, N. Kinomura, L. Fournes, J. Portier, and P. Hagenmuller
*Investigation of the K_2NiF_4-Type Layer Oxide Fluorides $Sr_2FeO_{3+x}F_{1-x}$ $(0 \leqq x \leqq 0.20)$
(a) Vol. **64**, No. 1, p. 261–274 (1981).
See also 5, p. 63; *18.4*, p. *281*; 22.8, p. 486.

B. P. Chandra
Mechanoluminescence and Piezoelectric Behaviour of Molecular Crystals
(a) Vol. **64**, No. 1, p. 395–405 (1981).
See also *12.2*, p. *139*; *20.3*, p. *325*; 22.9, p. 509.

J. P. Hauck, M. M. Beguwala, C. L. Hayes, and R. F. Palys
Laser Annealing of Silicon Wafers at 10.6 μm
(a) Vol. **64**, No. 1, p. 407–412 (1981).
See also *11*, p. *123*; 22.1.2, p. 411.

M. A. Shaikh
*Electrical Properties of Quinolinium $(TCNQ)_2$
(a) Vol. **64**, No. 1, p. K23–K25 (1981).
See also *15*, p. *247*; 22.9, p. 509.

M. Stordeur, H. T. Langhammer, H. Sobotta, and V. Riede
Valence Band Structure of $(Bi_{1-x}Sb_x)_2Te_3$ Single Crystals
(b) Vol. **104**, No. 2, p. 513–522 (1981).
See also *13.1*, p. *154*; 15, p. 248; 20.1, p. 307; 22.7, p. 483.

S. Porowski, L. Konczewcz, J. Kowalski, R. L. Aulombard, and J. L. Robert
*High Magnetic Field Observation of the Resonance Donor States of S in InSb
(b) Vol. **104**, No. 2, p. 657–666 (1981).
See also *13.4*, p. *173*; 22.2.3, p. 435.

M. Stordeur and H. T. Langhammer
*The Lorenz Number for Many-Valley Semiconductors on Special Transport Conditions
(b) Vol. **104**, No. 2, p. K121–K124 (1981).
See also *8*, p. *85*; 22, p. 393.

SH. A. GULAMOV, YU. A. KATULEVSKII, and R. A. MUMINOV
*Peculiarities of an Electromotive Force Arising in a Semiconductor in a Non-Homogeneous Magnetic Field (II)
(b) Vol. **105**, No. 1, p. 405–408 (1981).
See also 18, p. 261; 22, p. 394.

S. J. PEARTON
Magnetic Field Dependence of Defect State in GaAs
(b) Vol. **105**, No. 1, p. K19–K22 (1981).
See also *13.4*, p. *175*; 18, p. 261; 22.2.1, p. 428.

J. GARRIDO, J. L. CASTAÑO, and J. PIQUERAS
*Photoluminescence in Gallium Arsenide Irradiated with Thermal Neutrons
(a) Vol. **65**, No. 1, p. 103–106 (1981).
See also 11, p. 125; *20.3*, p. *327*; 22.2.1, p. 428.

V. K. MALYUTENKO, K. YU. GUGA, and YU. M. MALOZOVSKII
*Magnetoconcentration Effect at Auger Recombination of Current Carriers
(a) Vol. **65**, No. 1, p. 131–140 (1981).
See also 18, p. 261; 22.2.3, p. 436.

D. SHAW and H. R. MOGHADDAM
*A New Method for the Analysis of Thermally Stimulated Current Data
(a) Vol. **65**, No. 1, p. 167–176 (1981).
See also *13.4*, p. *175*.

B. GILLOT, R. M. BENLOUCIF, and A. ROUSSET
*Electrical Conductivity of Zinc–Iron Ferrites in Vacuum and in the Presence of Oxygen
(a) Vol. **65**, No. 1, p. 205–213 (1981).
See also 22.8.2, p. 504.

O. VALASSIADES, E. K. POLYCHRONIADIS, J. STOEMENOS, and N. A. ECONOMOU
*On the Disordered Nature of Single Crystalline $TlBiTe_2$ and Its Influence on Its Electrical Properties
(a) Vol. **65**, No. 1, p. 215–223 (1981).
See also 1.2, p. 17; *10.1*, p. *100*; 22.8, p. 487.

M. B. DUTT, R. BANERJEE, and A. K. BARUA
*Transport Properties of Lithium and Sodium Doped Nickel Oxide
(a) Vol. **65**, No. 1, p. 365–370 (1981).
See also *15*, p. *248*; 22.6, p. 477.

T. MOHAMMED-BRAHIM
*Defect Levels in HgI_2 and Illumination Effect
(a) Vol. **65**, No. 1, p. K1–K6 (1981).
See also *13.4*, p. *176*; 22.5, p. 459.

N. M. RAVINDRA, V. K. SRIVASTAVA, and B. MISHRA
*Temperature Dependent Electrical Conductivity of Natural Galena Crystals
(a) Vol. **65**, No. 1, p. K59–K61 (1981).
See also 13.1, p. 156; 22, p. 395.

Z. Januškevičius, N. Korech, A. Sakalas, and I. Tychina
*Electrical Properties of CdP_2 Single Crystals
(a) Vol. **65**, No. 2, p. K149–K151 (1981).
See also 13.4, p. 177; 22.3, p. 441.

T. Hejwowski and M. Subotowicz
The Mn^{2+} Ion Spectra in $Pb_xSn_{1-x}Te$ Alloys
(b) Vol. **106**, No. 1, p. 373–377 (1981).
See also *19*, p. *291*; 22, p. 397.

Y. Ishihara and I. Nakada
*Negative Magnetoresistance of GeSe in a Weak Magnetic Field
(b) Vol. **106**, No. 1, p. K27–K31 (1981).
See also 22, p. 397.

C. Scharager, P. Siffert, P. Höschl, P. Moravec, and M. Vaněček
Characterization of Germanium-Doped CdTe Crystals
(a) Vol. **66**, No. 1, p. 87–92 (1981).
See also *13.4*, p. *178*; 16, p. 254; 22.4.3, p. 454.

V. S. Shukla and D. M. Rowe
*Precipitation of Boron from Silicon–Germanium Alloy and Its Effect in the Thermoelectric Transport Properties
(a) Vol. **66**, No. 1, p. 243–248 (1981).
See also 10.2, p. 113; *15*, p. *249*; 22.1.1, p. 406; 22.1.2, p. 416.

J. Reichel
The Distinction of Several Different Kinds of Oxygen Donors in Silicon
(a) Vol. **66**, No. 1, p. 277–282 (1981).
See also *10.2*, p. *113*; 22.1.2, p. 416.

P. Höschl, P. Moravec, V. Prosser, A. Sakalas, and R. Jašinskaite
*Electrical Characterization of ZnTe Crystals Grown from In Solvent
(a) Vol. **66**, No. 1, p. K41–K42 (1981).
See also *13.4*, p. *178*; 22.4.3, p. 455.

S. S. Dyunaidov, N. A. Urmanov, and M. V. Gafurova
The Cadmium Levels in Silicon
(a) Vol. **66**, No 1, p. K79–K81 (1981).
See also *13.4*, p. *178*; 22.1.2, p. 417.

E. Litwin-Staszewska, W. Szymańska, and R. Piotrzkowski
*The Electron Mobility and Thermoelectric Power in InSb at Atmospheric and Hydrostatic Pressures
(b) Vol. **106**, No. 2, p. 551–559 (1981).
See also *15*, p. *249*; 22.2.3, p. 436.

J. F. Pouradier and G. Grimouille
*Growth of Semiconducting $CdF_2:Ho^{3+}$ Crystals and Study of Their Optical and Transport Properties
(b) Vol. **106**, No. 2, p. 657–665 (1981).
See also *3*, p. *55*; 13.4, p. 179; 20.1, p. 315; 22.5.3, p. 471.

R. W. A. Hillhouse and J. Woods
*The Evaluation of Thermally Stimulated Current and Thermoluminescence Curves in Rutile
(a) Vol. **67**, No. 1, p. 119–128 (1981).
See also *13.4*, p. *180*; 20.3, p. 332; 22.6, p. 479.

A. H. Abou El Ela, H. H. A. Labib, and N. Abdelmohsen
*Electrical Conductivity, Thermoelectric Power, and Thermal Conductivity of $CuTlSe_2$ in the Solid and Liquid States
(a) Vol. **67**, No. 1, p. K43–K48 (1981).
See also 8, p. 87; 15, p. 249; 22.8, p. 490.

B. B. Baizakov, R. A. Moominov, S. H. Shamirzaiev, and A. Ya. Yafasov
*Bias Current in Semiconductors during Overcharging of Impurity Centres
(a) Vol. **67**, No. 1, p. K59–K62 (1981).
See also *13.4*, p. *181*; 22, p. 398.

K. Sierański and J. Szatkowski
*Disorder Scattering in Mixed $Pb_{1-x}Sn_xTe$-Type Semiconductors
(b) Vol. **107**, No. 2, p. K87–K90 (1981).
See also 13.4, p. 181; 22.8, p. 490.

I. K. Bandrovskaya, V. Ya. Prokhorenko, V. G. Romanov, and B. I. Sokolovskii
*Optical and Thermoelectric Properties of Arsenic Selenide in Glassy and Liquid States
(b) Vol. **107**, No. 2, p. K95–K98 (1981).
See also 2, p. 50; 15, p. 249; *20.1*, p. *318*; 22, p. 398.

M. Matlak and J. Zieliński
*Electrical Resistivity and Insulator-to-Metal Transition for the Periodic Anderson Model
(b) Vol. **107**, No. 2, p. K133–K137 (1981).
See also *14.1*, p. *202*.

A. Priemuth
Electromigration of Cu in V_2VI_3 Compounds
(a) Vol. **67**, No. 2, p. 505–510 (1981).
See also *1*, p. *10*; *9*, p. *91*; 22.7, p. 483.

B. R. Nag and A. N. Chakravarti
*Generalized Einstein Relation – A Suggested Method for Experimental Verification
(a) Vol. **67**, No. 2, p. K113–K117 (1981).
See also 22, p. 399.

T. G. Kerimova, F. R. Adzalova, R. Kh. Nani, and V. Ya. Shteinshreiber
Interaction of Free Current Carriers with Longitudinal Optical Phonons in $CdIn_2Se_4$
(a) Vol. **67**, No. 2, p. K143–K146 (1981).
See also *6*, p. *80*; *20.1*, p. *318*; 22.8, p. 490.

E. Klier
*A Simple Method of Evaluation of TSC Measurements
(a) Vol. **67**, No. 2, p. K151–K154 (1981).
See also *13.4*, p. *182*; 22, p. 399.

I. P. Zvyagin
*AC Hopping Conductivity – Analytical Theory versus Numerical Calculation
(b) Vol. **108**, No. 2, p. K85–K88 (1981).
See also *2*, p. *51*.

G. Döding and R. Labusch
*Anisotropic Conductivity of CdS after Plastic Deformation and Conduction along Dislocations (II): Maesurements on a Microscopic Scale
(a) Vol. 68, No. 2, p. 469–476 (1981).
See also *10.2*, p. *119*; 22.4.1, p. 451.

J. Kodeš, J. Samek, and T. Šimko
*Lifetime Measurements on Electron Irradiated Silicon
(a) Vol. 68, No. 2, p. K177–K179 (1981).
See also *11*, p. *132*; 22.1.2, p. 424.

14.3.1 Films

Review Article

Original Papers and Short Notes

V. M. Efimov
*Study of Thin Silicon Nitride Films by Isothermal Depolarization on MNOS Structures
(a) Vol. **65**, No. 1, p. 177–187 (1981).
See also *13.4*, p. *176*; 14.3.3, p. 227; 22, p. 395.

R. O. Loutfy
*Effect of Temperature on the Dark and Photoconductivity of Metal-Free Phthalocyanine
(a) Vol. **65**, No. 2, p. 659–668 (1981).
See also *16*, p. *253*; 22.9, p. 510.

S. Uthanna and P. J. Reddy
*Electrical Properties of CdSe Thin Films
(a) Vol. **65**, No. 2, p. K113–K115 (1981).
See also *1.5*, p. *35*; 20.1, p. 312; 22.4.2, p. 453.

L. V. Rao, B. S. Naidu, and P. J. Reddy
*Electrical Conductivity and Thermopower Measurements on Silver Doped Tellurium Films
(a) Vol. **65**, No. 2, p. K135–K138 (1981).
See also 1.5, p. 35; *15*, p. *248*; 22.1.3, p. 425.

S. Sen and D. N. Bose
*Preparation and Electrical Properties of Ga_2Te_3 Thin Films
(a) Vol. **66**, No. 2, p. K117–K119 (1981).
See also 22.3, p. 442.

P. K. Bhatnagar, S. R. Dhariwal, and G. P. Srivastava
*Non-Uniformities in Ultra-Thin Oxide Layers and Their Effect on the Properties of Tunnel MOS, MOM, and SOS Devices
(a) Vol. **67**, No. 1, p. 305–311 (1981).
See also *14.3.3*, p. *229*.

Y. Colin, B. Fortin, and F. Raoult
*Resistance Variation of a Semiconducting Thin Film during a Thermal Desorption
(a) Vol. **67**, No. 2, p. 485–495 (1981).
See also 22.4.2, p. 453.

A. L. Dawar, S. K. Paradkar, P. Kumar, O. P. Taneja, and P. C. Mathur
*Electrical Transport Properties of Ag-Doped p-Type PbTe Epitaxial Thin Films
(a) Vol. **68**, No. 1, p. 227–232 (1981).
See also 22, p. 399.

S. J. Pearton and E. M. Lawson
*Laser Assisted Doping of n-GaAs from Sn Evaporated Films
(a) Vol. **68**, No. 1, p. K63–K66 (1981).
See also *20.2*, p. *322*; 22.2.1, p. 430.

G. Badics, I. A. Serbinov, Yu. D. Kalafati, and W. F. Fallmann
*Time Periodic Dissipative Structures in VO_2 Films
(a) Vol. **68**, No. 1, p. K101–K104 (1981).
See also 22.6, p. 480.

14.3.2 Surfaces and Interfaces

Original Papers and Short Notes

A. Szekeres, K. Kirov, and S. Alexandrova
*Plasma Photoresist Stripping in a Planar Reactor
(a) Vol. **63**, No. 1, p. 371–374 (1981).
See also 22.1.2, p. 409.

A. N. Chakravarti, A. K. Chowdhury, and K. P. Ghatak
*Oscillatory Electronic Heat Capacity in Inversion Layers on ⟨100⟩ Oriented p-Type Si in a Quantizing Magnetic Field
(a) Vol. **63**, No. 1, p. K97–K101 (1981).
See also *8*, p. *84*; 14.3.3, p. 225; 22.1.2, p. 409.

K. I. Kirov and K. B. Radev
*A Simple Charge-Based DLTS Technique
(a) Vol. **63**, No. 2, p. 711–716 (1981).
See also *13.3*, p. *165*; 22.1.2, p. 410.

W. Siegel, G. Kühnel, and E. Ziegler
*Electrical Properties of Grain Boundaries in n-Type and p-Type GaP
(a) Vol. **64**, No. 1, p. 249–259 (1981).
See also 1.4, p. 29; 22.2.2, p. 432.

J. Marczewski
*Low-Temperature Annealing of Surface States at the Si–SiO_2 Interfaces
(a) Vol. **64**, No. 2, p. K107–K110 (1981).
See also *13.3*, p. *165*; 22.1.2, p. 413.

S. N. Kozlov and N. L. Levshin
*On the Distribution Function of Activation Barriers Surmounted by Charge Carriers when Captured to Slow Surface States of a Semiconductor
(a) Vol. **64**, No. 2, p. K169–K172 (1981).
See also *13.3*, p. *165*; 22, p. 394.

V. L. Alperovich, A. F. Kravchenko, N. A. Pakhanov, and A. S. Terekhov
*Effect of Electron Heating by Surface Electric Fields on Oscillating Photoresponse Spectra of Gallium Arsenide–Metal Structures
(a) Vol. **65**, No. 1, p. 259–263 (1981).
See also *16*, p. *253*; 22.2.1, p. 429.

C. Donolato
*An Analytical Model of SEM and STEM Charge Collection Images of Dislocations in Thin Semiconductor Layers (I): Minority Carrier Generation, Diffusion, and Collection
(a) Vol. **65**, No. 2, p. 649–658 (1981).
See also *10.2*, p. *111*; 14.3.4, p. 234; 22.1.2, p. 415.

D. K. Ferry
*Charge Instabilities in Lateral Super-Lattices under Conditions of Population Inversion
(b) Vol. **106**, No. 1, p. 63–71 (1981).
See also *14.3.3*, p. *228*.

14.3.3 Devices. Junctions (Contact Problems See 14.3.4)

Review Articles

E. SPENKE and W. HEYWANG
Twenty Five Years of Semiconductor-Grade Silicon
(a) Vol. **64**, No. 1, p. 11–44 (1981).
See also *3*, p. *53*; 13.4, p. 168; *14.3*, p. *209*; 22.1.2, p. 408.

K. W. BÖER
*The CdS/Cu_2S Solar Cell (II): The Junction and Junction–Emitter Relation
(a) Vol. **66**, No. 1, p. 11–43 (1981).
See also 13.4, p. 168; 14.3.4, p. 232; *16*, p. *251*; 22.4.1, p. 447.

P. MÜLLER
Electric Relaxation Currents in High-Resistivity Solids
(a) Vol. **67**, No. 1, p. 11–60 (1981).
See also 14.3.4, p. 232; *14.4*, p. *237*; 22.3, p. 440; 22.5.1, p. 461; 22.5.2, p. 462.

Original Papers and Short Notes

H. LEMKE
On the Evidence of Recombination Centres in the Base Regions of Si Rectifiers
(a) Vol. **63**, No. 1, p. 127–136 (1981).
See also *13.4*, p. *169*; 22.1.2, p. 408.

R. K. GUPTA and QU. A. ALATIEF
*Schottky-Barrier Field-Effect Transistor with Static-Induction Principle (MESSIT)
(a) Vol. **63**, No. 1, p. 313–319 (1981).
See also 22.1.2, p. 408.

A. N. CHAKRAVARTI, A. K. CHOWDHURY, and K. P. GHATAK
Oscillatory Electronic Heat Capacity in Inversion Layers on ⟨100⟩ Oriented p-Type Si in a Quantizing Magnetic Field
(a) Vol. **63**, No. 1, p. K97–K101 (1981).
See also *8*, p. *84*; *14.3.2*, p. *223*; 22.1.2, p. 409.

J. GONG, K. M. VAN VLIET, A. D. SUTHERLAND, and E. R. CHENETTE
*Noise Measurements on Photo Avalanche Diodes
(a) Vol. **63**, No. 2, p. 445–460 (1981).
See also 16, p. 251; 22.1.2, p. 409.

H. HAEFNER, H. MORAWETZ, and G. OELGART
*Cathodoluminescence Measurements on Green Light-Emitting GaP Diodes
(a) Vol. **63**, No. 2, p. 495–500 (1981).
See also *20.3*, p. *324*; 22.2.2, p. 432.

D. TJAPKIN, V. MILANOVIĆ, and Ž. SPASOJEVIĆ
*The Einstein Relation for Two-Dimensional Strongly Inverted Layers
(a) Vol. **63**, No. 2, p. 737–742 (1981).
See also 22.1.2, p. 410.

R. S. Korde and M. S. Tyagi
*Breakdown Voltage Reduction of High Voltage PIN Diodes by a Process Induced Linear Defect
(a) Vol. **64**, No. 2, p. K89–K91 (1981).
See also 22.1.2, p. 413.

V. G. Baru and E. V. Grekov
Van der Waals Theory of the Phase Transition in Quasi-Two-Dimensional Electron Gas in Magnetic Semiconductors
(b) Vol. **105**, No. 1, p. 369–376 (1981).
See also *13*, p. *145*; 18.2, p. 270; 22, p. 394.

V. M. Efimov
Study of Thin Silicon Nitride Films by Isothermal Depolarization on MNOS Structures
(a) Vol. **65**, No. 1, p. 177–187 (1981).
See also *13.4*, p. *176*; *14.3.1*, p. *221*; 22, p. 395.

H. M. Przewłocki, S. Krawczyk, and A. Jakubowski
*A Simple Technique of Work Function Difference Determination in MOS Structure
(a) Vol. **65**, No. 1, p. 253–257 (1981).
See also 13.3, p. 165; 22.1.2, p. 414.

S. Cristoloveanu
*Magnetic Field and Surface Influences on Double Injection Phenomena in Semiconductors (II): The Magnetodiode Effect Theory for the Diffusion Regimes, Magnetodiode Experiments and Applications
(a) Vol. **65**, No. 1, p. 281–292 (1981).
See also 14.3.4, p. 234; *18*, p. *262*; 22.1.1, p. 405; 22.1.2, p. 414.

D. Sachelarie, M. Dragan, M. Stoica, and M. Sachelarie
Radiation Effects of Fast Neutrons in Silicon Diffused Detectors
(a) Vol. **65**, No. 1, p. 379–388 (1981).
See also *11*, p. *125*; *13.4*, p. *176*; 22.1.2, p. 414.

K. Nauka and J. Misiewicz
*Minority-Carrier Diffusion Length in Zn_3P_2
(a) Vol. **65**, No. 1, p. K95–K97 (1981).
See also 16, p. 253; 22.3, p. 441.

Z. S. Gribnikov, V. M. Ivastchenko, and V. V. Mitin
*Nonlocality of Carrier Multiplication in Semiconductor Depletion Layers
(b) Vol. **105**, No. 2, p. 451–459 (1981).
See also 13.4, p. 176; 22, p. 395.

G. Oelgart and G. Arnold
*Impurity Profile on GaP Diodes in the Neighbourhood of a p–n Junction
(a) Vol. **65**, No. 2, p. 609–616 (1981).
See also *10.2*, p. *111*; 22.2.2, p. 433.

H. Van Cong
*Heavy Doping Effects in Bipolar Silicon Transistors and p–n Junction Silicon Solar Cells
(a) Vol. **65**, No. 2, p. 683–693 (1981).
See also 13.1, p. 157; 22.1.2, p. 415.

P. SAXENA and P. C. MATHUR
*Intrinsic Noise in Avalanche Diodes (II): Experiment
(a) Vol. **66**, No. 2, p. 737–742 (1981).
See also 22.1.2, p. 417; 22.2.1, p. 430.

N. B. LUKYANCHIKOVA, T. M. PAVELKO, and G. S. PEKAR
*Physical Processes in Forward-Biased ZnS MIS Light-Emitting Diodes with a "Thick" Interfacial Layer
(a) Vol. **66**, No. 2, p. 749–759 (1981).
See also *20.3*, p. *331*; 22.4.1, p. 449.

B. A. KOMAROV and V. I. SOPRYAKOV
*Study of the Field Dependence for the Parameters of the A-Centre in Si by Capacitance Transient Spectroscopy
(a) Vol. **66**, No. 2, p. 783–791 (1981).
See also *13.4*, p. *180*; 14.3.4, p. 235; 22.1.2, p. 417.

E. A. AKOPYAN and A. SH. MEKHTIEV
*High Sensitive Fast-Response Photodetector
(a) Vol. **66**, No. 2, p. K135–K139 (1981).
See also *16*, p. *254*; 22.1.1, p. 406.

H. KLOSE and M. RIETH
Influence of Al Postimplantation on p–n Junctions in Gallium Phosphide
(a) Vol. **66**, No. 2, p. K153–K156 (1981).
See also *11*, p. *128*; 20.3, p. 331; 22.2.2, p. 433.

S. M. DAVIDSON, R. M. INNES, and S. M. LINDSAY
*The Sublinear Behavior of Electron-Beam and Photo-Induced Currents in a p–n Junction as a Test of Injection Levels
(b) Vol. **107**, No. 1, p. K9–K12 (1981).
See also 14.3.4, p. 235; 22.1.2, p. 418.

I. ZÓLOMY
*The Effect of Non-Zero Switch-Off Voltage upon the Transient of MISS Diodes
(a) Vol. **67**, No. 1, p. 69–74 (1981).
See also 14.3.4, p. 235; 22.1.2, p. 418.

P. K. BHATNAGAR, S. R. DHARIWAL, and G. P. SRIVASTAVA
*Non-Uniformities in Ultra-Thin Oxide Layers and Their Effect on the Properties of Tunnel MOS, MOM, and SOS Devices
(a) Vol. **67**, No. 1, p. 305–311 (1981).
See also *14.3.1*, p. *221*.

S. K. KRAWCZYK, A. JAKUBOWSKI, and M. ŻURAWSKA
*Effect of Temperature on Electrical and Photoelectrical Properties of Silicon MIS Solar Cells
(a) Vol. **67**, No. 1, p. 313–323 (1981).
See also *16*, p. *254*; 22.1.2, p. 419.

B. Balland, J. L. Paviot, B. de Cremoux, and P. Hirtz
*Sn–X Centres in GaAlAs Thin Layer Emitters
(a) Vol. 68, No. 2, p. 661–673 (1981).
See also 10.2, p. 120; *13.4*, p. *184*; 14.3.1, p. 222; 22.2.4, p. 439.

H. van Cong, S. Charar, and S. Brunet
*Minority-Hole Diffusion Coefficient in an n-Type Heavily Doped Semiconductor Region of Silicon Devices
(a) Vol. 68, No. 2, p. 697–702 (1981).
See also 22.1.2, p. 423.

M. S. Gilenko, P. M. Karageorgy-Alkalaev, and A. Yu. Leiderman
*On the Magnetosensitivity of Semiconductor Diode Structures with Strong Carrier Accumulation
(a) Vol. 68, No. 2, p. K165–K169 (1981).
See also 14.3.4, p. 236; 22.1.2, p. 424.

14.3.4 High-Field Phenomena, Space-Charge Effects, Inhomogeneities, Injected Carriers (Electroluminescence See 20.3; Junctions See 14.3.3)

V. M. LOMAKO and P. YA. STAROSTION
Simulation of Electrical Characteristics and Properties of Defect Clusters in Irradiated Semiconductors (I): Approximation of Electric Charge Density
(a) Vol. **63**, No. 2, p. 585–593 (1981).
See also *13.4*, p. *171*; 22.2.1, p. 427.

R. V. KONAKOVA, YU. A. TKHORIK, I. L. ZAITSEVSKII, P. KORDOŠ, M. MORVIC, and J. ČERVENÁK
*Avalanche Breakdown Peculiarities of $Al_xGa_{1-x}As$ IMPATT Diodes
(a) Vol. **63**, No. 2, p. K163–K166 (1981).
See also *14.3.3*, p. *226*; 22.2.4, p. 437.

S. SARKAR
*Second Harmonic Dependence of Domain Excess Voltage in GaAs
(a) Vol. **63**, No. 2, p. K171–K173 (1981).
See also 22.2.1, p. 427.

I. ZÓLOMY and A. ADÁN
Current Transients in Illuminated Polysilicon MIS Diodes
(a) Vol. **63**, No. 2, p. K179–K181 (1981).
See also *14.3.3*, p. *226*; 22.1.2, p. 410.

J. GODLEWSKI
*On the Interpretation of Unipolar Current Flow in Insulators with Exponential Trap Energy Distribution
(a) Vol. **63**, No. 2, p. K199–K202 (1981).
See also 16, p. 252; 22.4.1, p. 447.

N. G. BEBENIN
*The Critical Dynamics of Hot Electrons in a Ferromagnetic Semiconductor at $T = T_c$
(b) Vol. **104**, No. 1, p. K65–K70 (1981).
See also *13*, p. *143*; 18.2, p. 269; 22, p. 393.

A. M. BRODSKII and L. I. DAIKHIN
*On the Theory of Electroreflectance of Metals
(b) Vol. **104** No. 2, p. 415–426 (1981).
See also *20.1*, p. *306*; 21, p. 339.

F. STÖCKMANN
*An Exact Evaluation of Steady-State Space-Charge-Limited Currents for Arbitrary Trap Distributions
(a) Vol. **64**, No. 2, p. 475–483 (1981).
See also *13.4*, p. *174*; 22, p. 393.

J. GODLEWSKI
*Electrode-Limited Photoinjection in Aqueous Electrode–Organic Crystal Systems
(a) Vol. **64**, No. 2, p. 499–507 (1981).
See also *16*, p. *252*; 22.9, p. 509.

S. CRISTOLOVEANU
Magnetic Field and Surface Influences on Double Injection Phenomena in Semiconductors (I): The Magnetodiode Effect Theory for the Semiconductor and Insulator Regimes
(a) Vol. **64**, No. 2, p. 683–695 (1981).
See also *14.3.3*, p. *226*; *18*, p. *261*.

B. A. Komarov and V. I. Sopryakov
Capacitance Method Applied in Studying Defects in Heavily Doped Semiconductors
(a) Vol. **66**, No. 1, p. 139–147 (1981).
See also *13.4*, p. *178*; 14.3.3, p. 228; 22.1.2, p. 415.

Tran Chot
*A Modified Forward *I–U* Plot for Schottky Diodes with High Series Resistance
(a) Vol. **66**, No. 1, p. K43–K45 (1981).
See also *14.3.3*, p. *228*; 22.1.2, p. 416; 22.4.2, p. 453.

A. M. Berezhovskii, A. A. Ovchinnikov, and R. A. Suris
*A New Method for Detecting the Wannier-Star Ladder in a Semiconductor in a Strong Electric Field
(b) Vol. **106**, No. 2, p. 461–466 (1981).
See also 13.1, p. 158; *20.1*, p. *314*; 22, p. 397.

B. A. Komarov and V. I. Sopryakov
Study of the Field Dependence for the Parameters of the A-Centre in Si by Capacitance Transient Spectroscopy
(a) Vol. **66**, No. 2, p. 783–791 (1981).
See also *13.4*, p. *180*; *14.3.3*, p. *229*; 22.1.2, p. 417.

S. M. Davidson, R. M. Innes, and S. M. Lindsay
The Sublinear Behavior of Electron-Beam and Photo-Induced Currents in a p–n Junction as a Test of Injection Levels
(b) Vol. **107**, No. 1, p. K9–K12 (1981).
See also *14.3.3*, p. *229*; 22.1.2, p. 418.

I. Zólomy
The Effect of Non-Zero Switch-Off Voltage upon the Transient of MISS Diodes
(a) Vol. **67**, No. 1, p. 69–74 (1981).
See also *14.3.3*, p. *229*; 22.1.2, p. 418.

W. Moldenhauer, K.-H. Bäther, W. Brückner, D. Hinz, and D. Bühling
*Degradation Phenomena of ZnO Varistors
(a) Vol. **67**, No. 2, p. 533–542 (1981).
See also 22.4, p. 446.

X. Aymerich-Humet, F. Serra-Mestres, and J. Millán
*On the Resonant Braodening of the Trap Assisted Tunneling in MIM Structures
(a) Vol. **67**, No. 2, p. 597–602 (1981).
See also 13.4, p. 182; *14.3.3*, p. *230*; 22.6, p. 479.

N. N. Koren and T. E. Borisenko
*Space-Charge-Limited Currents in GeS Single-Crystal Layers
(a) Vol. **67**, No. 2, p. K147–K150 (1981).
See also 22, p. 399.

B. Olejníková
*The Influence of the Electric Field on the Electron Capture Coefficient of Screened Coulomb Centres
(b) Vol. **108**, No. 1, p. 79–85 (1981).
See also *13.4*, p. *182*; 22, p. 399.

14.4 Dielectrics

Review Article

P. Müller
*Electric Relaxation Currents in High-Resistivity Solids
(a) Vol. **67**, No. 1, p. 11–60 (1981).
See also 14.3.3, p. 225; 14.3.4, p. 232; 22.3, p. 440; 22.5.1, p. 461; 22.5.2, p. 462.

Original Papers and Short Notes

R. M. Hill
*Evaluation of Susceptibility Functions
(b) Vol. **103**, No. 1, p. 319–328 (1981).
See also 22.9, p. 507.

V. Patráš, J. Krempaský, G. Vlasák, and F. Hanic
*Superionic Conductivity in the System PrO_x–HfO_2
(a) Vol. **63**, No. 1, p. 79–83 (1981).
See also 22.6.1, p. 481.

H. Schäfer and H. W. Helberg
*Dielectric Properties of the One-Dimensional Conductors $(UDMTTF)_2BF_4$ and $(TMTTF)_2$-BF_4
(a) Vol. **63**, No. 1, p. 203–207 (1981).
See also 22.9, p. 507.

T. van Dijk and A. J. Burggraaf
*Grain Boundary Effects on Ionic Conductivity in Ceramic $Gd_xZr_{1-x}O_{2-(x/2)}$Solid Solutions
(a) Vol. **63**, No. 1, p. 229–240 (1981).
See also *1.4*, p. *28*; 22.6.1, p. 481.

I. L. Baginskii and E. G. Kostsov
Transient Single-Injection Currents in Dielectrics at Fast Trapping
(a) Vol. **63**, No. 1, p. 271–279 (1981).
See also *13.4*, p. *169*; *14.3.4*, p. *232*.

F. El-Kabbany, Y. Badr, and G. Said
*Pyroelectric and Dielectric Behaviour of Ammonium Nitrate at the Phase Transition II–I
(a) Vol. **63**, No. 1, p. 355–362 (1981).
See also *1.2*, p. *14*; 22.8.1, p. 493.

J. P. Jog, S. J. Walzade, and S. V. Bhoraskar
*Detection of Phase Transitions in Polymers by TSC Technique
(a) Vol. **63**, No. 1, p. K103–K106 (1981).
See also *1.2*, p. *14*; 22.9, p. 508.

S. Bozdemir
*An Ising Model Analysis of Dielectric Polarization (I): Analysis for the Dielectric Response of Infinite and Finite Dipolar Chains in Dielectrics
(b) Vol. **103**, No. 2, p. 459–470 (1981).

W. Gasser and E. Heiner
On the Peierls Transition in the Strong Correlation Limit of the Hubbard Chain
(b) Vol. **103**, No. 2, p. 659–664 (1981).
See also 6, p. 68; *13*, p. *142*; 22.9, p. 508.

A. V. Singh and J. Shanker
Effect of Three-Body Potential on the Dielectric Behaviour and Optic Mode Grüneisen Parameters of Alkali Halide Crystals
(b) Vol. **103**, No. 2, p. 733–738 (1981).
See also *4*, p. *57*; 6, p. 68; 12.2, p. 138; 22.5.2, p. 463.

S. Bozdemir
*An Ising Model Analysis of Dielectric Polarization (II): Computations of Dynamic Susceptibility and Comparison with Experimental Observations
(b) Vol. **104**, No. 1, p. 37–47 (1981).
See also 22.9, p. 508.

R. W. Munn
*External and Internal Strains in Piezoelectric Crystals
(b) Vol. **104**, No. 1, p. 325–330 (1981).
See also *12.2*, p. *139*.

D. R. Goyal, S. Walker, and K. K. Srivastava
*Errors Due to Lack of Contact in Measurements of Dielectric Relaxation Parameters for Solid Powders
(a) Vol. **64**, No. 1, p. 351–357 (1981).
See also 22.8, p. 486.

D. P. Yadav, H. N. Acharya, and K. V. Rao
*Effect of Quenching and X-Ray Irradiation on the Dielectric Properties of PbI_2 Single Crystals
(a) Vol. **64**, No. 1, p. 413–417 (1981).
See also 11, p. 123; 22.5, p. 458.

O. Litzman
*A Note on the Additional Boundary Conditions in Spatially Dispersive Media
(b) Vol. **104**, No. 2, p. K109–K111 (1981).
See also *20.1*, p. *308*.

E. M. Raspopova, M. Závětová, J. Krištofik, and B. T. Kolomiets
The Influence of Defects Generated during Densification on the Properties of the As–Se System
(a) Vol. **64**, No. 2, p. 457–460 (1981).
See also *2*, p. *44*; 10.2, p. 109; *20.1*, p. *308*; 22, p. 393.

B. K. Verma, O. P. Srivastava, and H. B. Lal
*Transport Studies of the Copper Solid Superionic Conductor $CuPb_3Br_7$
(a) Vol. **64**, No. 2, p. 467–473 (1981).
See also *15*, p. *248*; 22.8, p. 486.

V. P. Romanov and Yu. A. Chaplygin
*A Theoretical Model for the Stationary Distribution of Mobile Ions in a Double-Layer Insulator
(a) Vol. **64**, No. 2, p. 525–531 (1981).
See also *14.3.3*, p. *226*; 22, p. 394.

K. P. Bansal, S. Kumari, B. K. Das, and G. C. Jain
*Electrical Properties of Ceramic Lanthanum Chromite Doped with TiO_2
(a) Vol. **64**, No. 2, p. K125–K127 (1981).
See also 22.8.1, p. 496.

H. Kresse and B. Gajewska
*Dielectric Relaxation of the Liquid Crystalline S_E Phase
(a) Vol. **64**, No. 2, p. K161–K163 (1981).
See also *2*, p. *45*; 22.9, p. 509.

O. A. Grigorev and T. D. Shermergor
*The Scattering Coefficient and Velocity of Propagation of Electromagnetic Waves in Dielectric Polycrystals
(a) Vol. **65**, No. 1, p. 127–130 (1981).

H. Neumann, Yu. I. Tsipivka, and K. Unger
Influence of Dielectric Screening and Band Nonparabolicity on Ionized Impurity Scattering in Doped Semiconductors
(b) Vol. **105**, No. 2, p. 577–583 (1981).
See also 13.1, p. 156; *14.3*, p. *214*; 22.2.1, p. 429; 22.2.3, p. 436.

Yu. A. Gorokhovatskii, A. N. Gubkin, and A. I. Margulev
*Thermally Stimulated Depolarization for Quasi-Continuous Distribution of Relaxators According to Activation Energy and Frequency Factor
(a) Vol. **65**, No. 2, p. 485–490 (1981).
See also 13.4, p. 177; 22.6, p. 477.

J. Lecomte, J. P. Loup, M. Hervieu, and B. Raveau
*Non-Stoichiometry and Electrical Conductivity of Strontium Niobates with Perovskite Structure (I): Defect Structure of $Sr(Sr_{1/3}Nb_{2/3})O_3$
(a) Vol. **65**, No. 2, p. 743–752 (1981).
See also 1.3, p. 24; *10.2*, p. *112*; 22.8.1, p. 497.

R. Świetlik and A. Graja
*On the Dielectric Properties of Tetracyanoquinodimethane (TCNQ) Salts with N-Methyl Derivatives of Pyridine
(a) Vol. **65**, No. 2, p. K165–K168 (1981).
See also 22.9, p. 510.

Ch. Kokott and F. Fischer
*Dielectric Relaxation of Chalcogen–Vacancy Centres in Potassium Halide Crystals
(b) Vol. **106**, No. 1, p. 141–147 (1981).
See also *10.2*, p. *112*; 22.5.2, p. 465.

P. J. Chandler and E. Lilley
*Ionic Conductivity of Single Crystals of 6 NaCl · $CdCl_2$ Suzuki Phase
(a) Vol. **66**, No. 1, p. 183–187 (1981).
See also 22.5.4, p. 473.

G. LUTHER and H. ROEMER
*Electrical Conductivity of Superionic AgI between 18 and 26.5 GHz
(b) Vol. **106**, No. 2, p. 511–517 (1981).
See also 22.5.1, p. 461.

C. J. DELBECQ, S. A. MARSHALL, and S. SUSMAN
*Dielectric Properties of the Fast-Ion Conductor $Na_3Sc_2P_3O_{12}$
(b) Vol. **106**, No. 2, p. 731–734 (1981).
See also 22.8, p. 488.

J. LECOMTE, J. P. LOUP, M. HERVIEU, and B. RAVEAU
*Non-Stoichiometry and Electrical Conductivity of Strontium Noibates with Perovskite Structure (II): Ionic Conductivity of $Sr(Sr_{(1/3)+x}Nb_{(2/3)-x})O_{3-(3/2)x}$
(a) Vol. **66**, No. 2, p. 551–558 (1981).
See also *10.2*, p. *115*; 22.8.1, p. 499.

L. E. OLIVEIRA
*Local Field Effects and a Model Dielectric Response Matrix $\varepsilon_{G,G'}(\boldsymbol{q})$ for a Covalent Semiconductor
(b) Vol. **107**, No. 1, p. 255–265 (1981).
See also *13*, p. *147*; 22.1, p. 402.

E. IGUCHI, H. SAWATARI, and R. J. D. TILLEY
*Polarisabilities of Ions in Rutile (TiO_2) Calculated with Madelung Potentials Determined by Ewald's Method
(b) Vol. **107**, No. 1, p. K63–K66 (1981).
See also 22.6, p. 479.

H. KRESSE, S. DIELE, A. WIEGELEBEN, D. DEMUS, and B. GAJEWSKA
A New Liquid Crystalline Low Temperature Phase
(a) Vol. **67**, No. 1, p. K21–K23 (1981).
See also *2*, p. *49*; 22.9, p. 512.

N. N. SIROTA, L. A. BASHKIROV, E. Z. KATSNELSON, and E. F. SHAPOVALOVA
*The Effect of Small Quantities of Group IV and V Transition Elements on Dielectric Properties of Nickel–Zinc Ferrites
(a) Vol. **67**, No. 1, p. K67–K69 (1981).
See also 22.8.2, p. 506.

P. K. C. PILLAI, G. K. NARULA, A. K. TRIPATHI, and R. G. MENDIRATTA
*Thermally Stimulated Discharge Current Studies in Polarized Polypropylene
(a) Vol. **67**, No. 2, p. 649–654 (1981).
See also 22.9, p. 512.

H. JAIN and A. S. NOWICK
*Electrical Conduction in Paratellurite (TeO_2) Crystals
(a) Vol. **67**, No. 2, p. 701–707 (1981).
See also 10.2, p. 117; 22.6, p. 480.

J. WOLAK and Z. CZAPLA
*Phase Transition in $CsHSeO_4$ Single Crystal at 128 °C
(a) Vol. **67**, No. 2, p. K171–K174 (1981).
See also *1.2*, p. *19*; 22.8.1, p. 500.

G. Nierste and F. Fischer
*Thermally Stimulated Depolarization of NO_3^-, NO_2^-, and PO_2^- Centres
(b) Vol. **108**, No. 1, p. 269–274 (1981).
See also *10.2*, p. *117*; 22.5.2, p. 468.

S. Radhakrishnan, M. N. Kamalasanan, and P. C. Mehendru
*Surface Charge Decay Characteristics in Polyvinyl Acetate Films
(a) Vol. **68**, No. 1, p. 117–121 (1981).
See also 22.9, p. 512.

M. Böhm, S. Herrmann, and A. Scharmann
*Dipole Centres in $CaMoO_4$ Crystals
(a) Vol. **68**, No. 1, p. 129–134 (1981).
See also *10.2*, p. *118*; 20.1, p. 319; 22.8.1, p. 500.

V. P. Gupta, A. Gupta, and V. K. Srivastava
Electronic Properties of Gallium Chalcogenides (GaS, GaSe, and GaTe)
(b) Vol. **108**, No. 2, p. 323–327 (1981).
See also *4*, p. *61*; 22, p. 400.

A. C. Sharma and S. Auluck
*Small Wave Vector $\boldsymbol{q}$ and Frequency ω Dependent Longitudinal Dielectric Function for a Model Semiconductor
(b) Vol. **108**, No. 2, p. 587–600 (1981).
See also *13.1*, p. *161*; 22.1, p. 403; 22.2, p. 426; 22.4, p. 446.

14.4.1 Ferroelectrics

Original Papers and Short Notes

M. D. Fontana, G. E. Kugel, G. Metrat, and C. Carabatos
Long-Wavelength Phonons in the Different Phases of $KNbO_3$
(b) Vol. **103**, No. 1, p. 211 – 219 (1981).
See also *6*, p. *66*; *20.1*, p. *301*; 22.8.1, p. 493.

A. Krumins and P. Günter
Holographic Currents in Reduced $KNbO_3$ Crystals
(a) Vol. **63**, No. 1, p. K111 – K114 (1981).
Erratum
(a) Vol. **65**, No. 2, p. K189 (1981).
See also *16*, p. *251*; 22.8.1, p. 494.

W. Käppler and G. Arlt
Photovoltaic Effect in Pure and Modified Barium Titanate Ceramics
(a) Vol. **63**, No. 2, p. 475 – 480 (1981).
See also *16*, p. *252*; 22.8.1, p. 494.

G. Schmidt, H. Arndt, G. Borchardt, J. von Cieminski, T. Petzsche, K. Borman, A. Sternberg, A. Zirnite, and V. A. Isupov
*Induced Phase Transitions in Ferroelectrics with Diffuse Phase Transition
(a) Vol. **63**, No. 2, p. 501 – 510 (1981).
See also 22.8.1, p. 494.

M. K. Teng, M. Massot, M. R. Chaves, M. H. Amaral, S. Ziolkiewicz, and W. Young
*Tricritical Point Induced by Atomic Substitution in $Bi_xSb_{1-x}SI$
(a) Vol. **63**, No. 2, p. 605 – 615 (1981).
See also 22.8, p. 485.

J. Horváth and D. Kucharczyk
*Temperature Dependence of Lattice Parameters of $PbHPO_4$ and $PbDPO_4$ Single Crystals
(a) Vol. **63**, No. 2, p. 687 – 692 (1981).
See also *1.2*, p. *15*; 22.8.1, p. 494.

G. Spörl, S. Ullmann, W. Hässler, E. Hegenbarth, and B. Malike
*Thermal Conductivity of Tri-Sarcosine-Calcium-Chloride
(a) Vol. **63**, No. 2, p. K175 – K177 (1981).
See also *8*, p. *84*; 22.9, p. 508.

T. Kröber, S. Wartewig, and W. Windsch
*Electron Paramagnetic Resonance Studies of Ferroelectric Fe^{3+}-Doped Tris-Sarcosine Calcium Chloride Monocrystals
(b) Vol. **104**, No. 1, p. 177 – 184 (1981).
See also *19*, p. *287*; 22.9, p. 509.

W. Elflein, W. Käppler, and G. Arlt
*Drift Mobility in Insulating Barium Titanate Ceramics
(a) Vol. **64**, No. 1, p. K1 – K4 (1981).
See also 13.5.3, p. 196; 22.8.1, p. 495.

R. Le Bihan, M. Maussion, F. Lefacheux, E. Manghi, and M. C. Robert
*Observation of Ferroelectric Domains in Gel Grown $PbHPO_4$ Crystals
(a) Vol. **64**, No. 1, p. K5–K6 (1981).
See also 22.8.1, p. 495.

J. Kroupa and Z. Czapla
*Far Infrared Spectra of NH_4HSeO_4
(a) Vol. **64**, No. 1, p. K17–K18 (1981).
See also *20.1*, p. *306*; 22.8.1, p. 495.

R. Clemens, G. Luther, and H. E. Müser
*Dielectric Behaviour of Ferroelectric Barium Titanate in the Microwave Region
(a) Vol. **64**, No. 2, p. 637–645 (1981).
See also 22.8.1, p. 495.

A. Sternberg, P. Hlídek, and M. Zvára
*Electro-Optic Effect in PLZT Ceramics in the Infrared
(a) Vol. **64**, No. 2, p. K149–K153 (1981).
See also *20.1*, p. *309*; 22.8.1, p. 496.

A. Anderson, P. Gash, and F. Lüty
Lattice Dynamics of KCN and NaCN in the Anti-Ferroelectric Phase
(b) Vol. **105**, No. 1, p. 315–325 (1981).
See also *6*, p. *74*; *20.1*, p. *309*; 22, p. 394.

A. A. Mirzakhanyan, A. K. Petrosyan, and H. R. Asatryan
Angular Dependence of the Interdoublet EPR Spectrum of Fe^{3+} and Mn^{2+} in $Pb_5Ge_3O_{11}$ Single Crystals
(b) Vol. **105**, No. 1, p. K55–K59 (1981).
See also *19*, p. *290*; 22.8.1, p. 496.

G. Sorge, U. Straube, and N. R. Ivanov
*Elastic, Dilatational, and Electrostrictive Properties of TSCC Crystals
(a) Vol. **65**, No. 1, p. 189–196 (1981).
See also 1.2, p. 16; *12.2*, p. *140*; 22.8, p. 487.

L. A. Shebanov
X-Ray Temperature Study of Crystallographic Characteristics of Barium Titanate
(a) Vol. **65**, No. 1, p. 321–325 (1981).
See also *1.2*, p. *17*; 22.8.1, p. 496.

O. G. Vlokh, L. A. Lažko, and Ya. I. Shopa
*Electrooptic and Electrogyration Properties of the Solid Solutions on the Basis of Lead Germanate
(a) Vol. **65**, No. 1, p. 371–378 (1981).
See also *20*, p. *297*; 22.8.1, p. 496.

I. H. Ismailzade, R. M. Ismailov, and T. M. Stolpakova
*The Magnetoelectric Effect $(ME)_H$ in the Ferroelastics $NaFe(MoO_4)_2$ and $K_{0.8}Na_{0.2}Fe(MoO_4)_2$
(a) Vol. **65**, No. 1, p. K25–K28 (1981).
See also *18.4*, p. *282*; 22.8.1, p. 497.

V. P. Miroshkin, Ya. I. Panova, and V. V. Passynkov
*Dielectric Relaxation in Polycrystalline Ferrites
(a) Vol. **66**, No. 2, p. 779–782 (1981).
See also 22.8.2, p. 505.

K. Samudravijaya and B. K. Basu
*Dielectric Measurements in Potassium Dihydrogen Arsenate
(a) Vol. **66**, No. 2, p. K121–K124 (1981).
See also 22.8.1, p. 499.

E. Fischer, W. Hässler, E. Hegenbarth, and V. I. Fritsberg
*Amorphous Behaviour of PLZT Demonstrated by Thermal Conductivity Measurements
(a) Vol. **66**, No. 2, p. K169–K171 (1981).
See also *8*, p. *87*; 22.8.1, p. 499.

D. Wagner and D. Bäuerle
*Soft Modes in $SrTiO_3$: Zr, Nb, Ta
(b) Vol. **107**, No. 1, p. K39–K41 (1981).
See also *6*, p. *78*; 22.8.1, p. 499.

G. V. Liberts and V. Ya. Fritsberg
*SHG Investigations in the Paraelectric Phase of Perovskite Type Ferroelectrics
(a) Vol. **67**, No. 1, p. K81–K84 (1981).
See also 20, p. 298; 22.8.1, p. 500.

A. N. Vtyurin, V. P. Ermakov, B. I. Ostrovskii, and V. F. Shabanov
*Study of Optical Second Harmonic Generation in Ferroelectric Liquid Crystal
(b) Vol. **107**, No. 2, p. 397–402 (1981).
See also *2*, p. *49*; 20, p. 299; 22.9, p. 512.

V. E. Schneider and E. E. Tornau
*The Model of the Phase Transition in Squaric Acid
(b) Vol. **107**, No. 2, p. 491–496 (1981).
See also *1.2*, p. *19*; 22.9, p. 512.

F. Brehat, B. Wyncke, and A. Hadni
Far-Infrared Reflectivity Spectra of KH_2PO_4 between 300 and 400 K
(b) Vol. **107**, No. 2, p. 723–727 (1981).
See also *6*, p. *79*; *20.1*, p. *317*; 22.8.1, p. 500.

K. S. Aleksandrov, A. T. Anistratov, S. V. Melnikova, P. V. Klevtsov, A. I. Kruglik, and V. N. Voronov
*Ferroelectric Phase Transitions in $CsLiMoO_4$ Single Crystals
(a) Vol. **67**, No. 2, p. 377–380 (1981).
See also 1.2, p. 19; 22.8.1, p. 500.

Y. Uesu, J. Ogawa, N. Hanami, B. A. Strukov, and J. Kobayashi
*X-Ray Dilatometric Study of Phase Transitions of Ferroelectric $(NH_4)_2BeF_4$
(a) Vol. **67**, No. 2, p. 693–699 (1981).
See also *1.2*, p. *19*; 22.8, p. 490.

15. Thermoelectric and Thermomagnetic Properties

Original Papers and Short Notes

G. Simon and W. Eichler
*Galvanomagnetische und thermoelektrische Transportuntersuchungen an Sb_2Te_3
(Galvanomagnetic and Thermoelectric Transport Investigations on Sb_2Te_3)
(b) Vol. **103**, No. 1, p. 289–295 (1981).
See also *14.3*, p. *209*; 22.7, p. 483.

J. K. A. Amuzu
*The Effect of Tensile Stress on the Thermoelectric E.M.F. in Copper, Gold, and Silver
(a) Vol. **63**, No. 1, p. K7–K10 (1981).
See also 12.1, p. 134; 21.1, p. 352; 21.6, p. 384.

K. B. Sundaram and G. K. Bhagavat
Preparation and Properties of Indium Oxide Films
(a) Vol. **63**, No. 1, p. K15–K18 (1981).
See also 1.5, p. 34; *3*, p. *53*; *14.3.1*, p. *220*; 20.1, p. 303; 22.6, p. 474.

I. Gorczyca
Scattering on Short-Range Potentials in InSb. A Pseudopotential Calculation
(b) Vol. **103**, No. 2, p. 529–533 (1981).
See also *13.4*, p. *170*; 14.3, p. 210; 22.2.3, p. 435.

G. Srinivasan and C. M. Srivastava
Electrical Conductivity Mechanism in Zinc and Copper Substituted Magnetite
(b) Vol. **103**, No. 2, p. 665–671 (1981).
See also *14.3*, p. *210*; 22.8.2, p. 502.

P. Středa and P. Vašek
*Thermopower in Strong Magnetic Fields and the Mott Rule
(b) Vol. **103**, No. 2, p. K137–K140 (1981).
See also 14.1, p. 199; 21.7, p. 387.

J. Horák, L. Tichý, and P. Lošťák
Electrical Conductivity of BiTeI Crystals
(a) Vol. **63**, No. 2, p. 407–416 (1981).
See also 13.1, p. 153; *14.3*, p. *210*; 22.8, p. 485.

L. G. Khvostantsev and V. A. Sidorov
*High-Pressure Polymorphism of Antimony. Thermoelectric Properties and Electrical Resistance Studies
(a) Vol. **64**, No. 1, p. 379–384 (1981).
See also 1.2, p. 16; *14.1*, p. *199*; 21.7, p. 387.

M. A. Shaikh
*Electrical Properties of Quinolinium $(TCNQ)_2$
(a) Vol. **64**, No. 1, p. K23–K25 (1981).
See also *14.3*, p. *211*; 22.9, p. 509.

M. Stordeur, H. T. Langhammer, H. Sobotta, and V. Riede
*Valence Band Structure of $(Bi_{1-x}Sb_x)_2Te_3$ Single Crystals
(b) Vol. **104**, No. 2, p. 513–522 (1981).
See also *13.1*, p. *154*; **14.3**, p. 211; 20.1, p. 307; 22.7, p. 483.

R. D. Barnard
Some Remarks on Deviations from Matthiessen's Rule in Dilute Alloy Systems
(b) Vol. **104**, No. 2, p. 613–620 (1981).
See also 6, p. 72; 10.1, p. 99; 13.2, p. 163; *14.1*, p. *200*; 21.6, p. 384.

L. Tichý, M. Frumar, M. Kincl, J. Klikorka, and A. Tříska
*Mixed Scattering Mechanism of Free Current Carriers in $SnBi_4Te_7$ Single Crystals
(a) Vol. **64**, No. 2, p. 461–466 (1981).
See also *14.3*, p. *212*; 22.8, p. 486.

B. K. Verma, O. P. Srivastava, and H. B. Lal
*Transport Studies of the Copper Solid Superionic Conductor $CuPb_3Br_7$
(a) Vol. **64**, No. 2, p. 467–473 (1981).
See also *14.4*, p. *238*; 22.8, p. 486.

J. J. Dubowski
*Influence of Electron–Electron Scattering on Lorentz Number in Narrow-Gap $Cd_xHg_{1-x}Te$
(b) Vol. **105**, No. 1, p. 399–404 (1981).
See also *14.3*, p. *212*; 22.4.4, p. 456.

M. B. Dutt, R. Banerjee, and A. K. Barua
*Transport Properties of Lithium and Sodium Doped Nickel Oxide
(a) Vol. **65**, No. 1, p. 365–370 (1981).
See also *14.3*, p. *213*; 22.6, p. 477.

D. V. Gitsu, F. M. Muntyanu, and M. I. Onu
*Quantum Oscillations of Thermomagnetic Power and Magnetoresistance in Antimony–Bismuth Alloys
(b) Vol. **105**, No. 2, p. K95–K98 (1981).
See also *14.1*, p. *201*; 21.7, p. 387.

L. V. Rao, B. S. Naidu, and P. J. Reddy
*Electrical Conductivity and Thermopower Measurements on Silver Doped Tellurium Films
(a) Vol. **65**, No. 2, p. K135–K138 (1981).
See also 1.5, p. 35; *14.3.1*, p. *221*; 22.1.3, p. 425.

W. Heimke and V. I. Novikov
*On the Transport Properties of Chromium–Aluminium Alloys at Low Temperatures
(a) Vol. **65**, No. 2, p. K139–K142 (1981).
See also 8, p. 86; *14.1*, p. *201*; 21.1, p. 360.

K. Takita and G. Landwehr
*Very Large Phonon-Drag Thermoelectric Power of HgTe in Strong Magnetic Fields
(b) Vol. **106**, No. 1, p. 259–269 (1981).
See also *6*, p. *76*; 18, p. 263; 22.4.3, p. 454.

V. S. SHUKLA and D. M. ROWE
*Precipitation of Boron from Silicon-Germanium Alloy and Its Effect in the Thermoelectric Transport Properties
(a) Vol. **66**, No. 1, p. 243–248 (1981).
See also 10.2, p. 113; *14.3*, p. *215*; 22.1.1, p. 406; 22.1.2, p. 416.

V. MADHUSUDAN, S. G. SATHYANARAYAN, and G. SIVARAMA SASTRY
Electrical Conductivity and Stimulated Thermocurrent Studies in Dicalcium Barium Propionate Single Crystals
(a) Vol. **66**, No. 1, p. 371–376 (1981).
See also *14*, p. *198*; 22.9, p. 511.

E. LITWIN-STASZEWSKA, W. SZYMAŃSKA, and R. PIOTRZKOWSKI
*The Electron Mobility and Thermoelectric Power in InSb at Atmospheric and Hydrostatic Pressures
(b) Vol. **106**, No. 2, p. 551–559 (1981).
See also *14.3*, p. *215*; 22.2.3, p. 436.

H. OVERHOF and W. BEYER
Density of States in Amorphous Si:H
(b) Vol. **107**, No. 1, p. 207–213 (1981).
See also *13.2*, p. *163*; 14.3, p. 216; 22.1.2, p. 418.

A. H. ABOU EL ELA, H. H. A. LABIB, and N. ABDELMOHSEN
Electrical Conductivity, Thermoelectric Power, and Thermal Conductivity of $CuTlSe_2$ in the Solid and Liquid States
(a) Vol. **67**, No. 1, p. K43–K48 (1981).
See also 8, p. 87; *14.3*, p. *217*; 22.8, p. 490.

D. SEN and S. K. SARKAR
*Unified Study of Metals on Pseudopotential Theory: Lead
(b) Vol. **107**, No. 2, p. 559–567 (1981).
See also 2, p. 49; *6*, p. *79*; 12.1, p. 136; *14.1*, p. *202*; 21, p. 346.

M. IDREES, FARID A. KHWAJA, and M. S. K. RAZMI
A Comparative Study of the Pseudopotentials for Cu, Ag, and Au
(b) Vol. **107**, No. 2, p. 769–778 (1981).
See also *13*, p. *148*; *14.1*, p. *202*; 21.1, p. 364; 21.6, p. 386.

I. K. BANDROVSKAYA, V. YA. PROKHORENKO, V. G. ROMANOV, and B. I. SOKOLOVSKII
Optical and Thermoelectric Properties or Arsenic Selenide in Glassy and Liquid States
(b) Vol. **107**, No. 2, p. K95–K98 (1981).
See also 2, p. 50; *14.3*, p. *217*; *20.1*, p. *318*; 22, p. 398.

M. PAI and J. M. HONIG
*A Test for Electron Localization in Intrinsic Semiconductors
(b) Vol. **108**, No. 2, p. K79–K83 (1981).
See also *14.3*, p. *218*.

16. Photoconductivity. Photovoltaic Effects

Review Article

K. W. Böer
*The CdS/Cu_2S Solar Cell (II): The Junction and Junction–Emitter Relation
(a) Vol. **66**, No. 1, p. 11–43 (1981).
See also 13.4, p. 168; *14.3.3*, p. *225*; 14.3.4, p. 232; 22.4.1, p. 447.

Original Papers and Short Notes

Y. Maruyama and H. Bässler
*Mobility and Recombination Coefficient of Charge Carriers in Disordered Pentacene
(b) Vol. **103**, No. 1, p. 263–268 (1981).
See also 13.4, p. 168; 22.9, p. 507.

N. T. Bagraev, L. S. Vlasenko, A. A. Lebedev, I. A. Merkulov, and P. Yusupov
*Exchange Impurity Levels and Photoconductivity of Semiconductors
(b) Vol. **103**, No. 1, p. K51–K54 (1981).
See also *13.4*, p. *169*; 22.1.2, p. 408.

S. Kończak and M. Nowak
*The Estimation of Semiconductor Parameters Using Least Squares in Photomagnetoelectric Investigations
(a) Vol. **63**, No. 1, p. 305–311 (1981).
See also 22.1.1, p. 404; 22.2.3, p. 435.

Sh. M. Efendiev, V. E. Bagiev, A. Ch. Zeinally, V. A. Balashov, V. A. Lomonov, and A. A. Majer
Optical Properties of $Bi_{12}TiO_{20}$ Single Crystals
(a) Vol. **63**, No. 1, p. K19–K22 (1981).
See also *20.1*, p. *303*; 22.8, p. 484.

I. D. Joshi and D. Sharma
*Preparation and Properties of CdS Sintered Layers
(a) Vol. **63**, No. 1, p. K23–K26 (1981).
See also 22.4.1, p. 447.

A. Krumins and P. Günter
*Holographic Currents in Reduced $KNbO_3$ Crystals
(a) Vol. **63**, No. 1, p. K111–K114 (1981).
Erratum
(a) Vol. **65**, No. 2, p. K189 (1981).
See also 14.4.1, p. 242; 22.8.1, p. 494.

R. S. Lee, J. F. Merklin, C. D. Marrs, and M. H. Richter
*Impurity Photoconductivity of γ-Irradiated LiF
(b) Vol. **103**, No. 2, p. 605–611 (1981).
See also 10.2, p. 107; *20.1*, p. *303*; 22.5.2, p. 463.

J. Gong, K. M. van Vliet, A. D. Sutherland, and E. R. Chenette
Noise Measurements on Photo Avalanche Diodes
(a) Vol. **63**, No. 2, p. 445–460 (1981).
See also *14.3.3*, p. *225*; 22.1.2, p. 409.

L. V. CHERNYSH, V. A. KOVARSKII, and E. P. SINYAVSKII
Laser Induced Auger Recombination in Doped Semiconductors
(b) Vol. **105**, No. 1, p. K51–K53 (1981).
See also 6, p. 74; *13.4*, p. *175*; 22, p. 395.

V. L. ALPEROVICH, A. F. KRAVCHENKO, N. A. PAKHANOV, and A. S. TEREKHOV
*Effect of Electron Heating by Surface Electric Fields on Oscillating Photoresponse Spectra of Gallium Arsenide–Metal Structures
(a) Vol. **65**, No. 1, p. 259–263 (1981).
See also *14.3.2*, p. *223*; 22.2.1, p. 429.

N. E. KORSUNSKAYA, I. V. MARKEVICH, T. V. TORCHINSKAYA, and M. K. SHEINKMAN
*Erratum to: Photosensitivity Degradation Mechanism in CdS:Cu Single Crystals
(a) Vol. **65**, No. 1, p. 403 (1981).
See also *13.4*, p. *176*; 22.4.1, p. 448.

E. F. BAGIRZADE, B. G. TAGIEV, G. M. MAMEDOV, I. M. ASKEROV, and F. N. SULTANOV
*Interference Oscillations of Photocurrent in GaSe:Sn and $GeSe_{0.95}Te_{0.05}$:Sn Single Crystals
(a) Vol. **65**, No. 1, p. K55–K58 (1981).
See also 14.3.4, p. 234; *20.1*, p. *310*; 22.3, p. 441.

K. NAUKA and J. MISIEWICZ
Minority-Carrier Diffusion Length in Zn_3P_2
(a) Vol. **65**, No. 1, p. K95–K97 (1981).
See also *14.3.3*, p. *227*; 22.3, p. 441.

D. D. ELEY, S. KINNEAR, and M. R. WILLIS
*Real and Apparent Photoconduction in Single Crystals of Methyltriphenylphosphonium $(TCNQ)_2^-$
(b) Vol. **105**, No. 2, p. 677–683 (1981).
See also 22.9, p. 510.

S. KOŃCZAK, K. KOTEWICZ, and M. NOWAK
*High-Frequency Photomagnetoelectric Method for Determining Semiconductor Parameters
(a) Vol. **65**, No. 2, p. 447–451 (1981).
See also *18*, p. *262*; 22.1.2, p. 414.

I. R. DOROSH, YU. S. KUZMINOV, N. M. POLOZKOV, A M. PROKHOROV, V. V. OSIKO, N. V. TKACHENKO, V. V. VORONOV, and D. KH. NURLIGAREEV
Barium–Strontium Niobate Crystals for Optical Information Recording
(a) Vol. **65**, No. 2, p. 513–522 (1981).
See also *14.4.1*, p. *244*; *20.1*, p. *312*; 22.8.1, p. 497.

R. O. LOUTFY
*Effect of Temperature on the Dark and Photoconductivity of Metal-Free Phthalocyanine
(a) Vol. **65**, No. 2, p. 659–668 (1981).
See also *14.3.1*, p. *221*; 22.9, p. 510.

V. S. BALLANDOVICH, YU. M. TAIROV, and G. N. VIOLINA
*Acceptor Boron in α-SiC (6H): Investigation by the Photocapacitance Method
(a) Vol. **65**, No. 2, p. 709–716 (1981).
See also *13.4*, p. *177*; 20.3, p. 329; 22.3, p. 441.

G. A. Medvedkin and Yu. V. Rud
*The Parameters of Polarization Photosensitivity of Isotropic Semiconductors
(a) Vol. **67**, No. 1, p. 333–337 (1981).
See also 22, p. 398.

E. L. Frankevich, M. M. Tribel, I. A. Sokolik, L. I. Kolesnikova, and Yu. M. Stolovitski
*Magnetic Spin Phenomena in the Chlorophyll "a" Photoconductivity
(b) Vol. **107**, No. 2, p. 423–434 (1981).
See also 10.2, p. 116; 18, p. 263; 22.9, p. 512.

A. V. Voitsekhovskii and Yu. V. Lilenko
*Recombination Effects in Narrow-Gap Semiconductors p-$Hg_{1-x}Cd_xTe$
(a) Vol. **67**, No. 2, p. 381–386 (1981).
See also 13.4, p. 181; 22.4.4, p. 457.

J. Sinkkonen
*Photoconductivity of an Inhomogeneous Semiconductor
(a) Vol. **67**, No. 2, p. 555–564 (1981).
See also *13.4*, p. *182*; 22.1.3, p. 425; 22.4.1, p. 450.

D. D. Eley, S. Kinnear, and M. R. Willis
*Lifetime of Photogenerated Carriers in TCNQ Single Crystals
(b) Vol. **108**, No. 2, p. 373–376 (1981).
See also 22.9, p. 512.

V. K. Bashenov, A. G. Gontar, and A. G. Petukhov
*Excited States of a Shallow Acceptor in Semiconducting Diamond from Photoconductivity Spectra
(b) Vol. **108**, No. 2, p. K139–K142 (1981).
See also *13.4*, p. *184*; 22.1, p. 403.

17. Emission of Electrons and Ions

Original Papers and Short Notes

M. A. Blokhin, E. G. Orlova, and I. G. Schweitzer
*The Influence of the Environment of the Absorbing Atom on the Photoelectric Yield Spectra of Beryllium Compounds in the Soft X-Ray Region
(b) Vol. **103**, No. 1, p. 63–69 (1981).
See also *4*, p. *57*; 22.5, p. 458; 22.6, p. 474; 22.8, p. 484; 22.9, p. 507.

C. Ance and A. Amamou
Photoemission Spectra of the Ternary $Ga_{1-x}Al_xSb$ Alloy
(b) Vol. **103**, No. 2, p. 441–449 (1981).
See also *13.1*, p. *152*; 22.2.4, p. 437.

H.-J. Fitting
*Exoelectron Emission Due to Relaxation Processes at Surfaces
(a) Vol. **63**, No. 1, p. K47–K50 (1981).
See also 22.6, p. 474.

G. Margaritondo, A. D. Katnani, and F. Lévy
Electronic States of $CdIn_2Se_4$ and $ZnIn_2Se_4$: Role of the Cation Pseudovacancies
(b) Vol. **103**, No. 2, p. 725–731 (1981).
See also *10.2*, p. *107*; *13.1*, p. *153*; 22.8, p. 484.

V. Siegel and H.-H. Kirchner
*The Influence of the Ambient Atmosphere on the TSEE of Ceramic BeO
(a) Vol. **63**, No. 2, p. 481–485 (1981).
See also 22.6, p. 474.

M. Rösler and W. Brauer
*Theory of Secondary Electron Emission (I): General Theory for Nearly-Free-Electron Metals
(b) Vol. **104**, No. 1, p. 161–175 (1981).
See also 13.5.2, p. 194; 21, p. 338.

O. V. Farberovich
Band Structure and Electron–Electron Interaction in Samarium Monosulfide
(b) Vol. **104**, No. 1, p. 365–374 (1981).
See also *13.1*, p. *154*; 20.1, p. 305; 21.4, p. 380; 22, p. 392.

V. G. Kohn and M. V. Kovalchuk
*On the Theory of the External Photoeffect Accompanying X-Ray Diffraction in an Ideal Crystal with Disturbed Surface Layer
(a) Vol. **64**, No. 1, p. 359–366 (1981).
See also *1.6*, p. *38*; 22.1.2, p. 411.

M. Rösler and W. Brauer
*Theory of Secondary Electron Emission (II): Application to Aluminum
(b) Vol. **104**, No. 2, p. 575–587 (1981).
See also 13.5.2, p. 194; 21, p. 340.

E. P. Domashevskaya and V. A. Terekhov
d–s,p Resonance and Electronic Structure of Compounds Alloys and Solid Solutions
(b) Vol. **105**, No. 1, p. 121–127 (1981).
See also *13.1*, p. *155*; 20, p. 297; 21, p. 341; 21.1, p. 358.

P. Rennert
Multiple Scattering Theory in the Small Scattering Centre Approximation. Application to EXAFS, AES, XPS, MEED
(b) Vol. **105**, No. 1, p. 213–221 (1981).
See also 10, p. 93; *11*, p. *125*; 13.1, p. 155.

K. Uhlmann and F. Storbeck
*Relative Determination of L_3 Ionisation Cross Sections of Chromium, Iron, and Nickel
(a) Vol. **65**, No. 1, p. K37–K41 (1981).
See also 21.1, p. 359; 21.1.1, p. 371.

N. Schwentner, G. Martens, and H. W. Rudolf
Free Exciton Energy Transfer in Kr–Au Sandwiches
(b) Vol. **106**, No. 1, p. 183–192 (1981).
See also *13.5.1*, p. *190*; 23, p. 514.

E. A. Sozontov, M. V. Kruglov, and B. G. Zakharov
*Photoemission in Bragg Diffraction of X-Rays by Bicrystals
(a) Vol. **66**, No. 1, p. 303–310 (1981).
See also *1.1*, p. *12*; 22.1.1, p. 406; 22.2.1, p. 429; 22.2.4, p. 438.

E. P. Domashevskaya, L. N. Marshakova, V. A. Terekhov, A. N. Lukin, Ya. A. Ugai, V. I. Nefedov, and Ya. V. Salyn
Role of Noble Metal d-States in the Formation of the Electron Structure of Ternary Sulphides
(b) Vol. **106**, No. 2, p. 429–435 (1981).
See also *13.1*, p. *158*; 20, p. 298; 22.8, p. 488.

M. Wöhlecke and G. Borstel
Electron-Spin Polarization in Hexagonal Crystals Induced by Circulary Polarized Light
(b) Vol. **106**, No. 2, p. 593–599 (1981).
See also *13.1*, p. *159*.

J. Szuber
On the Correlation of Paramagnetic Properties and Electronic Structure of (111) Si Surfaces Baked at Low Temperature in High Vacuum
(a) Vol. **66**, No. 2, p. K141–K144 (1981).
See also *13.3*, p. *166*; 22.1.2, p. 418.

M. V. Kovalchuk, N. Hertel, M. K. Melkonyan, R. M. Imamov, and P. A. Aleksandrov
*Fluorescence Yield Curves under X-Ray Diffraction Conditions
(a) Vol. **66**, No. 2, p. K173–K177 (1981).
See also *20.3*, p. *331*; 22.1.1, p. 406; 22.1.2, p. 418.

I. Bartoš and F. Máca
*On Self-Deconvolutions in Auger Spectroscopy
(b) Vol. **107**, No. 2, p. 513–519 (1981).
See also 13.1, p. 160.

17.1 Field Emission Microscope Investigations

Original Papers

N. Rihon
*Band Bending in Field Emission Spectroscopy from the Zinc Oxide (0001) Polar Face
(a) Vol. **63**, No. 2, p. 617–624 (1981).
See also 22.4, p. 445.

H. C. Eaton
*Grain Boundary Dislocation Observations Using the Field Ion Microscope
(a) Vol. **65**, No. 2, p. 497–502 (1981).
See also *10.1*, p. *101*; 21.1, p. 359.

18. Magnetic Properties

Original Papers and Short Notes

Yu. E. Lozovik and D. R. Musin
*Stability Region of the Two-Dimensional Electron Crystal in Strong Magnetic Fields
(b) Vol. **104**, No. 2, p. 711–717 (1981).
See also *13*, p. *144*.

P. C. M. Gubbens, A. M. van der Kraan, and K. H. J. Buschow
*^{161}Dy Mössbauer Effect and Magnetic Properties of Some Amorphous and Crystalline Dy–Co Alloys
(a) Vol. **64**, No. 2, p. 657–663 (1981).
See also 2, p. 44; *5*, p. *63*; 21.1, p. 357; 21.4, p. 381.

S. Cristoloveanu
*Magnetic Field and Surface Influences on Double Injection Phenomena in Semiconductors (I): The Magnetodiode Effect Theory for the Semiconductor and Insulator Regimes
(a) Vol. **64**, No. 2, p. 683–695 (1981).
See also *14.3.3*, p. *226*; 14.3.4, p. 233.

D. Kostopoulos and J. Grammatikakis
*Disturbance in the State of Magnetite Due to Magnetic Prehistory at 78 K
(a) Vol. **64**, No. 2, p. K155–K159 (1981).
See also *14.3*, p. *212*; 22.8.2, p. 503.

B. Lorenz
Charge Order–Disorder Transition in the Extended Hubbard Model. Finite Bandwidth Effect
(b) Vol. **105**, No. 1, p. 233–241 (1981).
See also *13*, p. *145*; 22.6, p. 476.

Sh. A. Gulamov, Yu. A. Katulevskii, and R. A. Muminov
Pecularities of an Electromotive Force Arising in a Semiconductor in a Non-Homogeneous Magnetic Field (II)
(b) Vol. **105**, No. 1, p. 405–408 (1981).
See also *14.3*, p. *213*; 22, p. 394.

F. Tataru and I. Boskovitz
The Use of Magnetic Circular Dichroism Measurements in Determining the Reduced Matrix Elements of the Electric-Dipole Momentum Operator
(b) Vol. **105**, No. 1, p. K1–K4 (1981).
See also *13.4*, p. *175*; *20*, p. *297*; 22.8, p. 486.

S. J. Pearton
*Magnetic Field Dependence of Defect State in GaAs
(b) Vol. **105**, No. 1, p. K19–K22 (1981).
See also *13.4*, p. *175*; 14.3, p. 213; 22.2.1, p. 428.

V. K. Malyutenko, K. Yu. Guga, and Yu. M. Malozovskii
Magnetoconcentration Effect at Auger Recombination of Current Carriers
(a) Vol. **65**, No. 1, p. 131–140 (1981).
See also *14.3*, p. *213*; 22.2.3, p. 436.

K. Takita and G. Landwehr
Very Large Phonon-Drag Thermoelectric Power of HgTe in Strong Magnetic Fields
(b) Vol. **106**, No. 1, p. 259–269 (1981).
See also *6*, p. *76*; *15*, p. *248*; 22.4.3, p. 454.

T. Tsuboi
Assignment of $(Tl^+)_2$ Absorption Band in Alkali Halides by Magnetic Circular Dichroism Spectra
(b) Vol. **106**, No. 1, p. 301–307 (1981).
See also *13.4*, p. *178*; *20.1*, p. *313*; 22.5.2, p. 465.

A. V. Malakhovskii
*Influence of Crystal Field on the Magnetic Circular Birefringence
(b) Vol. **106**, No. 1, p. 327–336 (1981).
See also *20*, p. *298*.

A. Rauh and H. J. Schellnhuber
Diamagnetic Band Structure. Comparison of Second-Order Perturbation with a First-Principles Calculation
(b) Vol. **106**, No. 2, p. 537–544 (1981).
See also *13.1*, p. *158*.

E. Wieser, A. Handstein, J. Schneider, and K. Závěta
*Magnetic Structure of Amorphous $(Fe_{1-x}Mn_x)_{75}P_{15}C_{10}$ Alloys
(a) Vol. **66**, No. 2, p. 607–612 (1981).
See also 2, p. 48; 5, p. 65; 21.1, p. 362; 21.1.1, p. 373.

S. Ghosh, N. Apte, and V. K. Agarwal
Excitation of the Hybrid Mode Due to Modulation of an Intense Laser Beam in a Transversely Magnetised Semiconductor Plasma
(b) Vol. **107**, No. 1, p. 75–81 (1981).
See also *7*, p. *83*; *20*, p. *298*; 22, p. 398.

H. Heidrich
*Magnetic Susceptibility and Photomagnetic Measurements on ZnS:MnS under Low-Field Conditions with a SQUID Magnetometer
(a) Vol. **67**, No. 1, p. 163–170 (1981).
See also *20*, p. *298*; 22.4.4, p. 457.

E. L. Frankevich, M. M. Tribel, I. A. Sokolik, L. I. Kolesnikova, and Yu. M. Stolovitskii
Magnetic Spin Phenomena in the Chlorophyll "a" Photoconductivity
(b) Vol. **107**, No. 2, p. 423–434 (1981).
See also 10.2, p. 116; *16*, p. *255*; 22.9, p. 512.

A. del Moral
*A High-Order Magnetoeleastic Hamiltonian for the Conduction Electron Contribution in Cubic Rare-Earth Metals
(b) Vol. **108**, No. 1, p. 45–52 (1981).
See also 12.1, p. 137; 21.4, p. 382.

18.1 Paramagnetic Properties

Original Papers and Short Notes

RAMJIT SINGH, S. PRAKASH, and J. SINGH
*Dynamical Spin Susceptibility of 3d, 4d, and 5d Transition Metals
(b) Vol. **103**, No. 1, p. K65–K69 (1981).
See also 21.1, p. 351; 21.6, p. 384.

G. C. PSALTAKIS and M. G. COTTAM
Theory of Spin–Phonon Interactions in a System of Paramagnetic $S = 3/2$ Ions
(b) Vol. **103**, No. 2, p. 709–716 (1981).
See also *6*, p. *68*; 8, p. 84; *13.4*, p. *170*; 22.6, p. 474.

G. C. BARKER and J. W. TUCKER
Diagrammatic Perturbation Theory for a Spin–Phonon System
(b) Vol. **104**, No. 1, p. 313–323 (1981).
Erratum
(b) Vol. **108**, No. 2, p. 693 (1981).
See also *6*, p. *70*; *13.5*, p. *186*.

K. K. P. SRIVASTAVA and H. C. SINGH
Erratum to: Effect of Second-Order Spin–Orbit Coupling $\lambda' (\boldsymbol{L} \cdot \boldsymbol{S})^2$ on Electronic Spin Relaxation of Parametric Fe^{2+} in $CaCO_3$ and $CdCO_3$
(b) Vol. **104**, No. 1, p. 393 (1981).
See also *5*, p. *63*; 22.8, p. 485.

B. V. KARPENKO
*On the Magnetism in Narrow-Band Conductors
(b) Vol. **104**, No. 2, p. 407–413 (1981).
See also *13.1*, p. *154*; 21.4, p. 381.

J. KAPTURCZAK and Z. PAJĄK
*Paramagnetic Broadening of PMR Line in Ammonium Manganese Sulphate
(b) Vol. **104**, No. 2, p. 479–483 (1981).
See also *19*, p. *288*; 22.8.1, p. 495.

YU. G. SEMENOV and B. D. SHANINA
Exchange Interaction between Paramagnetic Centers and Valence Band Electrons in Semiconductors with Cubic Lattices
(b) Vol. **104**, No. 2, p. 631–639 (1981).
See also *13.1*, p. *154*; *13.4*, p. *173*; 22.4.2, p. 452; 22.4.3, p. 454.

D. K. DE
Electron Spin Resonance Study on Copper Salicylate Tetrahydrate (Blue Green) and Anhydrous Copper Salicylate (Brown)
(a) Vol. **64**, No. 2, p. 761–769 (1981).
See also *19*, p. *289*; 22.9, p. 509.

K. K. P. SRIVASTAVA and H. C. SINGH
*Paramagnetic Spin Relaxation Due to Spin–Spin Interaction Assisted by Spin–Phonon Interaction or Vice Versa (II)
(b) Vol. **105**, No. 1, p. 115–120 (1981).
See also *6*, p. *73*; 13, p. 144.

18.2 Ferromagnetic Properties

Original Papers and Short Notes

A. Bobák
*Exactly Solvable Model of Amorphous Ferromagnets
(b) Vol. **103**, No. 1, p. 101–105 (1981).
See also *2*, p. *40*.

V. Čápek
*Role of Magnetization in the Electron Quasiparticle Energy in the s–d Model of Ferromagnetic Semiconductors
(b) Vol. **103**, No. 1, p. 107–113 (1981).
See also *13.1*, p. *151*; 22, p. 390.

V. Szöcs, V. Čápek, and B. Velický
Simple Two-Band s–d Model for Ferromagnetic Semiconductors with Hybridization (II): Further Development and Optical Properties
(b) Vol. **103**, No. 1, p. 139–149 (1981).
See also *13.1*, p. *151*; 22, p. 390.

L. Palermo and X. A. da Silva
*Induced Magnetism via Conduction Electron Exchange. Application to $PrAl_2$
(b) Vol. **103**, No. 1, p. 419–422 (1981).
See also *13*, p. *142*; 21.4, p. 380.

W. Wójcik and K. Sokalski
*Pure Site Percolation Problem for Linear Ising Chain in the Molucular Field
(b) Vol. **103**, No. 1, p. K5–K8 (1981).
See also *13*, p. *142*.

E. P. Lyashenko and V. K. Raev
*On the Analysis of Current-Access Bubble Circuits
(a) Vol. **63**, No. 1, p. 55–61 (1981).

L. M. Sandratskii, R. F. Egorov, and A. A. Berdyshev
Energy Band Structure and Electronics Properties of NiAs Type Compounds (I): Ferromagnetic Manganese Arsenide
(b) Vol. **103**, No. 2, p. 511–518 (1981).
See also *13.1*, p. *152*; 21.1, p. 353.

R. Piasecki
*Anisotropic Ferromagnet with Two Spins per Site
(b) Vol. **103**, No. 2, p. 547–557 (1981).

M. J. Škrinjar, D. V. Kapor, and J. P. Šetrajčić
*Low-Temperature Poperties of an Anisotropic Heisenberg Ferromagnet
(b) Vol. **103**, No. 2, p. 559–566 (1981).

N. G. Bebenin
The Critical Dynamics of Hot Electrons in a Ferromagnetic Semiconductor at $T = T_c$
(b) Vol. **104**, No. 1, p. K65–K70 (1981).
See also *13*, p. *143*; *14.3.4*, p. *233*; 22, p. 393.

H. J. Blythe, F. Walz, and H. Kronmüller
*Magnetic Relaxation in Electron-Irradiated Ordered Ni_3Fe
(a) Vol. **64**, No. 1, p. 169–176 (1981).
See also 10.1, p. 98; *11*, p. *123*; 21.1, p. 355; 21.1.1, p. 369.

G. Le Caer and J. M. Dubois
*About the Asymmetries in Mössbauer Spectra of Magnetic Amorphous Transition Metal-Metalloid Alloys
(a) Vol. **64**, No. 1, p. 275–282 (1981).
See also 2, p. 43; *5*, p. *63*; 21.1.1, p. 369.

W. Bażela, A. Szytuła, J. Todorović, and A. Zieba
*Crystal and Magnetic Structure of the $NiMnGe_{1-x}Si_x$ System
(a) Vol. **64**, No. 1, p. 367–378 (1981).
See also 1.3, p. 23; *18.4*, p. *281*; 21.1, p. 355.

H. Oesterreicher and F. T. Parker
Erratum to: Aspects of Intrinsic Coercivity via Gd Analogs
(a) Vol. **64**, No. 1, p. 419 (1981).
See also 21.1, p. 356; 21.4, p. 381.

B. K. Ponomarev and V. G. Thiessen
*Temperature Dependence of the Baric Magnetization Coefficient of a Band Ferromagnet
(b) Vol. **104**, No. 2, p. 427–434 (1981).
See also 21.1, p. 356; 21.1.1, p. 369.

L. L. Gonçalves
*Isothermal and Frequency-Dependent Perpendicular Sucseptibilities of the One-Dimensional Ising Model with Nearest and Next-Nearest Neighbour Interactions
(b) Vol. **104**, No. 2, p. 491–496 (1981).
See also 18.4, p. 282.

Z. Bąk
*Heisenberg Ferromagnet with Anisotropic Biquadratic Exchange in a Transverse Magnetic Field
(b) Vol. **104**, No. 2, p. K125–K128 (1981).

S. Tyagi, J. Steinberg, A. E. Lord, Jr., and P. M. Anderson
*Acoustic Barkhausen Emission in Amorphous Metallic Alloys $Fe_{40}Ni_{40}P_{14}B_6$ and $Fe_{78}Mo_2B_{20}$
(a) Vol. **64**, No. 2, p. 443–448 (1981).
See also *2*, p. *44*; 21.1.1, p. 370.

L. Kraus, Z. Frait, and J. Schneider
*Ferromagnetic Resonance in Amorphous $(FeNi)_{80}P_{10}B_{10}$ Alloys (II): Resonance near the Curie Temperature
(a) Vol. **64**, No. 2, p. 449–456 (1981).
See also *2*, p. *44*; 19, p. 289; 21.1, p. 356; 21.1.1, p. 370.

J. Mwanje and E. M. Ayiera
Temperature-Dependent Resistivity of Ni-Base Alloys
(a) Vol. **65**, No. 1, p. K7—K8 (1981).
See also *14.1*, p. *200*; 18.1, p. 266; 21.1, p. 359.

V. E. Antonov, I. T. Belash, E. G. Ponyatovskii, V. G. Thiessen, and V. I. Shiryaev
*Magnetization of Iron Hydride
(a) Vol. **65**, No. 1, p. K43—K48 (1981).
See also 21.1.1, p. 371.

K. Hoselitz
*The Saturation Magnetization of Some Substituted Fe–Si–B Metallic Glasses
(a) Vol. **65**, No. 1, p. K91—K94 (1981).
See also *18.4*, p. *283*; 21.1.1, p. 371.

T. Kaneyoshi and I. P. Fittipaldi
*A Theory of an Amorphous Ferromagnet
(b) Vol. **105**, No. 2, p. 629—632 (1981).
See also *2*, p. *46*.

O. V. Farberovich and S. L. Vlasov
APW Research of Europium Monochalcogenides (I): Band Structure and Electron–Electron Interaction in Ferromagnetic Semiconductors
(b) Vol. **105**, No. 2, p. 755—768 (1981).
See also *13.1*, p. *156*; 20.1, p. 311; 22, p. 396.

E. Kolley and W. Kolley
Functional Integral Approach to the Electron–Magnon Interaction in Itinerant Ferromagnetism
(b) Vol. **105**, No. 2, p. K85—K89 (1981).
See also *13.5.4*, p. *197*; 21, p. 343.

V. K. Saxena
*Curie Temperature of Spin One Ferromagnets with Large Uniaxial Anisotropy
(b) Vol. **105**, No. 2, p. K125—K128 (1981).
See also *2*, p. *46*.

A. I. Mitsek, N. P. Kolmakova, and D. I. Sirota
*Metamagnetism of Two-Sublattice Uniaxial Antiferromagnets
(a) Vol. **65**, No. 2, p. 503—512 (1981).
See also *18.4*, p. *283*.

B. Lorenz
*Magnetism versus Charge Order in the Extended Hubbard Model with Ising-Like Exchange Interaction
(b) Vol. **106**, No. 1, p. K17—K21 (1981).
See also *13*, p. *146*; 18.4, p. 283.

W. Bażela and A. Szytuła
*Crystal and Magnetic Structure of the $NiMn_{1-t}Ti_tGe$ System
(a) Vol. **66**, No. 1, p. 45—52 (1981).
See also 1.1, p. 12; *18.4*, p. *283*; 22.8, p. 488.

Y. Umakoshi and H. Kronmüller
Observation of Dislocations in Deformed Iron Single Crystals by Means of High-Field Susceptibility Measurements
(a) Vol. **66**, No. 2, p. 509–520 (1981).
See also *10.1*, p. *101*; 21.1.1, p. 373.

A. Handstein, J. Kadlecová, and V. Gregor
Magnetoresistance of Amorphous $(Fe_xNi_{100-x})_{80}B_{10}P_{10}$ Alloys in Magnetic Fields up to 14.5 T
(a) Vol. **66**, No. 2, p. 657–663 (1981).
See also *2*, p. *48*; *14.1*, p. *202*; 21.1, p. 362; 21.1.1, p. 373.

R. Grössinger
*A Critical Examination of the Law of Approach to Saturation (I): Fit Procedure
(a) Vol. **66**, No. 2, p. 665–674 (1981).
See also 18.3, p. 279; 21.1, p. 362; 22.8.2, p. 505.

K. Watanabe
*The Permanent Magnet in Iron and Palladium Alloys "Pallamagnet"
(a) Vol. **66**, No. 2, p. 697–701 (1981).
See also 21.1.1, p. 374.

A. Zentko, M. Tinko, and P. Duhaj
*Effect of Neutron Irradiation on the Magnetic Properties of Amorphous $Fe_{47}Ni_{25}B_{18}Si_{10}$ Alloys
(a) Vol. **66**, No. 2, p. K125–K126 (1981).
See also *2*, p. *48*; 11, p. 128; 21.1.1, p. 374.

A. Z. Maksymowicz and W. Karaš
*Short Range Order and the Magnetic Properties of the Binary Alloys in the Ising Model
(b) Vol. **107**, No. 1, p. K1–K4 (1981).
See also 21.1, p. 363; 21.1.1, p. 374.

R. L. Jacobs
*Model Electronic Structure and Saturation Magnetisation of Some Substituted Fe–Si–B Metallic Glasses
(b) Vol. **107**, No. 1, p. K13–K17 (1981).
See also 2, p. 49; *13*, p. *147*; 21.1.1, p. 374.

V. E. Rode, B. Wurl, and J. Olszewski
*Magnetic Properties of Iron Alloys with Nickel and Ruthenium
(a) Vol. **67**, No. 1, p. 193–200 (1981).
See also *18.4*, p. *284*; 21.1, p. 363; 21.1.1, p. 374.

J. N. Chapman and J. P. Jakubovics
*Observation of Magnetic Domain Structures in Thin Foils at Low Temperatures by Transmission Electron Microscopy
(a) Vol. **67**, No. 1, p. K1–K3 (1981).
See also 21.4, p. 382.

J. Kamarád, J. Schneider, and Z. Arnold
*Pressure Dependence of Curie Temperature of Amorphous $Fe_{90}Zr_{10}$ Alloy
(a) Vol. **67**, No. 1, p. K85–K88 (1981).
See also *2*, p. *49*; 21.1.1, p. 375.

V. V. Valkov and S. G. Ovchinnikov
*Mixed Valence Phase Transition in Ferromagnets
(b) Vol. **108**, No. 2, p. 601–608 (1981).
See also 13, p. 150.

J. A. Tuszyński and H. Cofta
*Refined Model of Exchange Inversion in Spontaneous Metamagnetism
(b) Vol. **108**, No. 2, p. 617–622 (1981).
See also 18.4, p. 285; 22.8, p. 492.

D. K. De, A. K. Ghoshal, D. Pal and A. K. Pal
*EPR and Magnetic Studies on γ: bis (N-Methyl Salicylaldimine) Copper (II)
(b) Vol. **108**, No. 2, p. 645–652 (1981).
See also *19*, p. *294*; 22.9, p. 513.

Y. J. Wang and H. Kronmüller
*Effect of Hydrogen on the Coercive Field of Metallic Glass $Co_{75}Si_{15}B_{10}$
(a) Vol. **68**, No. 2, p. 499–504 (1981).
See also *2*, p. *51*; 21.1, p. 366.

Ch. Routsi and J. K. Yakinthos
*Magnetic Characteristic of Some TCu_2Si_2 Ternary Alloys (T = Nd, Dy, Ho, Er, Tm, Y, and La)
(a) Vol. **68**, No. 2, p. K153–K157 (1981).
See also *18.4*, p. *285*; 21.1, p. 367; 21.4, p. 383.

S. Vaskovskii, J. Schneider, and B. Springmann
*Low Field Magnetic Properties of Amorphous $Fe_{40}Ni_{40}P_{14}B_6$ Ribbons
(a) Vol. **68**, No. 2, p. K189–K193 (1981).
See also *2*, p. *52*; 21.1, p. 367; 21.1.1, p. 375.

18.2.1 Ferromagnetic Films

Original Papers and Short Notes

J. Duszewski
*Decay of Magnetic Fluctuations in Thin Films
(b) Vol. **103**, No. 1, p. 167 – 176 (1981).

K. Twarowski, H. K. Lachowicz, M. Gutowski, and H. Szymczak
*On the Origin of the Perpendicular Anisotropy and Magnetostriction in Amorphous RF Sputtered Gd–Co Films
(a) Vol. **63**, No. 1, p. 103 – 108 (1981).
See also *2*, p. *40*; 18.3, p. 278; 21.1, p. 352; 21.4, p. 380.

R. Żuberek
*Columnar Model of Magnetostriction in Amorphous Films
(a) Vol. **63**, No. 1, p. K51 – K54 (1981).
See also *2*, p. *41*; 18.3, p. 278; 21.1, p. 352; 21.4, p. 380.

J. Kaczér
*Bloch Wall Energy Including Cubic and Uniaxial Anisotropy
(a) Vol. **63**, No. 1, p. K87 – K92 (1981).

Diep-The-Hung
*Temperature-Dependent Surface Magnetization and Critical Temperature of Ferromagnetic Thin Films
(b) Vol. **103**, No. 2, p. 809 – 815 (1981).
See also 13.5.4, p. 197.

P. J. Schurer, A. H. Morrish, and M. J. Stavn
*Induced Perpendicular Anisotropy and Surface Crystallization in Amorphous $Fe_{78}B_{12}Si_{10}$ Ribbons
(a) Vol. **64**, No. 1, p. 343 – 349 (1981).
See also 2, p. 43; 5, p. 63; 21.1.1, p. 369.

O. Gebhardt and W. Richter
*Self-Biased Narrow Magnetoresistive Elements
(a) Vol. **64**, No. 2, p. 513 – 517 (1981).
See also *14.1*, p. *200*; 21.1, p. 357; 21.1.1, p. 370.

A. G. Shishkov, E. N. Ilyicheva, N. G. Kanavina, A. N. Ivanov, and A. M. Pirogova
*Magnetic and Electrical Properties of the Thin Film Magnetic Semiconductor $CdCr_2Se_4$
(a) Vol. **64**, No. 2, p. K93 – K97 (1981).
See also *14.3.1*, p. *220*; 22.8.2, p. 503.

V. I. Malyutin, V. E. Osukhovskii, Yu. D. Vorobiev, A. G. Shishkov, and V. V. Yudin
*Structure and Magnetic Properties of Etched NiFeCo Films
(a) Vol. **65**, No. 1, p. 45 – 52 (1981).
See also 21.1, p. 358; 21.1.1, p. 371.

O. Bostanjoglo and Th. Rosin
*New Reversal Process of Néel Walls Due to Nanosecond Field Pulses
(a) Vol. **65**, No. 2, p. K117–K119 (1981).
See also 21.1.1, p. 372.

O. Bostanjoglo and Th. Rosin
*Resonance Oscillations of Bloch Lines in Permalloy Films
(a) Vol. **66**, No. 1, p. K5–K10 (1981).
See also 21.1, p. 361; 21.1.1, p. 373.

I. Tomáš, B. N. D. Shija, and V. K. Srivastava
*To the Theory of 90° Stripe Domains in Films with Tetragonal Magnetic Anisotropy
(a) Vol. **67**, No. 1, p. 325–332 (1981).
See also 22.8.2, p. 505.

A. Duda
*Ellipticity Effects in Spin-Wave Spectrum of Thin Magnetic Films
(b) Vol. **107**, No. 2, p. 779–789 (1981).
See also *19*, p. *293*.

L. Vatskitchev
*Planar Hall Effect in Double Layer Magnetic Films with Unidirectional Anisotropy
(a) Vol. **67**, No. 2, p. 591–596 (1981).
See also 14.1, p. 203; *18.4*, p. *284*; 21.1, p. 364; 21.1.1, p. 375.

H. K. Lachowicz
*Remarks on the Extraordinary Hall Effect in Amorphous $Cd_{1-x}Co_x$ Sputtered Films
(a) Vol. **67**, No. 2, p. K131–K135 (1981).
See also *14.1*, p. *203*; 21.1, p. 365; 21.4, p. 382.

L. Wojtczak, A. Urbaniak-Kucharczyk, J. Mielnicki, and B. Mrygon
The Magnetic Contribution to Electrical Resistivity in Some Thin Ferromagnetic Films
(a) Vol. **68**, No. 2, p. 537–544 (1981).
See also *14.1*, p. *203*.

18.3 Ferrimagnetic Properties

Original Papers and Short Notes

A. Ślawska-Waniewska
Crystal Field of Cr^{3+} Ion in Spinel Lattice: The Effect of High Hydrostatic Pressure
(b) Vol. **104**, No. 2, p. 679–684 (1981).
See also *1*, p. *9*; *13*, p. *144*; 22.8.2, p. 503.

M. Wurlitzer and J. Franke
*A Model of the Photomagnetic Effect in YIG(Si)
(a) Vol. **64**, No. 2, p. 539–547 (1981).
See also *20*, p. *297*; 22.8.2, p. 503.

P. Muthukumarasamy, T. Nagarajan, and A. Narayanasamy
Mössbauer Study of the Ni–Cd Ferrite System
(a) Vol. **64**, No. 2, p. 747–754 (1981).
See also *5*, p. *63*; 22.8.2, p. 503.

C. M. Srivastava, O. Prakash, and R. Aiyar
*Magnetic Relaxation in Polycrystalline Garnets
(a) Vol. **64**, No. 2, p. 787–796 (1981).
See also 19, p. 289; 22.8.2, p. 503.

S. Ligenza, Z. Kucharski, and J. Suwalski
Mössbauer Study of the $3d^5$ Shell Deformation of Fe^{3+} Ions in $MnFe_2O_4$
(b) Vol. **105**, No. 1, p. 353–358 (1981).
See also *5*, p. *64*; 22.8.2, p. 503.

G. Dehe, J. Suwalski, E. Wieser, and R. Kabisch
The Cation Distribution in the Spinel System $Fe_{3-x}Ga_xO_4$
(a) Vol. **65**, No. 2, p. 669–675 (1981).
See also *5*, p. *64*; 22.8.2, p. 504.

C. M. Srivastava, S. N. Shringi, and M. V. Babu
Mössbauer Study of the Low-Temperature Phase of Magnetite
(a) Vol. **65**, No. 2, p. 731–735 (1981).
See also *5*, p. *65*; 22.8.2, p. 504.

M. B. Kurup
*Positron Lifetime Studies on the System $Fe_{3-x}Ti_xO_4$
(b) Vol. **106**, No. 1, p. K61–K65 (1981).
See also *18.1*, p. *266*; 22.8.2, p. 504.

R. G. Bowers
*The Exact Solution of the Curie-Weiss-Néel Model Ferrimagnet
(b) Vol. **106**, No. 2, p. 545–550 (1981).

R. Grössinger
A Critical Examination of the Law of Approach to Saturation (I): Fit Procedure
(a) Vol. **66**, No. 2, p. 665–674 (1981).
See also *18.2*, p. *273*; 21.1, p. 362; 22.8.2, p. 505.

H. V. Kiran, A. L. Shashimohan, D. K. Chakrabarty, and A. B. Biswas
*Structural and Magnetic Properties of Cooper–Nickel Ferrites
(a) Vol. **66**, No. 2, p. 743–747 (1981).
See also *1.3*, p. *25*; 22.8.2, p. 505.

18.4 Antiferromagnetic Properties

Original Papers and Short Notes

C. B. Cuden and R. Mota
Raman Scattering of Light in Antiferromagnetic Chromium
(b) Vol. **103**, No. 1, p. 51–62 (1981).
See also *6*, p. *66*; *20.1*, p. *301*; 21.1, p. 351.

S. A. Dyomin, V. V. Nemoshkalenko, O. N. Razumov, A. I. Ustinov, and K. V. Tchuistov
Mössbauer Study of Magnetic and Structure Transformations in Manganese–Cooper–Iron Alloy
(a) Vol. **63**, No. 1, p. K93–K96 (1981).
See also 1.2, p. 14; *5*, p. *62* ; 21.1, p. 353; 21.1.1, p. 368.

V. K. Saxena
*Spin-Glass Transition in a Heisenberg System with Competing Interactions
(b) Vol. **103**, No. 2, p. K99–K104 (1981).
See also 2, p. 41; *18.2*, p. *268*.

C. Tannous and A. Caillé
*Differential Susceptibility of the X-Y Model at the Spin-Peierls Transition in Presence of a Magnetic Field
(b) Vol. **103**, No. 2, p. K157–K161 (1981).

H. R. Salva and C. Fainstein
*Antiferromagnetic Resonance in $GdVO_4$
(a) Vol. **63**, No. 2, p. 417–421 (1981).
See also *19*, p. *287*; 22.8.1, p. 494.

L. M. Sandratskii, R. F. Egorov, and A. A. Berdyshev
Energy Band Structure and Electronic Properties of NiAs-Type Compounds (II): Antiferromagnetic Manganese Telluride
(b) Vol. **104**, No. 1, p. 103–107 (1981).
See also *13.1*, p. *153*; 22, p. 392.

F. Menil, N. Kinomura, L. Fournes, J. Portier, and P. Hagenmuller
*Investigation of the K_2NiF_4-Type Layer Oxide Fluorides $Sr_2FeO_{3+x}F_{1-x}$ $(0 \leqq x \leqq 0.20)$
(a) Vol. **64**, No. 1, p. 261–274 (1981).
See also 5, p. 63; *14.3*, p. *211*; 22.8, p. 486.

W. Bażela, A. Szytuła, J. Todorović, and A. Zięba
*Crystal and Magnetic Structure of the $NiMnGe_{1-n}Si_n$ System
(a) Vol. **64**, No. 1, p. 367–378 (1981).
See also 1.3, p. 23; *18.2*, p. *269*; 21.1, p. 355.

H. L. Alberts, L. de Koning, and S. J. Burger
The Effect of Hydrostatic Pressure on the Electrical Resistivity of Cr and Cr-Co Alloys at Room Temperature
(a) Vol. **64**, No. 1, p. K13–K15 (1981).
See also 12.1, p. 134; *14.1*, p. *200*; 21.1, p. 356.

J. Pietrzak, A. Maryanowska, and J. Leciejewicz
*Magnetic Ordering in $Pb(Fe_{0.5}Nb_{0.5})O_3$ at 4.2 K
(a) Vol. **65**, No. 1, p. K79–K82 (1981).
See also *14.4.1*, p. *244*; 22.8.1, p. 497.

K. Hoselitz
*The Saturation Magnetization of Some Substituted Fe–Si–B Metallic Glasses
(a) Vol. **65**, No. 1, p. K91–K94 (1981).
See also *18.2*, p. *271*; 21.1.1, p. 371.

W. Wegener, S. Hautecler, and G. Will
*Range of Magnetic Interactions in Some Anhydrous Sulphates of the 3d Transition Metals
(b) Vol. **105**, No. 2, p. 491–497 (1981).
See also *18.1*, p. *266*; 22.8.1, p. 497.

A. I. Mitsek, N. P. Kolmakova, and D. I. Sirota
*Metamagnetism of Two-Sublattice Uniaxial Antiferromagnets
(a) Vol. **65**, No. 2, p. 503–512 (1981).
See also *18.2*, p. *271*.

H. Gruber, E. Krautz, and H. P. Fritzer
Magnetic Susceptibility and Electrical Properties of the $\varkappa$-Phase $Mo_{17}O_{47}$ with Substitutions of Molybdenum
(a) Vol. **65**, No. 2, p. 589–594 (1981).
See also 14.1, p. 201; *14.3*, p. *214*; *18*, p. *262*; 22.6.1, p. 481.

G. K. Chepurnykh and S. V. Tarasenko
*Sound Interactions with Nuclear Subsystem in Antiferromagnets
(b) Vol. **106**, No. 1, p. 37–43 (1981).
See also 7, p. 82; *19*, p. *290*.

J. M. Viana and A. S. T. Pires
Proton Spin-Lattice Relaxation in TMMC
(b) Vol. **106**, No. 1, p. K1–K4 (1981).
See also *19*, p. *291*; 22.9, p. 511.

B. Lorenz
Magnetism versus Charge Order in the Extended Hubbard Model with Ising-Like Exchange Interaction
(b) Vol. **106**, No. 1, p. K17–K21 (1981).
See also *13*, p. *146*; 18.2, p. 271.

U. Lindner
On Solitary Solutions in a Modified Hubbard Chain
(b) Vol. **106**, No. 1, p. K77–K81 (1981).
See also *13.5*, p. *187*.

W. Bażela and A. Szytuła
*Crystal and Magnetic Structure of the $NiMn_{1-t}Ti_tGe$ System
(a) Vol. **66**, No. 1, p. 45–52 (1981).
See also 1.1, p. 12; *18.2*, p. *271*; 22.8, p. 488.

J. A. Tuszyński and H. Cofta
Refined Model of Exchange Inversion in Spontaneous Metamagnetism
(b) Vol. **108**, No. 2, p. 617–622 (1981).
See also *18.2*, p. *275*; 22.8, p. 492.

Ch. Routsi and J. K. Yakinthos
*Magnetic Characteristic of Some TCu_2Si_2 Ternary Alloys (T = Nd, Dy, Ho, Er, Im, Y, and La)
(a) Vol. **68**, No. 2, p. K153–K157 (1981).
See also *18.2*, p. *275*; 21.1, p. 367; 21.4, p. 383.

A. Maryanowska and J. Petrzak
Critical-Points Anomalies in EPR Linewidth of Pb $(Fe_{0.5}Nb_{0.5})O_3$
(a) Vol. **68**, No. 2, p. K185–K188 (1981).
See also *14.4.1*, p. *246*; *19*, p. *295*; 22.8.1, p. 501.

19. Magnetic Resonance

Review Articles

Original Papers and Short Notes

I. Bakonyi, L. Takács, and K. Tompa
*Dipole–Dipole Interaction and Short-Range Order in Amorphous Ni–P, Ni–Cu–P, and Ni–P–B Alloys
(b) Vol. **103**, No. 2, p. 489–497 (1981).
See also *2*, p. *41*; 21.1, p. 353.

V. A. Grazhulis, V. V. Kveder, and Yu. A. Osipyan
*Investigation of the Dislocation Spin System in Silicon as Model of One-Dimensional Spin Chains
(b) Vol. **103**, No. 2, p. 519–528 (1981).
See also *13.4*, p. *170*; 22.1.2, p. 409.

H. Möhwald and E. Erdle
*Guest Donor Arrangement in CT Crystals
(b) Vol. **103**, No. 2, p. 757–762 (1981).
See also *13.4*, p. *170*; 22.9, p. 508.

C. K. Sarkar
*Erratum to: On the Cyclotron Resonance Linewidth in Semiconductors at Low Temperatures
(b) Vol. **103**, No. 2, p. K179 (1981).
See also 6, p. 69; 22, p. 392.

H. R. Salva and C. Fainstein
*Antiferromagnetic Resonance in $GdVO_4$
(a) Vol. **63**, No. 2, p. 417–421 (1981).
See also *18.4*, p. *281*; 22.8.1, p. 494.

L. Kraus, Z. Frait, and J. Schneider
Ferromagnetic Resonance in Amorphous $(FeNi)_{80}P_{10}B_{10}$ Alloys (I): Alloys with High Curie Temperature
(a) Vol. **63**, No. 2, p. 669–675 (1981).
See also *2*, p. *42*; *18.2*, p. *268*; 21.1, p. 354; 21.1.1, p. 368.

J. Pietrzak and Z. Kruczyński
*EPR of Cr^{3+} in γ-Alum $NaAl(SO_4)_2 \cdot 12\,H_2O$
(a) Vol. **63**, No. 2, p. K143–K145 (1981).
See also 22.8, p. 485.

A. Ya. Dzyublik
*On the Theory of Double γ-Nuclear Magnetic Resonance
(b) Vol. **104**, No. 1, p. 81–88 (1981).
See also *5*, p. *63*.

T. Kröber, S. Wartewig, and W. Windsch
*Electron Paramagnetic Resonance Studies of Ferroelectric Fe^{3+}-Doped Tris-Sarcosine Calcium Chloride Monocrystals
(b) Vol. **104**, No. 1, p. 177–184 (1981).
See also *14.4.1*, p. *242*; 22.9, p. 509.

L. KRAUS, Z. FRAIT, and J. SCHNEIDER
Ferromagnetic Resonance in Amorphous $(FeNi)_{80}P_{10}B_{10}$ Alloys (II): Resonance near the Curie Temperature
(a) Vol. **64**, No. 2, p. 449–456 (1981).
See also *2*, p. *44*; *18.2*, p. *269*; 21.1, p. 356; 21.1.1, p. 370.

B. BORZECKA, S. F. SAGNOWSKI, and S. HODOROWICZ
*Proton Magnetic Resonance Study on Molecular Dynamics in $[Zn(H_2O)_6](ClO_4)_2$
(a) Vol. **64**, No. 2, p. 557–564 (1981).
See also 6, p. 73; 22.8, p. 486.

D. K. DE
*Electron Spin Resonance Study on Copper Salicylate Tetrahydrate (Blue Green) and Anhydrous Copper Salicylate (Brown)
(a) Vol. **64**, No. 2, p. 761–769 (1981).
See also 18.1, p. 265; 22.9, p. 509.

C. M. SRIVASTAVA, O. PRAKASH, and R. AIYAR
Magnetic Relaxation in Polycrystalline Garnets
(a) Vol. **64**, No. 2, p. 787–796 (1981).
See also *18.3*, p. *279*; 22.8.2, p. 503.

A. F. KHOKHLOV, A. I. MASHIN, and A. M. SATANIN
The Dependence of Magnetic Ordering Temperature in Amorphous Semiconductors on Paramagnetic Centre Concentration
(b) Vol. **105**, No. 1, p. 129–136 (1981).
See also 2, p. 45; 11, p. 124; *18.1*, p. *266*; *18.4*, p. *282*; 22.1.1, p. 405; 22.1.2, p. 413.

J. JAROSZ
*The Application of Kramers-Kronig Relations to the Analysis of the Resonance Lines of Metallic Samples in ESR Spectroscopy
(b) Vol. **105**, No. 1, p. 155–160 (1981).
See also 21, p. 341; 21.4, p. 381.

H. DERSCH, J. STUKE, and J. BEICHLER
*Electron Spin Resonance of Doped Glow-Discharge Amorphous Silicon
(b) Vol. **105**, No. 1, p. 265–274 (1981).
See also 2, p. 45; *13.4*, p. *175*; 22.1.2, p. 413.

B. ARUNA, Y. V. S. RAMAKRISHNA, and P. A. NARAYANA
*Electron Paramagnetic Resonance Studies of Cu^{2+} in $NH_4Cl_{1-x}Br_x$ Single Crystals
(b) Vol. **105**, No. 1, p. 347–351 (1981).
See also 22.5.4, p. 473.

V. SEEMAN, S. REIFMAN, and T. LEHTO
*A New Type V Centre in SrO
(b) Vol. **105**, No. 1, p. 409–412 (1981).
See also *10.2*, p. *110*; 22.6, p. 476.

W. Gehlhoff and K. H. Segsa
*Line Broadening and Line Displacement of EPR Transitions Produced by Inhomogeneous Plane Stresses in a Two-Layer System
(b) Vol. **106**, No. 1, p. 171 – 182 (1981).
See also *12.2*, p. *140*; 22.1.2, p. 415; 22.6, p. 477.

T. Hejwowski and M. Subotowicz
*The Mn^{2+} Ion Spectra in $Pb_xSn_{1-x}Te$ Alloys
(b) Vol. **106**, No. 1, p. 373 – 377 (1981).
See also 14.3, p. 215; 22, p. 397.

J. M. Viana and A. S. T. Pires
*Proton Spin-Lattice Relaxation in TMMC
(b) Vol. **106**, No. 1, p. K1 – K4 (1981).
See also 18.4, p. 283; 22.9, p. 511.

F. C. Montenegro and A. S. T. Pires
*Effects of the Anisotropy in the Positions of the Resonance Lines in EPR in TMMC
(b) Vol. **106**, No. 1, p. K13 – K15 (1981).
See also 22.9, p. 511.

D. Ballutaud, C. R'Kha, C. Sanchez, and J. Livage
*ESR of P_2O_5-Doped V_2O_5 Single Crystals
(a) Vol. **66**, No. 1, p. 271 – 276 (1981).
See also 22.6, p. 478.

I. A. Belobrova, I. P. Aleksandrova, and A. K. Moskalev
*^{81}Br NQR in the Incommensurate Phase of Rb_2ZnBr_4
(a) Vol. **66**, No. 1, p. K17 – K20 (1981).
See also *1.2*, p. *18*; 14.4.1, p. 244; 22.8, p. 488.

B. Cord, A. Hofstaetter, and A. Scharmann
*EPR Investigations on the Dynamic Behaviour of Hole Centres in $CaWO_4$
(b) Vol. **106**, No. 2, p. 499 – 504 (1981).
See also 11, p. 127; 22.8.1, p. 498.

W. Braune, U. Straubel, N. Kubicki, J. Lebech, and K. Saermark
Relaxation Time and Effective Masses for Charge Carriers in Semimetallic BiSb Alloys versus Alloy Composition
(b) Vol. **106**, No. 2, p. 587 – 592 (1981).
See also *13.5.2*, p. *194*; *13.5.4*, p. *197*; 21.7, p. 388.

J. Kuriata, N. Guskos, and T. Rewaj
*Temperature Dependence of Second-Order Spin-Lattice Coefficients for Eu^{2+} in MeF_2 Single Crystals
(b) Vol. **106**, No. 2, p. 607 – 612 (1981).
See also 22.5.3, p. 471.

M. Krupski and E. G. Sharoyan
*ESR Study of Cr^{3+} Complexes in α-Lithium Iodine Single Crystals under Hydrostatic Pressure
(b) Vol. **106**, No. 2, p. K95 – K98 (1981).
See also 22.8.1, p. 498.

J. Dzieslaty, D. Backs, and D. Mantwitz
*Photosensitive EPR Spectra of Pb-Doped ZnS Single Crystals
(b) Vol. **107**, No. 2, p. 553–557 (1981).
See also *13.4*, p. *181*; 22.4.1, p. 449.

K. Recker, F. Wallrafen, R. Büscher, and G. Lehmann
*Crystal Growth, EPR, and Luminescence of $BaZnF_4$:Mn
(b) Vol. **107**, No. 2, p. 699–706 (1981).
See also 3, p. 55; *20.3*, p. *333*; 22.5.4, p. 473.

A. Duda
*Ellipticity Effects in Spin-Wave Spectrum of Thin Magnetic Films
(b) Vol. **107**, No. 2, p. 779–789 (1981).
See also *18.2.1*, p. *277*.

N. G. Romanov, Yu. P. Veshchunov, V. A. Vetrov, and P. G. Baranov
*Application of Tunneling Recombination Afterglow for EPR Optical Detection of Recombining Centres in Ionic Crystals
(b) Vol. **107**, No. 2, p. K119–K124 (1981).
See also 10.2, p. 116; 22.5.2, p. 468; 22.5.3, p. 472.

S. F. Sagnowski and J. Ogar
*NMR Study of ^{23}Na Shielding Tensor in Sodium Bromate Single Crystal
(b) Vol. **107**, No. 2, p. K125–K128 (1981).
See also 22.8.1, p. 500.

M. Baran, A. A. Mirzakhanyan, E. G. Sharoyan, and H. Szymczak
*Zero Field Splitting of the Mo^{3+} Ground State in Corundum
(b) Vol. **107**, No. 2, p. K129–K132 (1981).
See also *13.4*, p. *181*; 22.6, p. 479.

M. M. Kucherov, A. V. Ponomarenko, and V. E. Zobov
*On Dipole-Zeeman Cross Relaxation in Multiple Pulse Spin Locking
(b) Vol. **107**, No. 2, p. K159–K162 (1981).
See also 22.5.3, p. 472.

J. Pietrzak, J. Kudyńska, and Z. Kruczyński
*Phase Transition Study in VO_2 Using the EPR Method
(a) Vol. **67**, No. 2, p. K175–K178 (1981).
See also 14.1, p. 203; 14.3, p. 218; 22.6, p. 480.

M. Corti, A. Rigamonti, and A. Magistris
*^{133}Cs Quadrupole Perturbed NMR Study of Jahn-Teller Phase Transitions in $CsCuCl_3$
(b) Vol. **108**, No. 1, p. 29–36 (1981).
See also *1.2*, p. *20*; 22.8, p. 490.

M. N. Aliev
*Theory of the Nuclear Magnetic Resonance Lineshape in Magnetically Diluted Crystals
(b) Vol. **108**, No. 1, p. 59–64 (1981).
See also 22, p. 399.

A. S. N. Rao, P. N. Murty, C. R. K. Murty, and T. Rs. Reddy
*EPR Study of Molecular Ordering in Liquid Crystal: NOBA
(a) Vol. 68, No. 2, p. 373–377 (1981).
See also *2*, p. *51*; 22.9, p. 513.

K. P. Hoffmann, H. Drulis, and B. Staliński
*Electron Paramagnetic Resonance of Er^{3+} Ions in Lutetium Hydrides
(a) Vol. 68, No. 2, p. 439–446 (1981).
See also 21.4, p. 383.

S. N. Karyagin
*On the Nature of Centers Responsible for the Spin-Dependent Recombination in Si and Ge
(a) Vol. 68, No. 2, p. K113–K116 (1981).
See also *10.2*, p. *120*; 22.1.1, p. 407; 22.1.2, p. 424.

J. Sthanapati, D. K. De, and A. K. Pal
*Studies on Phase Transitions in Cobalt and Ferrous Fluosilicates by Cu (II) EPR
(a) Vol. 68, No. 2, p. K123–K127 (1981).
See also *1.2*, p. *21*; 22.8, p. 492.

A. Maryanowska and J. Petrzak
*Critical-Points Anomalies in EPR Linewidth of $Pb(Fe_{0.5}Nb_{0.5})O_3$
(a) Vol. 68, No. 2, p. K185–K188 (1981).
See also *14.4.1*, p. *246*; 18.4, p. 285; 22.8.1, p. 501.

20. Optical Properties

Original Papers and Short Notes

N. S. Orlova
X-Ray Thermal Diffuse Scattering Measurements of the [110] Phonon Dispersion Curves of Indium Arsenide
(b) Vol. **103**, No. 1, p. 115–121 (1981).
See also *6*, p. *66*; 22.2.1, p. 427.

T. K. Hatwar, M. N. Ghatikar, B. D. Padalia, S. K. Malik, and R. Vijayaraghavan
*X-Ray Absorption Spectroscopic Study of Rare-Earth–Manganese Intermetallic Compounds
(b) Vol. **103**, No. 1, p. 159–166 (1981).
See also 21.1, p. 351; 21.4, p. 380.

Nguyen Ba An, Nguyen Van Hieu, Nguyen Toan Thang, and Nguyen Ai Viet
*Interband Electronic Raman Scattering of Polarized Light in CuCl
(b) Vol. **103**, No. 1, p. 205–210 (1981).
See also *13.1*, p. *151*; 13.5.1, p. 188; 22.5, p. 458.

V. Dose and H. Scheidt
Isochromat and SXAPS Measurements on Hafnium, Tantalum, and Tungsten
(b) Vol. **103**, No. 1, p. 247–252 (1981).
See also *13.1*, p. *151*; 21, p. 336.

M. Torfeh and H. Le Gall
*Theoretical Analysis of Hybrid Modes of Magnetooptical Wave-Guides
(a) Vol. **63**, No. 1, p. 247–258 (1981).
See also *18*, p. *260*; 22.8.2, p. 502.

M. Wurlitzer and J. Franke
*Some Experiments on Recombination in the Photomagnetic Effect in Si-Doped YIG
(a) Vol. **63**, No. 1, p. K11–K14 (1981).
See also *18.3*, p. *278*; 22.8.2, p. 502.

S. Guha and P. Sen
*Parametric Decay of Alfvén Waves in Piezoelectric Semiconductors
(b) Vol. **103**, No. 2, p. K131–K136 (1981).
See also *13*, p. *143*; 22.2.3, p. 435.

G. Dräger, O. Brümmer, V. Heera, G. Seifert, and P. Ziesche
*Polarized TiKβ Valence Band Spectra of TiO_2 (Rutile)
(b) Vol. **104**, No. 1, p. 219–226 (1981).
See also *13.1*, p. *154*; 22.6, p. 475.

K. S. Srivastava and O. K. Harsh
*Double Plasmon High Energy Satellites in the X-Ray Emission Spectra of Metals
(b) Vol. **104**, No. 1, p. K35–K38 (1981).
See also *13.5.2*, p. *194*; 21, p. 339; 21.1, p. 355; 21.2, p. 376.

V. V. Nemoshkalenko, V. N. Antonov, Vl. N. Antonov, and W. John
*Relativistic APW Calculation of Soft X-Ray Emission Spectra in Ta
(b) Vol. **104**, No. 1, p. K45 – K48 (1981).
See also *13.1*, p. *154*; 21, p. 339.

G. D. Priftis and J. Boviatsis
Negative Volume Plasmon Dispersion in Be for $k > k_F$
(b) Vol. **104**, No. 2, p. 673 – 677 (1981).
See also *13.5.2*, p. *194*; 21.3, p. 378.

M. Wurlitzer and J. Franke
*A Model of the Photomagnetic Effect in YIG(Si)
(a) Vol. **64**, No. 2, p. 539 – 547 (1981).
See also *18.3*, p. *279*; 22.8.2, p. 503.

E. P. Domashevskaya and V. A. Terekhov
d–s,p Resonance and Electronic Structure of Compounds, Alloys and Solid Solutions
(b) Vol. **105**, No. 1, p. 121 – 127 (1981).
See also *13.1*, p. *155*; 17, p. 257; 21, p. 341; 21.1, p. 358.

E. Tataru and I. Boskovitz
*The Use of Magnetic Circular Dichroism Measurements in Determining the Reduced Matrix Elements of the Electric-Dipole Momentum Operator
(b) Vol. **105**, No. 1, p. K1 – K4 (1981).
See also *13.4*, p. *175*; 18, p. 261; 22.8, p. 486.

L. Kaihola
*Anomalous Soft X-Ray Reflectivity of Borosilicate Glass by Successive Grazing Reflections
(a) Vol. **65**, No. 1, p. 163 – 166 (1981).
See also *2*, p. *45*; 22.6.1, p. 481.

O. G. Vlokh, L. A. Lažko, and Ya. I. Shopa
*Electrooptic and Electrogyration Properties of the Solid Solutions on the Basis of Lead Germanate
(a) Vol. **65**, No. 1, p. 371 – 378 (1981).
See also *14.4.1*, p. *243*; 22.8.1, p. 496.

M. Dobrowolska, W. Dobrowolski, R. R. Gałązka, and A. Mycielski
*Interband Magnetooptical Studies on the Semimagnetic Semiconductor $Hg_{1-x}Mn_xSe$
(b) Vol. **105**, No. 2, p. 477 – 484 (1981).
See also 13.1, p. 156; *18*, p. *262*; 22.4.4, p. 456.

P. Rennert
*Calculation of the Compton Profile of Beryllium
(b) Vol. **105**, No. 2, p. 567 – 575 (1981).
See also *13.1*, p. *156*; 13.2, p. 163; 21.3, p. 378.

S. N. Klymin and E. P. Pokatilov
*Non-Linear Optical Properties of the Degenerate Band Charge Carrier Gas
(b) Vol. **105**, No. 2, p. 641 – 648 (1981).
See also 22, p. 395.

A. N. Vtyurin, V. P. Ermakov, B. I. Ostrovskii, and V. F. Shabanov
Study of Optical Second Harmonic Generation in Ferroelectric Liquid Crystal
(b) Vol. **107**, No. 2, p. 397–402 (1981).
See also *2*, p. *49*; *14.4.1*, p. *245*; 22.9, p. 512.

F. A. Oliveira, M. G. Cottam, and D. R. Tilley
*Green Functions, Power Spectra, and Raman Scattering of Thin Film Polaritons
(b) Vol. **107**, No. 2, p. 737–747 (1981).
See also *6*, p. *79*; 13.5, p. 187; 22.5.2, p. 468.

Y. Aoyagi, Y. Segawa, S. Namba, T. Suhara, H. Nishihara, and H. Gamo
Dynamic Behavior of the Photodarkening Process in As_2S_3 Chalcogenide Glass
(a) Vol. **67**, No. 2, p. 669–676 (1981).
See also *2*, p. *50*; *10.2*, p. *117*; 22, p. 399.

C. Landron
*Interaction d'électrons avec des cibles amorphes ou polycristallines: Emission X caractéristique
(Interaction of Electrons with Amorphous and Polycrystalline Specimens: Characteristic X-Ray Emission)
(a) Vol. **67**, No. 2, p. 717–724 (1981).
See also *11*, p. *130*.

N. V. Ershov, A. L. Ageev, V. V. Vasin, and Yu. A. Babanov
*A New Interpretation of EXAFS Spectra in Real Space (II): A Comparison of the Regularization Technique with the Fourier Transformation Method
(b) Vol. **108**, No. 1, p. 103–111 (1981).
See also 21.1, p. 365.

K. Searmark, J. Lebech, W. Braune, and N. Kubicki
*Non-Local Calculations of the Dispersion of Electromagnetic Waves in Bismuth. Dispersion Relation for the Voigt Configuration: $\boldsymbol{q} \parallel \boldsymbol{n} \parallel \boldsymbol{C}_1$, $\boldsymbol{q} \perp \boldsymbol{B}$
(b) Vol. **108**, No. 1, p. 275–280 (1981).
See also *13*, p. *149*; 21.7, p. 389.

A. V. Tulupov
*Some Characteristic Features of the Kumakhov Radiation in an X-Ray Range
(b) Vol. **108**, No. 1, p. K11–K14 (1981).
See also *11*, p. *130*; 13, p. 149; 22.1.2, p. 421.

W. Grevendonk, L. van den Abeele, P. van den Keybus, B. Vanhuyse, and G. Ruymbeek
*Faraday Rotation in Quartz and Oxide Glasses
(b) Vol. **108**, No. 1, p. K53–K56 (1981).
See also 2, p. 50; 13.1, p. 160; *18*, p. *264*; 22.6, p. 480; 22.6.1, p. 482.

K. Harnisch, S. M. Neubacher, and O. Ennemoser
Optical Loss in Selenium Thin Films as a Measure of Crystallinity
(a) Vol. **68**, No. 1, p. K87–K91 (1981).
See also *1.5*, p. *37*; 3, p. 56; 22.1.3, p. 425.

A. N. Gusatinskii, M. A. Blokhin, P. Shumichen, M. A. Bunin, and S. S. Mikhailova
*X-Ray L_1 Spectra and XPS Investigation of the Electronic Structure of Indium Compounds
(b) Vol. **108**, No. 2, p. 405–410 (1981).
See also *13.1*, p. *161*; 21, p. 349; 22, p. 400; 22.2, p. 426; 22.3, p. 442.

20.1 Spectra. Optical Constants
(X-Ray Spectra See 20)

Review Articles

M. NARAYANA, V. S. SIVASANKAR, and S. RADHAKRISHNA
*Optical and Magnetic Resonance Studies of Transition Metal Ions and Complexes in Alkali Halides
(b) Vol. **105**, No. 1, p. 11–45 (1981).
See also 10.2, p. 106; 13.4, p. 168; *19*, p. *286*; 22.5.2, p. 462.

H. BÄSSLER
Localized States and Electronic Transport in Single Component Organic Solids with Diagonal Disorder
(b) Vol. **107**, No. 1, p. 9–54 (1981).
See also 2, p. 40; *13.4*, p. *168*; 13.5.1, p. 188; *14.3*, p. *209*; 14.3.4, p. 232; 20.3, p. 323; 22.9, p. 507.

Original Papers and Short Notes

C. B. CUDEN and R. MOTA
*Raman Scattering of Light in Antiferromagnetic Chromium
(b) Vol. **103**, No. 1, p. 51–62 (1981).
See also *6*, p. *66*; 18.4, p. 281; 21.1, p. 351.

YU. N. POLIVANOV and R. SH. SAYAKHOV
*Hyper-Raman Scattering from Infrared Active Modes in Calcite Crystals
(b) Vol. **103**, No. 1, p. 89–92 (1981).
See also *6*, p. *66*; 22.8.1, p. 493.

C. CARLONE, S. JANDL, and H. R. SHANKS
*Optical Phonons and Crystalline Symmetry of InSe
(b) Vol. **103**, No. 1, p. 123–130 (1981).
See also 1.3, p. 22; *6*, p. *66*; 22.3, p. 440.

G. B. DUBROVSKII and V. I. BELYAVSKII
*Selection Rules and Absorption Lineshapes for Electron Transitions between Minizones in a Superlattice
(b) Vol. **103**, No. 1, p. 131–137 (1981).
See also *13.1*, p. *151*; 22.3, p. 440.

S. B. S. SASTRY and S. SAPRU
*Thermoluminescence, Fluorescence, and Optical Absorption Studies on Europium-Doped Rubidium Chloride
(b) Vol. **103**, No. 1, p. 185–192 (1981).
See also 10.2, p. 106; *20.3*, p. *323*; 22.5.2, p. 462.

M. D. FONTANA, G. E. KUGEL, G. METRAT, and C. CARABATOS
*Long-Wavelength Phonons in the Different Phases of $KNbO_3$
(b) Vol. **103**, No. 1, p. 211–219 (1981).
See also *6*, p. *66*; 14.4.1, p. 242; 22.8.1, p. 493.

U. A. Jayasooriya, G. J. Kearley, S. F. A. Kettle, and H. J. Lauter
*A Comparison of Inelastic Neutron Scattering and Vibrational Spectral Data for Phase V of Ammonium Nitrate
(a) Vol. **63**, No. 1, p. 169–177 (1981).
See also *6*, p. *67*; 22.8.1, p. 493.

K. B. Sundaram and G. K. Bhagavat
Preparation and Properties of Indium Oxide Films
(a) Vol. **63**, No. 1, p. K15–K18 (1981).
See also 1.5, p. 34; *3*, p. *53*; *14.3.1*, p. *220*; 15, p. 247; 22.6, p. 474.

Sh. M. Efendiev, V. E. Bagiev, A. Ch. Zeinally, V. A. Balashov, V. A. Lomonov, and A. A. Majer
*Optical Properties of $Bi_{12}TiO_{20}$ Single Crystals
(a) Vol. **63**, No. 1, p. K19–K22 (1981).
See also 16, p. 251; 22.8, p. 484.

V. V. Sobolev, O. G. Maksimova, and S. G. Kroitoru
*Reflectivity Spectra and Band Structure of the ZnTe–CdTe System
(b) Vol. **103**, No. 2, p. 499–509 (1981).
See also *13.1*, p. *152*; 22.4.4, p. 456.

V. A. Tyagai†, O. V. Snitko, A. M. Evstigneev, and A. N. Krasiko
*Electroreflectance of InN Semimetallic Thin Films
(b) Vol. **103**, No. 2, p. 589–594 (1981).
See also *14.3.1*, p. *220*; 22.2, p. 426.

R. S. Lee, J. F. Merklin, C. D. Marrs, and M. H. Richter
*Impurity Photoconductivity of γ-Irradiated LiF
(b) Vol. **103**, No. 2, p. 605–611 (1981).
See also 10.2, p. 107; *16*, p. *251*; 22.5.2, p. 463.

T. Skettrup
*Urbach's Rule and Phase Fluctuations of the Transmitted Light
(b) Vol. **103**, No. 2, p. 613–621 (1981).
See also 6, p. 67; 13.5.1, p. 188; 22.4, p. 445.

T. Chattopadhyay and H. G. von Schnering
*Infrared Absorption Studies on Crystalline α-P_4S_3
(b) Vol. **103**, No. 2, p. 637–642 (1981).
See also *6*, p. *68*; 22, p. 391.

S. E. Kulkova, V. E. Egorushkin, I. N. Anokhina, and V. P. Fadin
*The Change of Electronic Structure and Optical Properties in FeCo at the Order–Discorder Transition
(b) Vol. **103**, No. 2, p. 653–658 (1981).
See also *13.1*, p. *152*; 21.1, p. 353; 21.1.1, p. 368.

A. R. Hassan
Phonon-Polaritons by Two-Photon Absorption in Solids
(b) Vol. **103**, No. 2, p. 739–744 (1981).
See also *6*, p. *68*; *13.5*, p. *186*; 22.2.2, p. 432.

P. Dub
*Reflection of Light on a Spatially Dispersive Crystal
(b) Vol. **104**, No. 1, p. 109–117 (1981).
See also *13.5.1*, p. *189*; 22.4.1, p. 447.

K. Asami, T. Naka, and M. Ishiguro
*Assignment of the Optical Absorption Bands in CsI:Tl and CsBr:Tl by Application of High Pressure
(b) Vol. **104**, No. 1, p. 145–149 (1981).
See also 1.2, p. 15; *13.4*, p. *171*; 22.5.2, p. 463.

A. Giessler, G. Schaack, and H. Bleif
*Phonon Frequencies and Dynamics of the OH^-Molecule in Sodium Hydroxide
(b) Vol. **104**, No. 1, p. 151–159 (1981).
See also 1.2, p. 15; *6*, p. *69*; 22, p. 392.

Th. Harbich and G. Mahler
The Electron-Hole Plasma in a Strong Electromagnetic Field. Optical Porperties in a Simplified Model
(b) Vol. **104**, No. 1, p. 185–191 (1981).
See also *13.1*, p. *153*; *13.5*, p. *186*; 22.2.1, p. 427.

H. H. Grosse, E. Jäger, and U. Telloke
*Phonon Modes and Infrared Absorption in the Coupled System Film–Substrate
(b) Vol. **104**, No. 1, p. 267–275 (1981).
See also *6*, p. *70*.

V. Riede, H. Neumann, F. Lévy, and H. Sobotta
*Two-Phonon Absorption Spectra in ε-GaSe
(b) Vol. **104**, No. 1, p. 277–282 (1981).
See also *6*, p. *70*; 22.3, p. 440.

O. V. Farberovich
Band Structure and Electron-Electron Interaction in Samarium Monosulfide
(b) Vol. **104**, No. 1, p. 365–374 (1981).
See also *13.1*, p. *154*; 17, p. 256; 21.4, p. 380; 22, p. 392.

K. Hübner, L. Schumann, A. Lehmann, H. H. Vajen, and G. Zuther†
*Detection of LO and TO Phonons in Amorphous SiO_2 Films by Oblique Incidence of IR Light
(b) Vol. **104**, No. 1, p. K1–K5 (1981).
See also 2, p. 42; *6*, p. *70*; 22.6, p. 475.

J. Pörsch and N. B. Manson
*Magnetic Circular Dichroism of the $^4A_{2g} \rightarrow {}^4T_{2g}$ Zero-Phonon Line in $MgO:Cr^{3+}$
(b) Vol. **104**, No. 1, p. K7–K10 (1981).
See also *13.4*, p. *171*; 22.6, p. 475.

P. Ketolainen and A. Vaittinen
*On the Production of Z_1 Centres in Irradiated Crystals
(b) Vol. **104**, No. 1, p. K53–K56 (1981).
See also *10.2*, p. *108*; *11*, p. *122*; 22.5.2, p. 463.

D. G. GALIMOV, K. D. TARZIMANOV, and N. R. YAFAEV
On the Nature of Colour Centres in Ultramarine and Blue Alkali Borate Glasses Containing Sulphur
(a) Vol. **64**, No. 1, p. 65–71 (1981).
See also *10.2*, p. *108*; 13.4, p. 172; 19, p. 288; 22.8.1, p. 495.

P. SRIDARAN, R. K. GARTIA, R. C. BHUNIYA, and V. V. RATNAM
Thermoluminescence of Z_2 Centers in X-Irradiated Ca-Doped KCl Crystals
(a) Vol. **64**, No. 1, p. 127–131 (1981).
See also 10.2, p. 108; *20.3*, p. *325*; 22.5.2, p. 463.

D. SCHALCH, A. SCHARMANN, and H. WEBER
Electrical Conduction in Reactively Sputtered Silicon Nitride Films at Medium Field Strengths
(a) Vol. **64**, No. 1, p. 159–167 (1981).
See also *2*, p. *42*; 13.4, p. 172; *14.3.3*, p. *226*; 22, p. 393.

T. KAWAKUBO
*Color Centers in Irradiated Durene Crystals
(a) Vol. **64**, No. 1, p. 303–310 (1981).
See also *10.2*, p. *108*; 11, p. 123; 22.9, p. 509.

J. KROUPA and Z. CZAPLA
*Far Infrared Spectra of NH_4HSeO_4
(a) Vol. **64**, No. 1, p. K17–K18 (1981).
See also *14.4.1*, p. *243*; 22.8.1, p. 495.

H. SOBOTTA, V. RIEDE, C. ASCHERON, V. GEIST, and D. OPPERMANN
*Infrared Absorption of Hydrogen in Proton-Implanted GaP
(a) Vol. **64**, No. 1, p. K77–K80 (1981).
See also 11, p. 123; 22.2.2, p. 432.

A. M. BRODSKII and L. I. DAIKHIN
*On the Theory of Electroreflectance of Metals
(b) Vol. **104**, No. 2, p. 415–426 (1981).
See also *14.3.4*, p. *233*; 21, p. 339.

P. SRIDARAN, R. C. BHUNIYA, and V. V. RATNAM
Optical and Thermal Stability of Ag Centres in RT X-Irradiated KCl:Ag Crystals
(b) Vol. **104**, No. 2, p. 459–468 (1981).
See also *10.2*, p. *109*; 20.3, p. 326; 22.5.2, p. 464.

E. S. KISELYOVA
*The Phenomenon of Self-Induced Transparency in the Range of the Proper Lattice Absorption
(b) Vol. **104**, No. 2, p. 497–505 (1981).
See also *6*, p. *71*.

R. TARANKO, E. TARANKO, and K. I. WYSOKIŃSKI
*Effect of Electron–Phonon Interaction on the Optical Conductivity of Alloys
(b) Vol. **104**, No. 2, p. 507–512 (1981).
See also 6, p. 71; 13, p. 144; 21, p. 340.

M. Stordeur, H. T. Langhammer, H. Sobotta, and V. Riede
Valence Band Structure of $(Bi_{1-x}Sb_x)_2Te_3$ Single Crystals
(b) Vol. **104**, No. 2, p. 513–522 (1981).
See also *13.1*, p. *154*; 14.3, p. 211; 15, p. 248; 22.7, p. 483.

E. Jahne, P. Kleinert, B. Kh. Bairamov, and V. V. Toporov
*Impurity Vibrational Properties of Ga in InP Crystals
(b) Vol. **104**, No. 2, p. 531–539 (1981).
See also *6*, p. *71*; 10.2, p. 109; 22.2.2, p. 432.

P. John, I. M. Odeh, M. J. K. Thomas, M. J. Tricker, and J. I. B. Wilson
*Studies of the Oscillator Strengths of Infrared Vibrational Modes in Glow-Discharge Hydrogenated Amorphous Silicon
(b) Vol. **104**, No. 2, p. 607–612 (1981).
See also *2*, p. *43*; 6, p. 72; 22.1.2, p. 412.

I. Yaeger, R. Shuker, and B. M. Wanklyn
*Optical Studies of Er^{3+} Ions in $Er_2Ge_2O_7$ Single Crystals
(b) Vol. **104**, No. 2, p. 621–625 (1981).
See also *13.4*, p. *173*; 22.8, p. 486.

G. Gevers, J. Pichon, and A. S. Barrière
*Optical Absorption Study of Nickel Fluoride Thin Films
(b) Vol. **104**, No. 2, p. 641–647 (1981).
See also 13.4, p. 173; 22.5, p. 458.

T. Tsuboi
*The Parity-Forbidden $4f^5 \rightarrow 4f^5$ Transition of the Sm^{3+} Ion in CaF_2
(b) Vol. **104**, No. 2, p. 697–701 (1981).
See also *13.4*, p. *173*; 22.5.3, p. 471.

A. R. Hassan and R. Raouf
*Polarization Dependence of Three-Photon Absorption in Solids
(b) Vol. **104**, No. 2, p. 703–709 (1981).
See also 13.1, p. 155; 22.5, p. 458.

M. Baleva and S. D. Stoyanov
*On the Optical Properties and Electronic Structure of Solid Solutions of $(Ag_{x/2}Pb_{1-x}Sb_{x/2}Te)$ Type
(b) Vol. **104**, No. 2, p. 737–741 (1981).
See also *13.1*, p. *155*; 22.8, p. 486.

M. Ya. Valakh, A. P. Litvinchuk, G. S. Pekar, and G. N. Polysskii
*Multiple-Phonon Resonant Raman Scattering in $Zn_{1-x}Cd_xSe$ Crystals
(b) Vol. **104**, No. 2, p. 743–750 (1981).
See also *6*, p. *72*; 13.5.1, p. 189; 22.4.4, p. 456.

I. Gregora, B. Velický, and M. Závětová
*Refractive Index of Crystalline and Amorphous GeS
(b) Vol. **104**, No. 2, p. K95–K100 (1981).
See also 2, p. 44; 22, p. 393.

O. Litzman
*A Note on the Additional Boundary Conditions in Spatially Dispersive Media
(b) Vol. **104**, No. 2, p. K109–K111 (1981).
See also *14.4*, p. *238*.

Sh. M. Efendiev, I. I. Khammadov, I. A. Velichko, E. A. Vinogradov, V. A. Yakovlev, A. Kh. Zeinally, and G. N. Zhizhin
*High-Frequency Vibrational Spectrum of $Gd_2(MoO_4)_3$
(b) Vol. **104**, No. 2, p. K129–K133 (1981).
See also *6*, p. *72*; 22.8.1, p. 495.

Sh. M. Efendiev, I. I. Khammadov, I. A. Velichko, E. A. Vinogradov, A. Kh. Zeinally, G. N. Zhizhin, and V. A. Yakovlev
*High-Frequency Surface Polariton Dispersion in $Gd_2(MoO_4)_3$
(b) Vol. **104**, No. 2, p. K135–K138 (1981).
See also *6*, p. *72*; 22.8.1, p. 495.

J. Provoost and G. Jacobs
On the Formation Time of Absorption Centres in KI and KBr with a Stimulated Raman Probe Beam
(b) Vol. **104**, No. 2, p. K149–K154 (1981).
See also *10.2*, p. *109*; 13.5.1, p. 189; 22.5.2, p. 464.

A. Olzewski, W. Wojdowski, and W. Nazarewicz
*Infrared Spectra of $Mn_xZn_{1-x}Te$ Mixed Crystals
(b) Vol. **104**, No. 2, p. K155–K157 (1981).
See also *6*, p. *72* ; 22.4.4, p. 456.

E. M. Raspopova, M. Závětová, J. Krištofik, and B. T. Kolomiets
*The Influence of Defects Generated during Densification on the Properties of the As–Se System
(a) Vol. **64**, No. 2, p. 457–460 (1981).
See also *2*, p. *44*; 10.2, p. 109; 14.4, p. 238; 22, p. 393:

M. Kitao, H. Akao, T. Ishikawa, and S. Yamada
*Influence of Copper Addition on Electrical and Optical Properties of Amorphous As_2Se_3
(a) Vol. **64**, No. 2, p. 493–498 (1981).
See also 2, p. 44; 13.4, p. 174; *14.3*, p. *212*; 22, p. 394.

W. D. Wagner and R. Claus
*Phase Matching for Difference Frequency Generation in the View of Spontaneous Raman Scattering by Hot Polaritons
(a) Vol. **64**, No. 2, p. 647–656 (1981).
See also *6*, p. *73*; 22.2.2, p. 432; 22.8.1, p. 496.

U. Dębska, M. Dietl, G. Grabecki, E. Janik, E. Kierzek-Pecold, and M. Klimkiewicz
*Energy Band Gaps of $Hg_xCd_yMn_zTe$ ($x + y + z = 1$) Epitaxial Layers
(a) Vol. **64**, No. 2, p. 707–713 (1981).
See also 3, p. 54; 13.1, p. 155; 22.4.4, p. 456.

V. DOLOCAN and F. IOVA
*Optical Properties of Bi_2O_3 Thin Films
(a) Vol. **64**, No. 2, p. 755–759 (1981).
See also 13.1, p. 155; 22.6, p. 476.

A. STERNBERG, P. HLÍDEK, and M. ZVÁRA
*Electro-Optic Effect in PLZT Ceramics in the Infrared
(a) Vol. **64**, No. 2, p. K149–K153 (1981).
See also *14.4.1*, p. *243*; 22.8.1, p. 496.

R. RAUCH and E. LIEBOLD
*VUV Absorption Spectra of Oxygenated CaF_2 Crystals
(a) Vol. **64**, No. 2, p. K165–K168 (1981).
See also *10.2*, p. *110*; 22.5.3, p. 471.

V. ANGELINI, M. CASALBONI, M. GRANDOLFO, and P. VECCHIA
*Exciton Thermoabsorption of $GaSe_{1-x}S_x$ Solid Solutions
(b) Vol. **105**, No. 1, p. 63–68 (1981).
See also *13.5.1*, p. *189*; 22.3.1, p. 444.

N. J. TRAPPENIERS, F. S. STIBBE, and J. L. RAO
*Optical Absorption Spectrum of Cr^{3+} Ions Doped in CsCl
(b) Vol. **105**, No. 1, p. 243–247 (1981).
See also *13.4*, p. *174*; 22.5.2, p. 464.

K. KAMBAS, J. SPYRIDELIS, and M. BALKANSKI
*Far Infrared and Raman Optical Study of α- and β-In_2S_3 Compounds
(b) Vol. **105**, No. 1, p. 291–296 (1981).
See also *6*, p. *73*; 22.3, p. 441.

A. ANDERSON, P. GASH, and F. LÜTY
*Lattice Dynamics of KCN and NaCN in the Anti-Ferroelectric Phase
(b) Vol. **105**, No. 1, p. 315–325 (1981).
See also *6*, p. *74*; 14.4.1, p. 243; 22, p. 394.

J. M. HVAM and A. BIVAS
*Erratum to: Optical Gain and Induced Absorption in High-Density Exciton System in CuCl
(b) Vol. **105**, No. 1, p. 439 (1981).
See also *13.5.1*, p. *189*; 22.5, p. 459.

YU. P. GNATENKO and P. A. SKUBENKO
*Exciton Absorption of GaSe Crystals in the Indirect Transition Region
(b) Vol. **105**, No. 1, p. K9–K12 (1981).
See also *13.5.1*, p. *189*; 22.3, p. 441.

J. MISIEWICZ and J. A. GAJ
*Optical Anisotropy in the Zn_3P_2 Band Gap Region
(b) Vol. **105**, No. 1, p. K23–K25 (1981).
See also 16, p. 252; 22.3, p. 441.

O. Goede, W. Heimbrodt, and R. Müller
*$CdS_{1-x}Te_x$ as Persistence-Type Semiconductor Mixed Crystals
(b) Vol. **105**, No. 2, p. 543–550 (1981).
See also *13.5.1*, p. *190*; 20.3, p. 328; 22.4.4, p. 457.

G. Jungk
*On the Experimental Determination of Many-Particle Effects at Interband Transitions
(b) Vol. **105**, No. 2, p. 551–556 (1981).
See also *13.1*, p. *156*; 13.5.1, p. 190; 22.2.1, p. 429; 22.2.2, p. 433.

O. V. Farberovich and S. V. Vlasov
APW Research of Europium Monochalcogenides (I): Band Structures and Electron–Electron Interaction in Ferromagnetic Semiconductors
(b) Vol. **105**, No. 2, p. 755–768 (1981).
See also *13.1*, p. *156*; 18.2, p. 271; 22, p. 396.

V. S. Gorelik, B. S. Umarov, and L. G. Reznik
*The Peculiarity of the A_1-Type Polariton in Lithium-Niobate Crystal
(b) Vol. **105**, No. 2, p. K73–K75 (1981).
See also *6*, p. *75*; 22.8.1, p. 497.

K. M. Häusler
*Infrared Spectra of Magnetic Excitations in CoF_2
(b) Vol. **105**, No. 2, p. K81–K83 (1981).
See also *13.5.4*, p. *197*; 22.5, p. 459.

V. V. Artamonov, M. Ya. Valakh, and M. P. Lisitsa
*Interlayer Interaction in Phonon Spectra of ε-GaSe Crystals
(b) Vol. **105**, No. 2, p. K103–K106 (1981).
See also *6*, p. *75*; 22.3, p. 441.

P. Bussemer and A. Taubert
*To the Asymmetry of the Absorption Spectrum at Localized Centres
(b) Vol. **105**, No. 2, p. K107–K110 (1981).
See also 6, p. 75; *13.4*, p. *177*.

I. V. Bodnar, L. V. Golubev, V. G. Plotnichenko, and E. A. Smolyaninova
*Raman Scattering in $CuGaSe_2$
(b) Vol. **105**, No. 2, p. K111–K114 (1981).
See also *6*, p. *75*; 22.8, p. 487.

I. V. Bodnar, B. V. Korzun, and A. I. Lukomskii
*Composition Dependence of the Band Gap of $CuInS_{2x}Se_{2(1-x)}$
(b) Vol. **105**, No. 2, p. K143–K145 (1981).
See also *13.1*, p. *157*; 22.8, p. 487.

O. S. Josyulu and J. Sobhanadri
*The Far-Infrared Spectra of Some Mixed Cobalt Zinc and Magnesium Zinc Ferrites
(a) Vol. **65**, No. 2, p. 479–483 (1981).
See also 22.8.2, p. 504.

M. GIEHLER
*The Effect of Short-Range Order on the Vibrational Spectra of 3:2 Coordinated Chalcogenide Glasses Analysis of the Vibrational Spectra by Using Pyramidal Structural Units
(b) Vol. **106**, No. 1, p. 193–205 (1981).
See also 2, p. 47; 4, p. 59; *6*, p. *76*; 22, p. 396.

F. BREHAT, B. WYNCKE, and A. HADNI
*Lattice Mode Anharmonicity in Sodium Nitrate
(b) Vol. **106**, No. 1, p. 207–214 (1981).
See also 1.2, p. 17; *6*, p. *76*; 22.8.1, p. 498.

K. SAMINADAYAR, N. MAGNEA, J. L. PAUTRAT, and B. PAJOT
*Infrared Absorption by Ag, Cu, and Au Acceptors in ZnTe
(b) Vol. **106**, No. 1, p. 215–222 (1981).
See also *13.4*, p. *178*; 20.3, p. 329; 22.4.3, p. 454.

R. K. SINGH and K. P. SINGH
Free-Carrier Doping and Many-Body Effects in Crystal Dynamics of Lead Sulphide Semiconductors
(b) Vol. **106**, No. 1, p. 229–236 (1981).
See also *6*, p. *76*; 22, p. 396.

T. SAUDER, B. STÉBÉ, and M. CERTIER
*Acceptor Bound Exciton States in High-Purity CuCl
(b) Vol. **106**, No. 1, p. 279–286 (1981).
See also 13.4, p. 178; *13.5.1*, p. *190*; 20.3, p. 329; 22.5, p. 459.

S. SAPRU and S. B. S. SASTRY
*Optical and Polarized Luminescence Studies on $RbI:Pb^{2+}$ at Liquid Helium Temperatures
(b) Vol. **106**, No. 1, p. 295–300 (1981).
See also 10.2, p. 112; *20.3*, p. *329*; 22.5.2, p. 465.

T. TSUBOI
*Assignment of $(Tl^{+})_2$ Absorption Bands in Alkali Halides by Magnetic Circular Dichroism Spectra
(b) Vol. **106**, No. 1, p. 301–307 (1981).
See also *13.4*, p. *178*; 18, p. 263; 22.5.2, p. 465.

Z. V. POPOVIĆ and H. J. STOLZ
*Infrared and Raman Spectra of Germanium Dichalcogenides (I): GeS_2
(b) Vol. **106**, No. 1, p. 337–348 (1981).
See also *6*, p. *76*; 22, p. 396.

C. TRALLERO GINER, I. G. LANG, and S. T. PAVLOV
*Theory of Multiphonon Resonance Raman Scattering as a Function of Temperature in Polar Semiconductors
(b) Vol. **106**, No. 1, p. 349–357 (1981).
See also *6*, p. *76*; 13.5.1, p. 190; 22, p. 396.

N. M. GASANLY, B. M. DZHAVADOV, A. S. RAGIMOV, V. I. TAGIROV, and R. E. GUSEINOV
*Long-Wave Optical Phonons in In_6S_7 Layer Crystals
(b) Vol. **106**, No. 1, p. K47–K51 (1981).
See also *6*, p. *77*; 22.3, p. 442.

YU. P. GNATENKO, Z. D. KOVALYUK, and P. A. SKUBENKO
*Influence of Ni Impurity on the Absorption Spectrum of Layered GaSe Crystals
(b) Vol. **106**, No. 2, p. 621–626 (1981).
See also *13.4*, p. *179*; 22.3, p. 442.

V. D. ISKRA
*Phonon Assisted Interband Optical Transitions in Semiconductors Substaining a Coulomb Random Field
(b) Vol. **106**, No. 2, p. 627–633 (1981).
See also 6, p. 77; *13.1*, p. *159*; 22, p. 397.

M. J. GELTEN and L. A. BOSCH
*A New Method to Determine the Coupled Mode Parameters of a Plasmon–Multiphonon System
(b) Vol. **106**, No. 2, p. 635–645 (1981).
See also *6*, p. *77*; *13.5.2*, p. *195*; 22, p. 397.

J. F. POURADIER and G. GRIMOUILLE
Growth of Semiconducting CdF_2:Ho^{3+} Crystals and Study of Their Optical and Transport Properties
(b) Vol. **106**, No. 2, p. 657–665 (1981).
See also *3*, p. *55*; 13.4, p. 179; *14.3*, p. *215*; 22.5.3, p. 471.

S. MOCHIZUKI and M. SATOH
*Infrared Spectra of Single-Crystal NiO at High Temperatures
(b) Vol. **106**, No. 2, p. 667–674 (1981).
See also 6, p. 77; 22.6, p. 478.

R. V. JIMENEZ, F. LÜTY, M. SIU LI, and M. DE SOUZA
*Paraelastic Alignment and Electric Dipole Relaxation Behavior of Off-Center Ag^+ Defects in RbI
(b) Vol. **106**, No. 2, p. 683–692 (1981).
See also *10.2*, p. *114*; 22.5.2, p. 467.

B. GHOSH and R. K. MUKHERJEE
*Electronic Absorption Spectrum of Rb_2MnCl_4 Single Crystals
(b) Vol. **106**, No. 2, p. 699–706 (1981).
See also 22.5, p. 459.

B. KUBIER, H. BOUDRIOT, K. DEUS, R. GRÜNDLER, H. ZSCHEILE, and E. BUHRIG
*Plasmon Data, Dielectric Function, and Optical Constants of $CdGeP_2$ from Electron Energy Loss Measurements between 4.5 and 35 eV
(b) Vol. **106**, No. 2, p. K103–K106 (1981).
See also *13.5.2*, p. *195*; 22.8, p. 488.

YU. M. FEDOROV, A. A. LEKSIKOV, A. E. AKSYONOV, and I. S. EDELMAN
*Magnetic Linear Dichroism and Birefringence of $FeBO_3$ in the $^6A_{1g} \rightarrow {}^4T_{1g}$ Transition Region
(b) Vol. **106**, No. 2, p. K127–K130 (1981).
See also 6, p. 78; 13.5.1, p. 191; 22.8.1, p. 498.

S. Mochizuki
*Polarized Infrared Absorption in Thin-Film Cerium Dioxide
(b) Vol. **107**, No. 1, p. K53–K56 (1981).
See also *6*, p. *78*; 22.6, p. 479.

F. El-Kabbany, G. Said, Y. Badr, and S. Taha
Infrared Investigation of the Phase Transition in K_2SO_4
(a) Vol. **67**, No. 1, p. 339–345 (1981).
See also *1.2*, p. *18*; 6, p. 78; 22.8.1, p. 499.

V. D. Akhmetov, V. V. Bolotov, and L. S. Smirnov
*Optical Properties of Defects Produced at Annealing of Neutron-Induced Clusters in Silicon
(a) Vol. **67**, No. 1, p. K33–K37 (1981).
See also 11, p. 129; *13.4*, p. *181*; 22.1.2, p. 419.

P. F. Robusto and R. Braunstein
*Optical Measurements of the Surface Plasmon of Copper
(b) Vol. **107**, No. 2, p. 443–449 (1981).
See also *13.5.2*, p. *195*; 21.1, p. 364.

R. Kink, A. Lôhmus, and M. Selg
Self-Trapping and Hot Luminescence of Excitons in Rare Gas Solids
(b) Vol. **107**, No. 2, p. 479–490 (1981).
See also *13.5.1*, p. *192*; *20.3*, p. *332*; 23, p. 514.

Doan Nhat Quang, B. Esser, and R. Keiper
*On the Theory of Resonance Raman Scattering in Heavily Doped Degenerate Semiconductors
(b) Vol. **107**, No. 2, p. 535–542 (1981).
See also *6*, p. *79*; 22, p. 398.

R. Zimmermann, M. Rösler, and V. M. Asnin
*Direct Absorption inside the Electron–Hole Liquid in Ge: Warping Effects and Final State Interactions
(b) Vol. **107**, No. 2, p. 579–590 (1981).
See also *13.5.1*, p. *192*; 13.5.2, p. 195; 22.1.1, p. 407.

A. C. Mezger and C. Jaccard
Electron Transfer within F Center Pairs in KCl from Optical Emission and Absorption
(b) Vol. **107**, No. 2, p. 689–697 (1981).
See also *13.4*, p. *181*; *20.3*, p. *332*; 22.5.2, p. 468.

F. Brehat, B. Wyncke, and A. Hadni
*Far-Infrared Reflectivity Spectra of KH_2PO_4 between 300 and 400 K
(b) Vol. **107**, No. 2, p. 723–727 (1981).
See also *6*, p. *79*; 14.4.1, p. 245; 22.8.1, p. 500.

G. Debiais
*Fonctions optiques d'une couche mince de sulfure de mercure
(Optical Function of a Thin Film of Mercury Sulfide)
(b) Vol. **107**, No. 2, p. 729–736 (1981).
See also 6, p. 79; 22.4.1, p. 450.

Z. V. Popović and H. J. Stolz
*Infrared and Raman Spectra of Germanium Dichalcogenides (II): $GeSe_2$
(b) Vol. **108**, No. 1, p. 153–163 (1981).
See also *6*, p. *80*; 22, p. 399.

S. R. Kuindersma, P. R. Boudewijn, and C. Haas
*Near-Infrared d–d Transitions of NiI_2, CdI_2:Ni^{2+}, and CoI_2
(b) Vol. **108**, No. 1, p. 187–194 (1981).
See also 1.2, p. 20; 22.5, p. 460.

S. Asano, N. Yamashita, Y. Nakao, and Y. Matsushima
Luminescence and Paramagnetic Resonance of the Mn^{2+} Ion in MgS Phosphor
(b) Vol. **108**, No. 1, p. 229–239 (1981).
See also *13.4*, p. *182*; 19, p. 294; 20.3, p. 333; 22.4.1, p. 450.

T. Chattopadhyay and H. G. von Schnering
*Temperature Dependence of the Infrared Absorption Spectrum of P_4S_3
(b) Vol. **108**, No. 1, p. 241–244 (1981).
See also *6*, p. *80*; 22, p. 399.

D. N. Popov, N. M. Nancheva, M. R. Minev, and K. I. Terzijsky
Positron Annihilation in Z_1-Centers in Additively Coloured KCl:Ca
(b) Vol. **108**, No. 1, p. K7–K9 (1981).
See also *10.2*, p. *117*; 22.5.2, p. 468.

C. Rincón, J. González, and G. Sanchez Perez
*Optical Absorption of $CuInSe_2$ in Bulk Single Crystal
(b) Vol. **108**, No. 1, p. K19–K22 (1981).
See also 22.8, p. 490.

K. E. Peiponen and T. Jääskeläinen
*Simple Formulas for the Complex Refractive Index of Gaussian Type
(b) Vol. **108**, No. 1, p. K71–K73 (1981).
See also 22.5.2, p. 469.

E. A. Radyabov and A. I. Nepomnyachikh
Magnesium Color Centers at 3.5 and 5.0 eV in Lithium Fluoride
(a) Vol. **68**, No. 1, p. 77–81 (1981).
See also *10.2*, p. *118*; 20.3, p. 333; 22.5.2, p. 469.

M. Böhm, S. Herrmann, and A. Scharmann
Dipole Centres in $CaMoO_4$ Crystals
(a) Vol. **68**, No. 1, p. 129–134 (1981).
See also *10.2*, p. *118*; *14.4*, p. *241*; 22.8.1, p. 500.

W. Kowalsky, A. Schlachetzki, and F. Fiedler
*Near-Band-Gap Absorption of InGaAsP at 1.3 μm Wavelength
(a) Vol. **68**, No. 1, p. 153–158 (1981).
See also 22.2.4, p. 439.

Bhanwar Singh and K. L. Chopra
*A Novel Bi-Layer System for High Resolution Optical Imaging
(a) Vol. **68**, No. 1, p. K5–K9 (1981).
See also *2*, p. *50*; 22.1.1, p. 407; 22.1.3, p. 425.

V. V. Kind, V. P. Fedorov, and A. V. Korshunov
*The Manifestation of Intermolecular Hydrogen Bond in a Binary Naphthalene-β-Naphthol Mixture
(b) Vol. **108**, No. 2, p. K143–K144 (1981).
See also *4*, p. *61*; 22.9, p. 513.

M. Paić and V. Paić
*Diffuse Reflectance Spectra and Polymorphism of Mercury (II) Iodide
(a) Vol. **68**, No. 2, p. 447–455 (1981).
See also 1, p. 10; 13.5.1, p. 193; *20.3*, p. *334*; 22.5, p. 460.

C. A. Hogarth and M. Y. Nadeem
*The Optical Absorption Edge in Amorphous Thin Films of Germania and of Germania with Barium Oxide
(a) Vol. **68**, No. 2, p. K181–K184 (1981).
See also *2*, p. *52*; 22.6.1, p. 482.

20.2 Lasers

Original Papers and Short Notes

20.3 Luminescence
(X-Ray Spectra See 20)

Review Article

H. BÄSSLER
Localized States and Electronic Transport in Single Component Organic Solids with Diagonal Disorder
(b) Vol. **107**, No. 1, p. 9–54 (1981).
See also 2, p. 40; *13.4*, p. *168*; 13.5.1, p. 188; *14.3*, p. *209*; 14.3.4, p. 232; 20.1, p. 301; 22.9, p. 507.

Original Papers and Short Notes

S. B. S. SASTRY and S. SAPRU
*Thermoluminescence, Fluorescence, and Optical Absorption Studies on Europium-Doped Rubidium Chloride
(b) Vol. **103**, No. 1, p. 185–192 (1981).
See also 10.2, p. 106; *20.1*, p. *301*; 22.5.2, p. 462.

H. J. JODL and F. BOLDUAN
Laser Induced Fluorescence and Raman Studies with KI:NO_2^-
(b) Vol. **103**, No. 1, p. 297–310 (1981).
See also *6*, p. *66*; *10.2*, p. *106*; 20.1, p. 302; 22.5.2, p. 462.

J. BIELLMANN, B. MEYER, and C. SCHWAB
*Anisotropy of Exciton Polaritons and Polariton–Phonon Interaction in 2H-PbI_2 Crystals (I): Structure of the Fundamental Excitonic Emission
(b) Vol. **103**, No. 1, p. 361–365 (1981).
See also *13.5.1*, p. *188*; 20.1, p. 302; 22.5, p. 458.

J. COLLET, M. PUGNET, A. CORNET, M. BROUSSEAU, B. S. RAZBIRIN, and G. V. MIKHAILOV
*Picosecond Spectroscopy of CdSe at High Excitation
(b) Vol. **103**, No. 1, p. 367–379 (1981).
See also *13.5.1*, p. *188*; 22.4.2, p. 452.

D. GOEDE and R. MÜLLER
*Calorimetric Measurements of Luminescence Quantum Efficiency in Crystals
(b) Vol. **103**, No. 1, p. K75–K79 (1981).
See also 22.4.1, p. 447.

T. IKARI, S. SHIGETOMI, Y. KOGA, and S. SHIGETOMI
*Photoluminescence of Zn Doped InSe Single Crystals
(b) Vol. **103**, No. 1, p. K81–K83 (1981).
See also 22.3, p. 440.

A. M. RASTOGI and S. L. MOR
*Studies on the Electroluminophor $(Ca_xSr_{1-x})S$:Cu, Nd
(a) Vol. **63**, No. 1, p. 75–78 (1981).
See also 22.4.4, p. 456.

M. J. J. Lammers, G. Blasse, and D. S. Robertson
*The Luminescence of Cadmium Tungstate ($CdWO_4$)
(a) Vol. **63**, No. 2, p. 569–572 (1981).
See also 22.8.1, p. 494.

V. V. Mürk, A. I. Kuznetsov, and B. R. Namozov
*Kinetics of Intrinsic Luminescence and Energy Transfer in Third Group Metal Oxides
(a) Vol. **63**, No. 2, p. K131–K135 (1981).
See also 22.6, p. 475.

A. C. van der Steen and L. T. F. Dijcks
*The Luminescence Properties of Alkaline-Earth Oxides Activated with $6s^2$ Ions
(b) Vol. **104**, No. 1, p. 283–292 (1981).
See also 6, p. 70; 13.4, p. 171; 22.6, p. 475.

H. Yoshida, H. Saito, and S. Shionoya
*Picosecond Relaxation Processes of High Density Electron–Hole Plasma in CdSe
(b) Vol. **104**, No. 1, p. 331–340 (1981).
See also 13.5, p. 186; 22.4.2, p. 452.

K. Hamada, T. Taguchi, and S. Fujita
*Annealing Behaviour of Oxygen-Bound Exciton in Electron-Irradiated ZnTe
(b) Vol. **104**, No. 1, p. K27–K30 (1981).
See also 11, p. 122; *13.5.1*, p. *189*; 22.4.3, p. 454.

G. Blasse
*Some Remarks on the Ce^{3+} and Bi^{3+} Luminescence in Alkaline-Earth Chalcogenides
(b) Vol. **104**, No. 1, p. K57–K59 (1981).
See also *13.4*, p. *172*; 22.4.1, p. 447; 22.6, p. 475.

P. Sridaran, R. K. Gartia, R. C. Bhuniya, and V. V. Ratnam
*Thermoluminescence of Z_2 Centers in X-Irradiated Ca-Doped KCl Crystals
(a) Vol. **64**, No. 1, p. 127–131 (1981).
See also 10.2, p. 108; 20.1, p. 306; 22.5.2, p. 463.

Hj. Bernhardt and R. Schnell
*Modulation-Spectroscopic Investigations of Defect Centres in $PbMoO_4$ Crystals
(a) Vol. **64**, No. 1, p. 207–214 (1981).
See also 10.2, p. 108; *16*, p. *252*; 22.8.1, p. 495.

L. N. Tripathi, V. K. Dubey, B. R. Chaubey, and C. P. Mishra
*Dielectric Dependence of EL Brightness of Powdered Phosphor and Particle Size Distribution
(a) Vol. **64**, No. 1, p. 297–302 (1981).
See also 22.4.1, p. 447.

B. P. Chandra
*Mechanoluminescence and Piezoelectric Behaviour of Molecular Crystals
(a) Vol. **64**, No. 1, p. 395–405 (1981).
See also *12.2*, p. *139*; 14.3, p. 211; 22.9, p. 509.

F. VOILLOT, J. BARRAU, M. BROUSSEAU, and J. C. BRABANT
*The Fine Structure of the 0.84 eV No-Phonon Luminescence in GaAs:Cr
(a) Vol. **64**, No. 1, p. K39–K41 (1981).
See also 22.2.1, p. 428.

N. A. DROZDOV, A. A. PATRIN, and V. D. TKACHEV
*Modification of the Dislocation Luminescence Spectrum by Oxygen Atmospheres in Silicon
(a) Vol. **64**, No. 1, p. K63–K65 (1981).
See also *10.2*, p. *108*; 22.1.2, p. 412.

P. SRIDARAN, R. C. BHUNIYA, and V. V. RATNAM
Optical and Thermal Stability of Ag Centres in RT X-Irradiated KCl:Ag Crystals
(b) Vol. **104**, No. 2, p. 459–468 (1981).
See also *10.2*, p. *109*; 20.1, p. 306; 22.5.2, p. 464.

P. A. KULIS, M. J. SPRINGIS, I. A. TALE, V. S. VAINER, and J. A. VALBIS
Impurity-Associated Colour Centres in Mg- and Ca-Doped Al_2O_3 Single Crystals
(b) Vol. **104**, No. 2, p. 719–725 (1981).
See also *10.2*, p. *109*; 19, p. 288; 22.6, p. 476.

F. THUSELT
Erratum to: On the Phase Separation of Carriers in Nitrogen-Doped GaP
(b) Vol. **104**, No. 2, p. 752 (1981).
See also *13.5.1*, p. *189*; 22.2.2, p. 432.

A. H. BUTH and G. BLASSE
*Luminescence and Energy Transfer in Yttrium Niobate ($YNbO_4$)
(a) Vol. **64**, No. 2, p. 669–676 (1981).
See also 22.8.1, p. 496.

N. B. LUKYANCHIKOVA, T. M. PAVELKO, G. S. PEKAR, N. N. TKACHENKO, and M. K. SHEINKMAN
*Mechanism of Electroluminescence Excitation in Forward-Biased MS and MIS Light-Emitting Diodes Based on Wide-Band-Gap II–VI Compounds
(a) Vol. **64**, No. 2, p. 697–706 (1981).
See also *14.3.3*, p. *226*; 22.4.1, p. 448.

K. D. GLINCHUK, K. LUKAT, and A. V. PROKHOROVICH
*Non-Trivial Excitation Dependence of the Kinetics of the Extrinsic Luminescence Decay in GaAs
(a) Vol. **64**, No. 2, p. K139–K143 (1981).
See also *13.4*, p. *174*; 22.2.1, p. 428.

R. K. SINHA and M. L. MUKHERJEE
*On the Kinetics of Phosphorescence and High-Temperature Glow Peak in CaF_2:Pr Single Crystals
(b) Vol. **105**, No. 1, p. 69–75 (1981).
See also *13.4*, p. *174*; 22.5.3, p. 471.

G. A. SOLER, J.-P. AICARDI, and J.-P. LEYRIS
*Red Photoluminescence of α-HgS Single Crystals
(b) Vol. **105**, No. 1, p. 249–255 (1981).
See also 16, p. 252; 22.4.1, p. 448.

S. Asano et N. Yamashita
*Luminescence et interaction phonon–électron dans le luminophore MgS: Bi^{3+} (I): Experience générale
(Luminescence and Electron–Phonon Interaction in the MgS: Bi^{3+} Phosphor (I): Experiments)
(b) Vol. **105**, No. 1, p. 305–310 (1981).
See also *6*, p. *74*; 13.4, p. 175; 22.4.1, p. 448; 22.6, p. 476.

B. V. R. Chowdari and S. Selvasekarapandian
*Thermoluminescence of Heavy Metal Ion Doped Cesium Halides X-Irradiated at Low Temperature
(b) Vol. **105**, No. 1, p. 377–383 (1981).
See also 10.2, p. 110; 22.5.2, p. 464.

J. Garrido, J. L. Castaño, and J. Piqueras
*Photoluminescence in Gallium Arsenide Irradiated with Thermal Neutrons
(a) Vol. **65**, No. 1, p. 103–106 (1981).
See also 11, p. 125; *14.3*, p. *213*; 22.2.1, p. 428.

R. Jayaprakash and V. V. Ratnam
*On the Thermoluminescence of Sb-Doped Calcium Fluorophosphate
(a) Vol. **65**, No. 1, p. 147–152 (1981).
See also *10.2*, p. *110*; 22.8.1, p. 496.

K. Wandel, G. Wurbs, R. Bindemannn, A. Kirsch, and R. Schwabe
*Minority Carrier Lifetime in VPE-GaP: N, Te at Various Excitation Levels
(a) Vol. **65**, No. 1, p. 233–243 (1981).
See also *13.4*, p. *176*; 22.2.2, p. 433.

M. Sharon, R. R. Pradhananga, and C. M. Sunta
*Thermoluminescence Glow Curves of Ca^{2+} and Sr^{2+} Doped LiCl Single Crystals
(a) Vol. **65**, No. 1, p. K67–K70 (1981).
See also *10.2*, p. *111*; 22.5.2, p. 465.

S. P. Kuila, R. K. Gartia, and V. V. Ratnam
*Emission Processes in Thermoluminescence of Li-Doped KCl Single Crystals
(a) Vol. **65**, No. 1, p. K71–K74 (1981).
See also *10.2*, p. *111*; 22.5.2, p. 465.

V. Tabakova and K. Kynev
*Absorption Spectra of Lead Centers in Barium Chloride
(a) Vol. **65**, No. 1, p. K75–K78 (1981).
See also *13.4*, p. *176*; 22.5.3, p. 471.

S. Takata, T. Minami, H. Nanto, and T. Kawamura
*Temperature Dependence of Electro- and Photoluminescence in ZnO Single Crystals
(a) Vol. **65**, No. 1, p. K83–K86 (1981).
See also 22.4, p. 445.

D. Gwinner and R. Labusch
*Photoluminescence in Plastically Twisted Silicon
(a) Vol. **65**, No. 1, p. K99–K101 (1981).
See also 10.2, p. 111; 22.1.2, p. 414.

F. R. KESSLER and J. W. MANGELSDORF
*Free Carrier Compression and Galvanomagnetic Luminescence of Germanium at High Lorentz Fields
(b) Vol. **105**, No. 2, p. 525–535 (1981).
See also 14.3, p. 214; 22.1.1, p. 405.

O. GOEDE, W. HEIMBRODT, and R. MÜLLER
$CdS_{1-x}Te_x$ as Persistence-Type Semiconductor Mixed Crystals
(b) Vol. **105**, No. 2, p. 543–550 (1981).
See also *13.5.1*, p. *190*; *20.1*, p. *311*; 22.4.4, p. 457.

S. ASANO
*Luminescence et interaction phonon–électron dans le luminophore $MgS:Bi^{3+}$ (II): Théorie sur le mélange entre les etats $^3T_{1u}$ et $^3A_{1u}$
(Luminescence and Electron–Phonon Interaction in the $MgS:Bi^{3+}$ Phosphor (II): Theory of $^3T_{1u}$ and $^3A_{1u}$ Mixing)
(b) Vol. **105**, No. 2, p. 613–621 (1981).
See also *6*, p. *74*; 13.4, p. 177; 22.4.1, p. 448.

B. M. ASHKINADZE
Role of Free Electrons and Holes in the Exciton Condensation Process in Semiconductors
(b) Vol. **105**, No. 2, p. 633–640 (1981).
See also *13.5.1*, p. *190*; 22.1.1, p. 405.

R. J. SAYER, R. H. PRINCE, and W. W. DULEY
*Site Effects on N and O Luminescence in Electron-Irradiated Neon Films
(b) Vol. **105**, No. 2, p. 727–738 (1981).
See also 10.2, p. 111; 23, p. 514.

B. TRIEMAN and N. KRISTIANPOLLER
*Defects Induced in $RbMgF_3$ Crystals by X- and UV-Irradiation
(b) Vol. **105**, No. 2, p. 739–746 (1981).
See also *11*, p. *126*; 22.5, p. 459.

V. F. KOVALENKO, G. P. PEKA, and L. G. SHEPEL
*Photoluminescent Methods for Determination of Graded Band-Gap Semiconductor Parameters (I): Analysis of Methods
(a) Vol. **65**, No. 2, p. 529–534 (1981).
See also 13.1, p. 157; 13.3, p. 165; 22.2.4, p. 437.

T. C. LESLIE and J. W. ALLEN
*Thermal Quenching of Photoluminescence in ZnS:Mn and ZnSe:Mn
(a) Vol. **65**, No. 2, p. 545–551 (1981).
See also 13.4, p. 177; 22.4.1, p. 448; 22.4.2, p. 452.

O. E. AKSENOV, E. D. ALUKER, D. J. LUSIS, and I. P. MEZINA
*Luminescence of Activated Alkali Halide Crystals at Excitation in the Fundamental Absorption Region
(a) Vol. **65**, No. 2, p. 617–622 (1981).
See also *10.2*, p. *111*; 22.5.2, p. 465.

V. S. Ballandovich, Yu. M. Tairov, and G. N. Violina
Acceptor Boron in α-SiC (6H): Investigation by the Photocapacitance Method
(a) Vol. **65**, No. 2, p. 709–716 (1981).
See also *13.4*, p. *177*; *16*, p. *253*; 22.3, p. 441.

R. P. Khare and M. Khare
*A New Method of Analysis of Traps Taking Part in the Electroluminescence Process
(a) Vol. **65**, No. 2, p. 725–729 (1981).
See also *13.4*, p. *177*; 22.6, p. 477.

H. Nanto, T. Minami, and S. Takata
*Photoluminescence in Sputtered ZnO Thin Films
(a) Vol. **65**, No. 2, p. K131–K134 (1981).
See also 22.4, p. 445.

P. Swarup, R. K. Jain, S. N. Verma, Sh. Charan, and D. M. Tandle
Preparation and Properties of GaAsSb and GaAlAsSb Epi-Layers
(a) Vol. **65**, No. 2, p. K183–K187 (1981).
See also *3*, p. *54*; *20.1*, p. *312*; 22.2.4 p. 437.

Sh. A. Vakhidov, E. M. Ibragimova, I. Nuritdinov, A. F. Rakov, and G. I. Ikramov
Self-Trapped Particles in Complex Oxide Crystals
(b) Vol. **106**, No. 1, p. 31–35 (1981).
See also *11*, p. *126*; 13.5.1, p. 190; 19, p. 290; 20.1, p. 312; 22.6, p. 477; 22.8, p. 487.

I. B. Ermolovich, V. V. Gorbunov, and G. I. Matvievskaya
*Properties of the R_n and IR-1 Luminescence Bands in Cadmium Sulfide Irradiated with Thermal Neutrons
(b) Vol. **106**, No. 1, p. 45–53 (1981).
See also *11*, p. *126*; 13.4, p. 178; 16, p. 254; 22.4.1, p. 448.

K. Saminadayar, N. Magnea, J. L. Pautrat, and B. Pajot
Infrared Absorption by Ag, Cu, and Au Acceptors in ZnTe
(b) Vol. **106**, No. 1, p. 215–222 (1981).
See also *13.4*, p. *178* ; *20.1*, p. *313*; 22.4.3, p. 454.

R. J. Sayer, R. H. Prince, and W. W. Duley
*Emission Spectra of O and N Atoms and the Structure of Solid N_2/Ne Mixtures
(b) Vol. **106**, No. 1, p. 249–258 (1981).
See also *2*, p. *47*; 23, p. 514.

T. Sauder, B. Stébé, and M. Certier
Acceptor Bound Exciton States in High-Purity CuCl
(b) Vol. **106**, No. 1, p. 279–286 (1981).
See also 13.4, p. 178; *13.5.1*, p. *190*; *20.1*, p. *313*; 22.5, p. 459.

S. Sapru and S. B. S. Sastry
*Optical and Polarized Luminescence Studies on RbI:Pb^{2+} at Liquid Helium Temperatures
(b) Vol. **106**, No. 1, p. 295–300 (1981).
See also 10.2, p. 112; *20.1*, p. *313*; 22.5.2, p. 465.

V. Hizhnyakov, T. Soovik, and S. Zazubovich
*Origin of Radiative Transitions from Metastable Minima of the Excited State of $Sn^{2+}v_c^-$ Centres in Alkali Halides
(b) Vol. **106**, No. 2, p. 581–586 (1981).
See also *13.4*, p. *179*; 22.5.2, p. 466.

C. W. M. Timmermans and G. Blasse
*On the Luminescence of $Cs_3Bi_2Br_9$ Single Crystals
(b) Vol. **106**, No. 2, p. 647–655 (1981).
See also 13.5.1, p. 191; 22.8, p. 488.

O. Sild and I. Sildos
*Zeeman Effect in Low-Temperature Luminescence Spectra of Molecular Ions O_2^- and S_2^- in Alkali Halide Crystals
(b) Vol. **106**, No. 2, p. 675–682 (1981).
See also *13.4*, p. *179*; 22.5.2, p. 466.

N. B. Lukyanchikova, T. M. Pavelko, and G. S. Pekar
*Physical Processes in Forward-Biased ZnS MIS Light-Emitting Diodes with a "Thick" Interfacial Layer
(a) Vol. **66**, No. 2, p. 749–759 (1981).
See also *14.3.3*, p. *229*; 22.4.1, p. 449.

N. Singh, G. L. Marwaha, and V. K. Mathur
*Luminescence Centres and Charge Compensation in CaS Phosphors
(a) Vol. **66**, No. 2, p. 761–765 (1981).
See also 13.4, p. 179; 22, p. 397.

S. V. Moharil
*A New Method for Obtaining Trapping Parameters from Glow Curves
(a) Vol. **66**, No. 2, p. 767–772 (1981).
See also *13.4*, p. *179*; 22.5.2, p. 467.

H. Klose and M. Rieth
Influence of Al Postimplantation on p–n Junctions in Gallium Phosphide
(a) Vol. **66**, No. 2, p. K153–K156 (1981).
See also *11*, p. *128*; 14.3.3, p. 229; 22.2.2, p. 433.

M. V. Kovalchuk, N. Hertel, M. K. Melkonyan, R. M. Imamov, and P. A. Aleksandrov
*Fluorescence Yield Curves under X-Ray Diffraction Conditions
(a) Vol. **66**, No. 2, p. K173–K177 (1981).
See also *17*, p. *257*; 22.1.1, p. 406; 22.1.2, p. 418.

R. Kuhnert, R. Helbig, and K. Hümmer
*Photoluminescence of ZnO Single Crystals. Influence of *k*-Linear Interaction and Non-Analytic Long-Range Exchange on the Emission Spectrum of the Free A-Exciton
(b) Vol. **107**, No. 1, p. 83–92 (1981).
See also *13.5.1*, p. *191*; 22.4, p. 445.

K. Recker, F. Wallrafen, R. Büscher, and G. Lehmann
*Crystal Growth, EPR, and Luminescence of $BaZnF_4$:Mn
(b) Vol. **107**, No. 2, p. 699–706 (1981).
See also 3, p. 55; *19*, p. *293*; 22.5.4, p. 473.

R. Baumert and J. Gutowski
Excitation Mechanisms and Excited States of the I_1 Bound Exciton in CdS
(b) Vol. **107**, No. 2, p. 707–714 (1981).
See also *13.5.1*, p. *192*; 22.4.1, p. 450.

P. Kukk and A. Erm
*The Structure of Radiative and Non-Radiative Recombination Centres in Activated CdS Phosphors
(a) Vol. **67**, No. 2, p. 395–406 (1981).
See also *10.2*, p. *117*; 13.4, p. 182; 22.4.1, p. 450.

V.-P. Tanninen, M. Oikkonen, and T. O. Tuomi
X-Ray Diffraction Study of Thin Electroluminescent ZnS Films Grown by Atomic Layer Epitaxy
(a) Vol. **67**, No. 2, p. 573–583 (1981).
See also *1.5*, p. *37*; 22.4.1, p. 450.

B. Esser
On the Lifetime of Excited States in One-Dimensional Disordered Systems
(b) Vol. **108**, No. 1, p. 195–204 (1981).
See also *2*, p. *50*; *13.4*, p. *182*; 22, p. 399.

S. Asano, N. Yamashita, Y. Nakao, and Y. Matsushima
Luminescence and Paramagnetic Resonance of the Mn^{2+} Ion in MgS Phosphor
(b) Vol. **108**, No. 1, p. 229–239 (1981).
See also *13.4*, p. *182*; 19, p. 294; 20.1, p. 319; 22.4.1, p. 450.

S. Modesti, A. Frova, J. L. Staehli, M. Guzzi, and M. Capizzi
Electron–Hole Plasma Expansion in Direct-Gap $GaAs_{1-x}P_x$
(b) Vol. **108**, No. 1, p. 281–288 (1981).
See also *13.5*, p. *187*; 22.2.4, p. 438.

E. A. Radyabov and A. I. Nepomnyachikh
Neutral Magnesium Atoms on Anion Sites in LiF
(b) Vol. **108**, No. 1, p. K75–K78 (1981).
See also *10.2*, p. *118*; 22.5.2, p. 469.

E. A. Radyabov and A. I. Nepomnyachikh
Magnesium Color Centers at 3.5 and 5.0 eV in Lithium Fluoride
(a) Vol. **68**, No. 1, p. 77–81 (1981).
See also *10.2*, p. *118*; 20.1, p. 319; 22.5.2, p. 469.

H. L. Oczkowski
*Thermoluminescence in Conductive n-Type ZnSe Crystals
(a) Vol. **68**, No. 1, p. 199–207 (1981).
See also *13.4*, p. *183*; 22.4.2, p. 453.

N. S. Minaev and A. V. Mudryi
Thermally-Induced Defects in Silicon Containing Oxygen and Carbon
(a) Vol. 68, No. 2, p. 561–565 (1981).
See also *10.2*, p. *120*; *13.4*, p. *184*; 22.1.2, p. 423.

H. Berek and P. Kirsten
*Observation of Recombination-Enhanced Defect Reactions (REDR) at Low Temperature by Means of Cathodoluminescence and Photoluminescence in GaP
(a) Vol. 68, No. 2, p. K203–K207 (1981).
See also 11, p. 132; 22.2.2, p. 434.

B. P. Chandra, A. K. Jaiswal, T. R. Chandraker, and B. R. Kaza
*Mechanoluminescence and Electroluminescence of Non-Photoluminescent $Mn(Ph_3PO)_2Cl_2$ Crystals
(a) Vol. 68, No. 2, p. K207–K210 (1981).
See also 22.8, p. 492.

21. Metals. Alloys

Review Articles

Original Papers and Short Notes

V. Navrátil, P. Lukáč, V. I. Startsev, and V. P. Soldatov
The Concentration Dependence of the Critical Resolved Shear Stress of Cd–Zn Single Crystals at Very Low Temperatures
(a) Vol. **63**, No. 1, p. K73–K75 (1981).
See also *10.1*, p. *96*; *12.1*, p. *134*; 21.1, p. 352.

G. Dlubek, O. Brümmer, and B. Möser
A Positron Annihilation Study of Quenched-in Defects in Diluted Al–Mg Alloys
(a) Vol. **63**, No. 1, p. K115–K118 (1981).
See also *10.1*, p. *96*; 21.3, p. 378.

A. P. Zhernov and T. A. Mamedov
Theory of the Static Atomic Displacements in Ordinary Metals with Isoelectric Impurity Atoms
(b) Vol. **103**, No. 2, p. 477–488 (1981).
See also *10.1*, p. *97*; *13*, p. *142*.

H. V. Sharma and I. Singh
Phase Transition in the Extended Falicov-Kimball Model with Finite f-Bandwidth
(b) Vol. **103**, No. 2, p. 629–635 (1981).
See also *13.1*, p. *152*.

A. M. Oleś and K. A. Chao
Hund's-Rule Effect in Degenerate Anderson Model
(b) Vol. **103**, No. 2, p. 769–774 (1981).
See also *13.4*, p. *170*.

V. Paidar
Local Density of Electronic States at the Core of the Screw Dislocation in B.C.C. Lattice
(b) Vol. **103**, No. 2, p. K93–K97 (1981).
See also *10.1*, p. *97*; *13.4*, p. *171*.

O. Simmich
The Influence of the Gradient Free Energy on the Movement of a Flat Interface
(a) Vol. **63**, No. 2, p. 539–545 (1981).
See also *1.3*, p. *22*; *10.1*, p. *97*.

E. Bonetti, E. Evangelista, and P. Gondi
Experimental Observations on Dislocation Link Distribution and Supersaturation during Creep of Al
(a) Vol. **63**, No. 2, p. 645–653 (1981).
See also *10.1*, p. *97*; 12.1, p. 134.

P. Lobotka and I. Vávra
TEM in-Situ Observation of Electromigration Damage in Al–Cu Strips (II): Superimposed AC and DC Stressing
(a) Vol. **63**, No. 2, p. 655–661 (1981).
See also 1.5, p. 34; *9*, p. *89*; 10.1, p. 97; *14.1*, p. *199*; 21.1, p. 354.

K. S. Srivastava and O. K. Harsh
Double Plasmon High Energy Satellites in the X-Ray Emission Spectra of Metals
(b) Vol. **104**, No. 1, p. K35–K38 (1981).
See also *13.5.2*, p. *194*; *20*, p. *296*; 21.1, p. 355; 21.2, p. 376.

H. Hermann
A Non-Central Potential for Modelling the Structure of Metallic Glasses
(b) Vol. **104**, No. 1, p. K39–K43 (1981).
See also *2*, p. *42*.

V. V. Nemoshkalenko, V. N. Antonov, Vl. N. Antonov, and W. John
Relativistic APW Calculation of Soft X-Ray Emission Spectra in Ta
(b) Vol. **104**, No. 1, p. K45–K48 (1981).
See also *13.1*, p. *154*; *20*, p. *297*.

S. B. Shrivastava, N. Shrivastava, and V. K. Ojha
Positron Annihilation Rates and the Enhancement Factor in Metals
(b) Vol. **104**, No. 1, p. K61–K64 (1981).
Erratum
(b) Vol. **105**, No. 1, p. K71 (1981).
See also *13.2*, p. *163*.

A. J. Pedraza and D. Fainstein-Pedraza
Analytic Approximations for Describing the Irradiation Growth of Zirconium-Base Alloys (II)
(a) Vol. **64**, No. 1, p. 53–63 (1981).
See also *10.1*, p. *98*; *11*, p. *122*.

V. Valvoda
X-Ray Diffraction Study of Debye Temperature and Charge Distribution in Tantalum Monocarbide
(a) Vol. **64**, No. 1, p. 133–142 (1981).
See also *1.3*, p. *23*; 6, p. 71.

H.-U. Habermeier
Structure and Superconducting Properties of Nb_3Ge Prepared in a UHV System
(a) Vol. **64**, No. 1, p. 195–206 (1981).
See also *1.3*, p. *23*; 10.1, p. 98; *14.2*, p. *205*.

D. Hildebrandt, R. Manns, and H. Düsterhöft
The Mean Kinetic Energy of Sputtered Atoms as a Function of the Angle of Emission
(a) Vol. **64**, No. 1, p. K27–K30 (1981).
See also *11*, p. *123*.

W. Zahn
Inductance of Small Window Type Tunnel Junctions
(a) Vol. **64**, No. 1, p. K43–K47 (1981).
See also *14.2*, p. *206*; 21.4, p. 381.

A. M. Brodskii and L. I. Daihkin
On the Theory of Electroreflectance of Metals
(b) Vol. **104**, No. 2, p. 415–426 (1981).
See also *14.3.4*, p. *233*; *20.1*, p. *306*.

N. C. Halder
A Comparison between the Theoretical and Experimental Electronic Density of States of Liquid Metals
(b) Vol. **104**, No. 2, p. 469–478 (1981).
See also *2*, p. *43*; *13*, p. *144*; 21.1, p. 356.

R. Taranko, E. Taranko, and K. I. Wysokiński
Effect of Electron–Phonon Interaction on the Optical Conductivity of Alloys
(b) Vol. **104**, No. 2, p. 507–512 (1981).
See also 6, p. 71; 13, p. 144; *20.1*, p. *306*.

M. Rösler and W. Brauer
Theory of Secondary Electron Emission (II): Application to Aluminum
(b) Vol. **104**, No. 2, p. 575–587 (1981).
See also 13.5.2, p. 194; *17*, p. *256*.

P. Mascher, L. Breitenhuber, and W. Puff
Vacancy Formation Enthalpy in Cadmium by Positron Lifetime Measurements
(b) Vol. **104**, No. 2, p. 601–605 (1981).
See also *10.1*, p. *99*.

H. P. Aubauer and P. L. Rossiter
Effects of Microstructure Scale on the Temperature Coefficient of Electrical Resistivity
(b) Vol. **104**, No. 2, p. 685–689 (1981).
See also *14.1*, p. *200*.

Farid A. Khwaja and M. S. K. Razmi
A Statistical-Pseudopotential Approach to the Calculation of Atomic Short-Range Order in Solid Solutions
(b) Vol. **104**, No. 2, p. 727–735 (1981).
See also *1.3*, p. *23*; *13*, p. *144*; 21.1, p. 356; 21.1.1, p. 370.

H. C. Freyhardt, J. Goebbels, K. Lüders, and J. Reichelt
Erratum to: NMR Investigations on ^{9}Be in Amorphous Be–Nb–Zr Alloys
(b) Vol. **104**, No. 2, p. K171 (1981).
See also 2, p. 44; 14.2, p. 206; *19*, p. *288*.

W. Siebke
Considerations on the Bulk Modulus of Pure Metals
(a) Vol. **64**, No. 2, p. 577–583 (1981).
See also *4*, p. *58*; *12.1*, p. *135*.

D. Lepski and P. Burck
Percipitate Structure Analysis by Diffraction Contrast Image Matching for a Metastable Semicoherent Carbide Phase in Molybdenum (I): Generating Inelastic Strain
(a) Vol. **64**, No. 2, p. 625–636 (1981).
See also *1.2*, p. *16*; *1.4*, p. *29*; 10.1, p. 99; 12.1, p. 135.

V. V. Kirsanov, Yu. S. Pyatiletov, and O. G. Tyupkina
Irradiation Creep Due to Dislocation Glide
(a) Vol. **64**, No. 2, p. 735–740 (1981).
See also 10.1, p. 99; *11*, p. *124*; *12.1*, p. *135*.

P. D. Pathak and R. J. Desai
Thermal Properties of Some H.C.P. Metals (II): Debye-Waller Factors of Zinc, Cadmium and Magnesium
(a) Vol. **64**, No. 2, p. 741–745 (1981).
See also 6, p. 73; *8*, p. *85*; 21.1, p. 357; 21.3, p. 378.

J. Sosnowski
The Superconducting Properties of $Nb_{75}Ge_{25-x}Al_x$ A15-Type Alloys Annealed at 800 °C for One Week
(a) Vol. **64**, No. 2, p. 771–775 (1981).
See also *14.2*, p. *206*.

V. L. Arbuzov, S. M. Klotsman, and I. E. Podchinenov
Pseudocorrelated Recombination of Interstitials during State II of Annealing of Irradiated Metals
(a) Vol. **64**, No. 2, p. K111–K114 (1981).
See also *10.1*, p. *99*; *11*, p. *124*.

B. B. Sahay, P. L. Srivastava, and N. R. Mitra
Temperature Dependence of Electrical Resistivity of Zn, Cd, Hg, Al, In, Tl, and Pb by Nonlocal Harmonic Model Potential
(b) Vol. **105**, No. 1, p. 93–98 (1981).
See also 2, p. 45; *14.1*, p. *200*; 21.1, p. 358.

K. Masuda
Lattice Vibrations around a Vacancy in B.C.C. Transition Metals
(b) Vol. **105**, No. 1, p. 107–113 (1981).
See also *6*, p. *73*; *10.1*, p. *100*; 21.1.1, p. 371.

E. P. Domashevskaya and V. A. Terekhov
d–s,p Resonance and Electronic Structure of Compounds, Alloys and Solid Solutions
(b) Vol. **105**, No. 1, p. 121–127 (1981).
See also *13.1*, p. *155*; 17, p. 257; 20, p. 297; 21.1, p. 358.

A. Kiejna
Simple Non-Local Calculation of Jellium Metal Surface Properties
(b) Vol. **105**, No. 1, p. 147–153 (1981).
See also *13.3*, p. *165*.

J. Jarosz
The Application of Kramers-Kronig Relations to the Analysis of the Resonance Lines of Metallic Samples in ESR Spectroscopy
(b) Vol. **105**, No. 1, p. 155–160 (1981).
See also *19*, p. *289*; 21.4, p. 381.

N. A. Stolwijk
Correlation Factors for Diffusion in Binary Alloys with Short-Range Order
(b) Vol. **105**, No. 1, p. 223–232 (1981).
See also *9*, p. *90*; 10.1, p. 100.

K. ELK
On the Electrical Conductivity of the Periodic Anderson Model
(b) Vol. **105**, No. 2, p. 507–513 (1981).
Erratum
(b) Vol. **107**, No. 1, p. 387 (1981).
See also *13*, p. *145*; *14.1*, p. *201*.

L. A. GEVORGIAN and N. A. KORKHMAZIAN
On the Theory of Radiation of Non-Relativistic Electrons Grazing into Metal Targets
(b) Vol. **105**, No. 2, p. 623–628 (1981).
See also *11*, p. *126*; 13, p. 145.

S. P. MHASKAR and P. K. NAIK
Effect of Surface Relaxation on Point Defects in the Vicinity of W (100) Surfaces
(b) Vol. **105**, No. 2, p. 685–689 (1981).
See also *1.6*, p. *38*; *10.1*, p. *100*.

P. F. TUA and G. D. MAHAN
Variational Hartree-Fock Calculations of the Ground State Energy of Metallic Hydrogen
(b) Vol. **105**, No. 2, p. 769–775 (1981).
See also *13*, p. *145*.

E. KOLLEY and W. KOLLEY
Functional Integral Approach to the Electron–Magnon Interaction in Itinerant Ferromagnetism
(b) Vol. **105**, No. 2, p. K85–K89 (1981).
See also *13.5.4*, p. *197*; 18.2, p. 271.

D. SCHILLING, K. LÄUGER, B. STRÖBEL, and H. E. BÖMMEL
Angular Dependence of the Nuclear Quadrupole Relaxation Rate in Tantalum Due to Diffusing Hydrogen, Studied by Nuclear Acoustic Resonance (NAR)
(b) Vol. **105**, No. 2, p. K129–K133 (1981).
See also *9*, p. *90*; *19*, p. *290*.

Cz. MARUCHA and J. RAFAŁOWICZ
The Aluminium Thermal Conductivity and Deviation from Matthiessen's Rule
(a) Vol. **65**, No. 2, p. 523–527 (1981).
See also *8*, p. *85*.

H. S. CHEN, Y. WASEDA, and K. T. AUST
Structure of $Pt_{75}P_{25}$ Glass
(a) Vol. **65**, No. 2, p. 695–700 (1981).
See also *2*, p. *47*.

I. VÁVRA and P. LOBOTKA
TEM In-Situ Observation of Electromigration in Al Stripes with Quasi-Bamboo Structure
(a) Vol. **65**, No. 2, p. K107–K108 (1981).
See also *9* p. *90*; *10.1*, p. *101*.

H. WAGNER
Influence of Lattice Defects on the Superconducting Behaviour of Implanted PdH_x Thin Film Layers
(a) Vol. **65**, No. 2, p. K169–K174 (1981).
See also 11, p. 126; *14.2*, p. *206*.

K. Tsukuma, M. Shimada, M. Koizumi, and H. Kawamura
Superconductive Compound $Nb_5Se_2S_2$ with Ti_5Te_4 Structure
(a) Vol. **65**, No. 2, p. K179–K182 (1981).
See also *3*, p. *54*; *14.2*, p. *206*.

M. Matlak and J. Zieliński
On the Electrical Conductivity and the Density of States for the Narrow-Band Periodic Anderson Model in the Symmetric Case
(b) Vol. **106**, No. 1, p. 73–77 (1981).
See also *13.1*, p. *157*; *14.1*, p. *201*.

L. B. Dubovskii
Density of States in Multilayer Sandwiches
(b) Vol. **106**, No. 1, p. 315–325 (1981).
See also *13*, p. *146*; 14.1, p. 201; 14.2, p. 206; 21.7, p. 388.

R. Ramji Rao and A. Rajput
Lattice Dynamics and Some Related Thermal Properties of Body-Centred Cubic Metals
(b) Vol. **106**, No. 1, p. 393–406 (1981).
See also *6*, p. *77*; 12.1, p. 136; 21.1.1, p. 372; 21.2, p. 376.

J. S. Brown and A. H. M. Lopez-Escobar
A Muffin-Tin Calculation of the Electronic d-Bandwidth of F.C.C. Palladium
(b) Vol. **106**, No. 1, p. K33–K38 (1981).
See also *13.1*, p. *158*.

V. P. Skripov, V. P. Koverda, and V. N. Skokov
Size Effect on Melting of Small Particles
(a) Vol. **66**, No. 1, p. 109–118 (1981).
See also *1.5*, p. *35*; 1.6, p. 38; 2, p. 47; 8, p. 86.

E. Stracke and Ch. Herzig
Formation of Slowly Diffusing Gold Defects in Thermomigration Experiments of Gold in Lead
(a) Vol. **66**, No. 1, p. 189–197 (1981).
See also *9*, p. *91*; *10.1*, p. *101*.

A. Altshuler, S. N. Bezryadin, Yu. Kh. Vekilov, and V. D. Verner
On the Pseudopotential Calculation from First Principles
(b) Vol. **106**, No. 2, p. 447–459 (1981).
See also *13*, p. *146*; 21.1, p. 361; 21.2, p. 376; 22.1, p. 402; 22.1.1, p. 406; 22.1.2, p. 417.

M. Glugla, H. G. Severin, and G. Sicking
Low-Temperature Heat Capacities of Pd–Al Alloys at Low Al Contents
(b) Vol. **106**, No. 2, p. 505–509 (1981).
See also *8*, p. *86*; 14.2, p. 207.

J. Sołtys
X-Ray Diffraction Research of the Order–Disorder Transitions in the Ternary Heusler Alloys B_2MnAl (B = Cu, Ni, Co, Pd, Pt)
(a) Vol. **66**, No. 2, p. 485–491 (1981).
See also *1.2*, p. *18*; *1.3*, p. *24*; 21.1, p. 361; 21.6, p. 385.

B. Gueffroy and H. Löffler
On the Reversion Behaviour of an Al-Zn(4.5 at%)-Mg(2.0 at%) Alloy
(a) Vol. **66**, No. 2, p. 585–590 (1981).
See also *10.1*, p. *102*; 21.1, p. 362; 21.3, p. 378.

J. Beauvillain
Quantitative Approach to Electron Microscopic Contrasts
(a) Vol. **66**, No. 2, p. 675–682 (1981).
See also *1.1*, p. *12*; *1.4*, p. *30*.

A. Planes, J. L. Macqueron, M. Morin, and G. Guenin
Study of Martensitic Transformation of Cu–Zn–Al Alloy by Coupled Enthalpy and Acoustic Emission Measurements
(a) Vol. **66**, No. 2, p. 717–724 (1981).
See also *1.2*, p. *18*; *7*, p. *83*; 21.1, p. 362.

S. Ikeno
Dislocation Behaviour in Niobium Single Crystals with [110] and [100] Tensile Axis at Low Temperatures
(a) Vol. **66**, No. 2, p. K97–K100 (1981).
See also *10.1*, p. *103*.

S. P. Mhaskar and P. K. Naik
Helium Desorption from a Tungsten Single Crystal
(b) Vol. **107**, No. 1, p. 99–103 (1981).
See also *1.6*, p. *39*; *10.1*, p. *102*.

S. C. Jain and C. M. Kachhava
Pressure Dependence of Superconductivity in Non-Transition Metals
(b) Vol. **107**, No. 1, p. 139–145 (1981).
See also *14.2*, p. *207*.

Ch. Högfors, J. Kubát, and M. Rigdahl
A Cooperative Model of Stress Relaxation in Solids
(b) Vol. **107**, No. 1, p. 147–155 (1981).
See also *12.1*, p. *136*; *12.2*, p. *140*; 13.5, p. 187; 22.5.2, p. 467; 22.9, p. 511.

D. Ploumbidis and E. Böss
Nuclear Magnetic Resonance in Liquid In–Ga Alloys
(b) Vol. **107**, No. 1, p. 225–231 (1981).
See also 2, p. 48; *19*, p. *292*.

R. N. Khanna and R. P. S. Rathore
Shell Models for B.C.C. Transition Metals
(b) Vol. **107**, No. 1, p. 267–274 (1981).
See also *4*, p. *60*.

B. M. Bolotov, A. I. Polyakov, and S. P. Senshin
On the NMR Line Shape in Metals
(b) Vol. **107**, No. 1, p. K5–K7 (1981).
See also *19*, p. *292*; 21.1, p. 363.

E. B. LEIKO and A. LUFT
A Combined Study of the Dynamical Properties of Dislocation in Molybdenum Single Crystals by Etch Pitting and Interference Contrast Techniques
(a) Vol. 67, No. 2, p. 435–442 (1981).
See also *10.1*, p. *103*.

O. NISHIKAWA, M. NACHI, K. KURIHARA, Y. TSUNASHIMA, and Y. HARA
Atom-Probe Study of Ga–Mo and Sn–Ir Interfaces
(a) Vol. 67, No. 2, p. 461–468 (1981).
See also *1.3*, p. *26*; *1.6*, p. *39*.

F. DE LIMA and W. BENOIT
Internal Friction Study of Dislocation Mobility in Deformed Niobium
(a) Vol. 67, No. 2, p. 565–572 (1981).
See also *10.1*, p. *103*; *12.1*, p. *137*.

R. VETTER, R. H. J. FASTENAU, and M. I. BASKES
Atomistic Computer Calculation of the Dilatation Caused by a $\frac{1}{2}$ ⟨111⟩ {110} Edge Dislocation in Molybdenum
(a) Vol. 67, No. 2, p. 585–589 (1981).
See also *10.1*, p. *103*.

H. M. LEDBETTER, J. F. LABRECQUE, and J. L. DAHNKE
Propagation Errors in Elastic-Constant Inversion
(a) Vol. 67, No. 2, p. 643–647 (1981).
See also *12.1*, p. *137*; 21.1, p. 364; 21.6, p. 386.

E. A. ABOU-SAIF, M. ABD-RABO, M. M. EZZO, A. A. GAMIL, and A. A. MOHAMED
Effect of Heat Treatment on the Structure and Mechanical Properties of Aluminium–Tin Alloys
(a) Vol. 67, No. 2, p. K101–K105 (1981).
See also *1.3*, p. *26*; *12.1*, p. *137*.

R. BONNET, G. MARCON, and M. DUPEUX
Comment on "An Approach to the Anisotropic Elastic Fields Related to a Semi-Coherent or Coherent Faceted Rod"
(a) Vol. 67, No. 2, p. K137–K142 (1981).
See also *10.1*, p. *103*; *12.1*, p. *137*.

A. I. LANDA, V. E. PANIN, and M. F. ZHOROVKOV
The Pseudopotential Method and Thermodynamics of Eutectic Systems
(b) Vol. 108, No. 1, p. 113–122 (1981).
See also *1.2*, p. *20*; *2*, p. *50*; 8, p. 88; 13, p. 149; 21.1, p. 365.

P. K. ACHARYYA, A. K. TEWARI, S. K. ROY, and D. L. BHATTACHARYA
Effect of Anharmonicity on the Second-Order Doppler Shift for a Vacancy-Trapped Mössbauer Impurity Atom in Metallic Solid Solution
(b) Vol. 108, No. 1, p. 211–220 (1981).
See also *5*, p. *65*; 10.1, p. 103.

A. N. Gusatinskii, M. A. Blokhin, P. Shumichen, M. A. Bunin, and S. S. Mikhailova
X-Ray L_1Spectra and XPS Investigation of the Electronic Structure of Indium Compounds
(b) Vol. 108, No. 2, p. 405–410 (1981).
See also *13.1*, p. *161*; *20*, p. *299*; 22, p. 400; 22.2, p. 426; 22.3, p. 442.

M. Surma, S. Maciejewski, R. Kubiak, and M. Wołcyrz
Electronic Structure of the $Cd_{65}Hg_{35}$ Alloy and Its Temperature Dependence near the Phase Transition
(b) Vol. 108, No. 2, p. 475–482 (1981).
See also 1.2, p. 21; *13.1*, p. *161*.

K. S. Srivastava, M. Husain, P. Gupta, A. K. Srivastava, V. Kumar, K. Sinha, and S. Singh
Shift in the Binding Energy of the Inner Electrons Due to Chemical Combination
(b) Vol. 108, No. 2, p. 575–579 (1981).
See also *4*, p. *61*; *20*, p. *300*.

A. M. Brodskii, L. I. Daikhin, and M. I. Urbakh
On the Polarization Dependence of Photoemission from Metals
(b) Vol. 108, No. 2, p. 653–662 (1981).
See also 13.5.2, p. 195; *17*, p. *258*.

K. Kranjc, D. Kunstelj, and V. Marinković
Parallel Moiré Effect in Electron Micrographs of Crystal Sandwiches (II): Comparison of Theory and Experiment
(a) Vol. 68, No. 2, p. 363–372 (1981).
See also *1.4*, p. *32*; 1.5, p. 37; 21.1, p. 366; 21.7, p. 389; 22.4.3, p. 455.

C. R. Pichard, C. R. Tellier, L. Ouarbya, and A. J. Tosser
Perturbations in Transport Properties of Thin Metal Films Due to Energy-Dependent Relaxation Time
(a) Vol. 68, No. 2, p. 477–482 (1981).
See also *14.1*, p. *203*; *15*, p. *250*.

R. Ramlau and H. Löffler
Guinier-Preston Zones in an Al-3 at% Zn Alloy
(a) Vol. 68, No. 2, p. 531–536 (1981).
See also *1.3*, p. *27*; *10.1*, p. *104*; 21.1, p. 367.

M. Tokuda, J. Kratochvíl, and Y. Ohashi
Mechanism of Induced Plastic Anisotropy on Polycristalline Metals
(a) Vol. 68, No. 2, p. 629–636 (1981).
See also *1.4*, p. *32*; *10.1*, p. *105*; 12.1, p. 137.

F. C. Lovey, R. Rapacioli, and M. Chandrasekaran
Precipitates with an Incommensurate Structure in β-Cu–Zn–Al
(a) Vol. 68, No. 2, p. K105–K108 (1981).
See also *1.3*, p. *27*; 21.1, p. 367.

Y. Waseda and K. T. Jacob
Refinement of a Correlation between Isothermal Compressibility and Surface Tension of Liquid Metals
(a) Vol. 68, No. 2, p. K117–K122 (1981).
See also *2*, p. *51*; *12.1*, p. *137*.

21.1 Transition Metals (Group Sc to Zn) and Their Alloys

Review Article

H. R. KHAN and K. LÜDERS
Nuclear Magnetic Resonance on Amorphous Alloys
(b) Vol. **108**, No. 1, p. 9–18 (1981).
See also *2*, p. *40*; 14.2, p. 205; *19*, p. *286*; 21, p. 336; 21.3, p. 378.

Original Papers and Short Notes

C. B. CUDEN and R. MOTA
Raman Scattering of Light in Antiferromagnetic Chromium
(b) Vol. **103**, No. 1, p. 51–62 (1981).
See also *6*, p. *66*; 18.4, p. 281; *20.1*, p. *301*.

T. K. HATWAR, M. N. GHATIKAR, B. D. PADALIA, S. K. MALIK, and R. VIJAYARAGHAVAN
X-Ray Absorption Spectroscopic Study of Rare-Earth–Manganese Intermetallic Compounds
(b) Vol. **103**, No. 1, p. 159–166 (1981).
See also *20*, p. *296*; 21.4, p. 380.

I. A. GARIFULLIN, G. G. KHALIULLIN, E. G. KHARAKHASHYAN, T. O. FARZAN, and E. F. KUKOVITZKII
Effect of Uniaxial Stresses on the EPR Spectra of the Er^{3+} Ions in Cu
(b) Vol. **103**, No. 1, p. K37–K39 (1981).
See also 12.1, p. 134; *19*, p. *286*.

RAMJIT SINGH, S. PRAKASH, and J. SINGH
Dynamical Spin Susceptibility of 3d, 4d, and 5d Transition Metals
(b) Vol. **103**, No. 1, p. K65–K69 (1981).
See also *18.1*, p. *265*; 21.6, p. 384.

D. KISHORE, UMENDRA, and K. C. LAL
Effect of Pressure on Second Order Doppler Shift
(b) Vol. **103**, No. 1, p. K91–K92 (1981).
See also *5*, p. *62*; *12.1*, p. *134*; 21.6, p. 384.

S. SPOONER, H. R. CHILD, F. H. HSU, E. R. VANCE, and J. H. SMITH
Small-Angle Neutron Scattering and Positron Annihilation in Mn–Cu Alloys
(a) Vol. **63**, No. 1, p. 31–34 (1981).
See also *1.3*, p. *22*.

J. SOŁTYS and R. KOZUBSKI
A Simple Model of the Order–Disorder Phase Transitions in Ternary Alloys and Its Application to Several Selected Heusler Alloys
(a) Vol. **63**, No. 1, p. 35–44 (1981).
See also *1.3*, p. *22*; 21, p. 336.

N. F. Fedorov and C. Heiden
Some Observations on the Influence of Oxygen on the Resistive Behavior of Superconducting Vanadium Foils
(a) Vol. **63**, No. 1, p. K83 – K86 (1981).
See also *14.2*, p. *205*.

S. A. Dyomin, V. V. Nemoshkalenko, O. N. Razumov, A. I. Ustinov, and K. V. Tchuistov
Mössbauer Study of Magnetic and Structure Transformations in Manganese–Copper–Iron Alloy
(a) Vol. **63**, No. 1, p. K93 – K96 (1981).
See also 1.2, p. 14; *5*, p. *62*; 18.4, p. 281; 21.1.1, p. 368.

I. Bakonyi, L. Takács, and K. Tompa
Dipole–Dipole Interaction and Short-Range Order in Amorphous Ni–P, Ni–Cu–P, and Ni–P–B Alloys
(b) Vol. **103**, No. 2, p. 489 – 497 (1981).
See also *2*, p. *41*; *19*, p. *287*.

L. M. Sandratskii, R. F. Egorov, and A. A. Berdyshev
Energy Band Structure and Electronic Properties of NiAs Type Compounds (I): Ferromagnetic Manganese Arsenide
(b) Vol. **103**, No. 2, p. 511 – 518 (1981).
See also *13.1*, p. *152*; 18.2, p. 267.

S. E. Kulkova, V. E. Egorushkin, I. N. Anokhina, and V. P. Fadin
The Change of Electronic Structure and Optical Properties in FeCo at the Order–Disorder Transition
(b) Vol. **103**, No. 2, p. 653 – 658 (1981).
See also *13.1*, p. *152*; *20.1*, p. *303*; 21.1.1, p. 368.

E. Kolley, W. Kolley, and A. Beckmann
Transition Temperature of Superconducting Alloys with Random Transfer Integrals
(b) Vol. **103**, No. 2, p. 775 – 781 (1981).
See also 13, p. 143; *14.2*, p. *205*.

J. Sołtys
Order–Disorder Phase Transitions in Ternary Alloys $Cu_{3-x}Mn_xAl$
(a) Vol. **63**, No. 2, p. 401 – 406 (1981).
See also *1.2*, p. *15*; 1.3, p. 22.

E. Wieser, J. Henke, M. Müller, and C. Cruz
Connection between Structure and Magnetic Properties of a Magnetically Semi-Permanent Fe–Ni–Al–Ti Alloy
(a) Vol. **63**, No. 2, p. 487 – 494 (1981).
See also *1.3*, p. *22*; 5, p. 62; *18.2*, p. *268*; 21.1.1, p. 368.

J. D. Stephenson
Synchrotron Radiation Tropography of (100)Co–8 wt% Fe Binary Alloy
(a) Vol. **63**, No. 2, p. 557 – 568 (1981).
See also *18.2*, p. *268*.

N. N. SIROTA and T. E. ZHABKO
X-Ray Study of the Anisotropy of Thermal Properties in Titanium
(a) Vol. **63**, No. 2, p. K211–K215 (1981).
See also 6, p. 69; *8*, p. *84*.

A. A. TUKHFATULLIN, V. E. PANIN, V. I. SIMAKOV, V. S. DEMIDENKO, and A. P. KALYANOV
Electron Structure and the Nature of Co_3Ti Phase Formation
(b) Vol. **104**, No. 2, p. 69–74 (1981).
See also 1.2, p. 15; *13.1*, p. *153*.

K. S. SRIVASTAVA and O. K. HARSH
Double Plasmon High Energy Satellites in the X-Ray Emission Spectra of Metals
(b) Vol. **104**, No. 1, p. K35–K38 (1981).
See also *13.5.2*, p. *194*; *20*, p. *296*; 21, p. 339; 21.2, p. 376.

K. BERNDT, C. WENZEL, T. DÖRRE, T. HERZEL, and A. NEUMANN
Electron-Energy Loss Spectra of Chromium in the Reflection Mode
(b) Vol. **104**, No. 1, p. K49–K52 (1981).
See also *13.5.2*, p. *194*.

J. M. ROLAND, X. QUILLARD and A. MOREAU
Study of the α-Phase of Cu–Al Alloys. Investigations of the Near-Equilibrium Disordered State
(a) Vol. **64**, No. 1, p. 45–51 (1981).
See also *1.3*, p. *22*; 14.1, p. 199.

Y. K. WU, J. Z. LIANG, and K. H. KUO
Micro Inversion Domains Formed during the Crystallization of the Amorphous Alloy $Fe_{40}Ni_{40}P_{14}B_6$
(a) Vol. **64**, No. 1, p. 113–120 (1981).
See also *1.4*, p. *28*; *2*, p. *42*; 21.1.1, p. 369.

H. J. BLYTHE, F. WALZ, and H. KRONMÜLLER
Magnetic Relaxation in Electron-Irradiated Ordered Ni_3Fe
(a) Vol. **64**, No. 1, p. 169–176 (1981).
See also 10.1, p. 98; *11*, p. *123*; *18.2*, p. *269*; 21.1.1, p. 369.

W. GUST, M. B. HINTZ, A. LODDING, H. ODELIUS, and B. PREDEL
Impurity Diffusion of Al in Ni Single Crystals Studied by Secondary Ion Mass Spectrometry (SIMS)
(a) Vol. **64**, No. 1, p. 187–194 (1981).
See also *9*, p. *89*.

F. X. KAYSER and C. STASSIS
The Elastic Constants of Ni_3Al at 0 and 23.5 °C
(a) Vol. **64**, No. 1, p. 335–342 (1981).
See also *12.1*, p. *134*.

W. BAŻELA, A. SZYTUŁA, J. TODOROVIĆ, and A. ZIĘBA
Crystal and Magnetic Structure of the $NiMnGe_{1-n}Si_n$ System
(a) Vol. **64**, No. 1, p. 367–378 (1981).
See also 1.3, p. 23; *18.2*, p. *269*; *18.4*, p. *281*.

O. Gebhardt and W. Richter
Self-Biased Narrow Magnetoresistive Elements
(a) Vol. **64**, No. 2, p. 513–517 (1981).
See also *14.1*, p. *200*; *18.2.1*, p. *276*; 21.1.1, p. 370.

T. Reichert, R. Meier-Hirmer, and H. Küpfer
Annealing Behaviour of the Critical Current Density and the Superconducting Parameters T_c, $\varkappa$, and H_{c2} in a Neutron-Irradiated V_3Si Single Crystal
(a) Vol. **64**, No. 2, p. 585–592 (1981).
See also **11**, p. **124**; *14.2*, p. *206*.

V. E. Rode, S. A. Finkelberg, B. Wurl, and A. I. Lyalin
Magnetic Structure of Cr–Fe Alloys in the Transition Range from Ferromagnetism to Antiferromagnetism
(a) Vol. **64** No. 2 p. 603–610 (1981).
See also *18.2*, p. *270*; *18.4*, p. *282*; 21.1.1 p. 370.

H.-J. Burmeister and C. Holste
Change of Activation Area during Cyclic Deformation (II): Quantitative Interpretation with a Model of Heterogeneous Plastic Deformation
(a) Vol. **64**, No. 2, p. 611–624 (1981).
See also *10.1*, p. *99*.

P. C. M. Gubbens, A. M. van der Kraan, and K. H. J. Buschow
^{161}Dy Mössbauer Effect and Magnetic Properties of Some Amorphous and Crystalline Dy–Co Alloys
(a) Vol. **64**, No. 2, p. 657–663 (1981).
See also **2**, p. **44**; *5*, p. *63*; *18*, p. *261*; 21.4, p. 381.

P. D. Pathak and R. J. Desai
Thermal Properties of Some H.C.P. Metals (II): Debye-Waller Factors of Zinc, Cadmium and Magnesium
(a) Vol. **64**, No. 2, p. 741–745 (1981).
See also **6**, p. **73**; *8*, p. *85*; 21, p. 341; 21.3, p. 378.

I. Yu. Bezotosnyi and I. N. Nikolaev
Effect of Pressure on the Hyperfine Magnetic Fields at ^{57}Fe Nuclei in $Pd_{0.95}Fe_{0.05}$ and $Fe_{0.65}Ni_{0.08}Mn_{0.27}$ Alloys
(a) Vol. **64**, No. 2, p. K129–K131 (1981).
See also 12.1, p. 135; *18.2*, p. *270*; *18.4*, p. *282*; 21.1.1, p. 370.

A. Vehanen, P. Hautojärvi, and V. S. Mikhalenkov
Vacancy Clusters in Deformed CuAs Alloy
(a) Vol. **64**, No. 2, p. K145–K148 (1981).
See also *10.1*, p. *99*; 12.1, p. 135.

A. R. McGurn and D. M. Esterling
Liquid Metal Pair-Correlation Functions from Empirical Pair Potentials
(b) Vol. **105**, No. 1, p. 47–53 (1981).
See also *2*, p. *45*; *4*, p. *58*; 21.1.1, p. 371; 21.6, p. 385.

J. MWANJE and E. M. AYIERA
Temperature-Dependent Resistivity of Ni-Base Alloys
(a) Vol. **65**, No. 1, p. K7–K8 (1981).
See also *14.1*, p. *200*; 18.1, p. 266; 18.2, p. 271.

K. UHLMANN and F. STORBECK
Relative Determination of L_3 Ionisation Cross Sections of Chromium, Iron, and Nickel
(a) Vol. **65**, No. 1, p. K37–K41 (1981).
See also *17*, p. *257*; 21.1.1, p. 371.

S. A. MAHMOUD, M. A. SEMARY, and Z. FARID
Internal Friction Spectra of Some Al–Cu Alloys
(a) Vol. **65**, No. 1, p. K63–K66 (1981).
See also *10.1*, p. *100*; *12.1*, p. *135*; 21, p. 342.

A. K. AGYEI
Theory of Second-Harmonic Generation by Soft Ultrasonic Shear Waves in Transforming A-15 Compounds
(b) Vol. **105**, No. 2, p. 469–476 (1981).
See also 6, p. 74; *7*, p. *82*.

K. N. KHANNA
Wigner-Seitz Radius Dependent Model Potential
(b) Vol. **105**, No. 2, p. 485–490 (1981).
See also 1, p. 9; 4, p. 59; *13*, p. *145*; 14.1, p. 200; 21, p. 342; 21.2, p. 376; 21.3, p. 378.

P. VALENTA, K. MAIER, H. KRONMÜLLER, and K. FREITAG
Self-Diffusion of Phosphorus in the Amorphous Alloy $Fe_{40}Ni_{40}P_{14}B_6$
(b) Vol. **105**, No. 2, p. 537–542 (1981).
See also 2, p. 46; *9*, p. *90*; 21.1.1, p. 371.

K. MASUDA
On the Vibrational Spectra for Atoms at B.C.C. Transition Metal Surfaces: Effects of Surface Relaxation and Reconstruction
(b) Vol. **105**, No. 2, p. K119–K124 (1981).
See also *1.6*, p. *38*; *6*, p. *75*.

G. VAN TENDELOO and S. AMELINCKX
Electron Diffraction and High Resolution Electron Microscopy Study of Ordering in the Gold–Manganese System (II): Square Island Structures Derived from the Au_4Mn Structure ($Au_{4-x}Mn_1$)
(a) Vol. **65**, No. 2, p. 431–446 (1981).
See also *1.1*, p. *11*; 1.3, p. 24; *1.4*, p. *30*; 21.6, p. 385.

A. A. PREDVODITELEV, G. I. NICHUGOVSKII, and V. I. VESELOV
Simulation of Dislocation Motion through a Dislocation Forest
(a) Vol. **65**, No. 2, p. 469–478 (1981).
See also *10.1*, p. *100*; *10.2*. p. *111*; 22.5.2, p. 465.

H. C. EATON
Grain Boundary Dislocation Observations Using the Field Ion Microscope
(a) Vol. **65**, No. 2, p. 497–502 (1981).
See also *10.1*, p. *101*; *17.1*, p. *259*.

O. Bostanjoglo and Th. Rosin
Resonance Oscillations of Bloch Lines in Permalloy Films
(a) Vol. **66**, No. 1, p. K5–K10 (1981).
See also *18.2.1*, p. *277*; 21.1.1, p. 373.

A. H. Abou El Ela and S. Mahmoud
Some Optical Properties of Thin Chromium Films
(a) Vol. **66**, No. 1, p. K11–K16 (1981).
See also *20.1*, p. *314*.

M. R. Nagy, M. S. Sakr, and R. Kamel
Effect of Transformation on the Steady Creep Characteristics of Zn–Al Alloys
(a) Vol. **66**, No. 1, p. K25–K29 (1981).
See also *12.1*, p. *136*.

T. Maeda, T. Soumura, and S. Saito
Comparison of the Wall-Pinning Model Theory with Magnetic Damping Experiments
(a) Vol. **66**, No. 1, p. K51–K54 (1981).
See also *18.2*, p. *272*; 21.1.1, p. 373.

I. G. Salisbury
Comments on Weak-Beam Contrast of Stacking Faults in Transmission Electron Microscopy
(a) Vol. **66**, No. 1, p. K63–K66 (1981).
See also *1.4*, p. *30*; 21.1.1, p. 373; 22.1.2, p. 416.

F. Ogiba, W. H. Schreiner, J. V. Kunzler, and D. E. Brandão
Electrical Resistivity of the Ni_2MnSn Heusler Alloy
(a) Vol. **66**, No. 1, p. K83–K86 (1981).
See also *14.1*, p. *202*.

A. Altshuler, S. N. Bezryadin, Yu. Kh. Vekilov, and V. D. Verner
On the Pseudopotential Calculation from First Principles
(b) Vol. **106**, No. 2, p. 447–459 (1981).
See also *13*, p. *146*; 21, p. 344; 21.2, p. 376; 22.1, p. 402; 22.1.1, p. 406; 22.1.2, p. 417.

M. Fähnle
Micromagnetic Theory of Phase Transitions in Inhomogeneous Ferromagnets (IV): Dynamics of the Spin System
(b) Vol. **106**, No. 2, p. 519–528 (1981).
See also *18.2*, p. *272*.

H. M. Ledbetter
*Elastic Constants of Polycrystalline Copper at Low Temperatures. Relationship to Single-Crystal Elastic Constants
(a) Vol. **66**, No. 2, p. 477–484 (1981).
See also *12.1*, p. *136*.

J. Sołtys
X-Ray Diffraction Research of the Order–Disorder Transitions in the Ternary Heusler Alloys B_2MnAl (B = Cu, Ni, Co, Pd, Pt)
(a) Vol. **66**, No. 2, p. 485–491 (1981).
See also *1.2*, p. *18*; *1.3*, p. *24*; 21, p. 344; 21.6, p. 385.

A. Z. MAKSYMOWICZ and W. KARAŠ
Short Range Order and the Magnetic Properties of the Binary Alloys in the Ising Model
(b) Vol. **107**, No. 1, p. K1–K4 (1981).
See also *18.2*, p. *273*; 21.1.1, p. 374.

B. M. BOLOTOV, A. I. POLYAKOV, and S. P. SENSHIN
On the NMR Line Shape in Metals
(b) Vol. **107**, No. 1, p. K5–K7 (1981).
See also *19*, p. *292*; 21, p. 345.

T. SOMA, H. SATOH, and H. MATSUO
Pressure–Volume Relations and Elastic Constants under Pressure of Cu, Ag, and Au
(b) Vol. **107**, No. 1, p. K25–K29 (1981).
See also *12.1*, p. *136*; 21.6, p. 386.

S. F. DUBININ, YU. A. IZYUMOV, S. K. SIDOROV, V. N. SYROMYATNIKOV, and S. F. TEPLOUCHOV
The Role of the Incommensurable Atomic Structure in the Mechanism of the Martensite Transformation in the Invar Alloys
(a) Vol. **67**, No. 1, p. 75–82 (1981).
See also *1.2*, p. *18*; *1.3*, p. *25*; 21.1.1, p. 374.

V. E. RODE, B. WURL, and J. OLSZEWSKI
Magnetic Properties of Iron Alloys with Nickel and Ruthenium
(a) Vol. **67**, No. 1, p. 193–200 (1981).
See also *18.2*, p. *273*; *18.4*, p. *284*; 21.1.1, p. 374.

G. VAN TENDELOO, F. J. A. DEN BROEDER, S. AMELINCKX, R. DE RIDDER, J. VAN LANDUYT, and H. J. VAN DAAL
Microstructure of $Cr_{100-x}Al_x$ Alloys (10 at% $\lessgtr x \lessgtr$ 33 at%) Studied by Means of Transmissions Electron Microscopy and Diffraction (I): Microstructure of the β-Phase
(a) Vol. **67**, No. 1, p. 217–232 (1981).
See also *1.3*, p. *25*; *1.4*, p. *31*.

F. J. A. DEN BROEDER, G. VAN TENDELOO, S. AMELINCKX, J. HORNSTRA, R. DE RIDDER, J. VAN LANDUYT, and H. J. VAN DAAL
Microstructure of $Cr_{100-x}Al_x$ Alloys (10 at% $\lessgtr x \lessgtr$ 33 at%) Studied by Means of Transmission Electron Microscopy and Diffraction (II): Discovery of a New Phase
(a) Vol. **67**, No. 1, p. 233–248 (1981).
See also *1.3*, p. *25*; *1.4*, p. *31*.

H. TAKAHASHI and K. URBAN
A Study of the Point Defect Mobility in Ag–Cu by High-Voltage Electron Microscopy
(a) Vol. **67**, No. 1, p. 347–359 (1981).
See also *9*, p. *91*; *10.1*, p. *102*; 11, p. 128; 21.6, p. 386.

H. VAN SWIJGENHOVEN, J. VANOPPEN, and L. M. STALS
Vitrification of Recrystallized $Fe_{40}Ni_{38}Mo_4B_{18}$ during 5 keV Ar^+-Sputtering
(a) Vol. **67**, No. 1, p. K5–K9 (1981).
Erratum
(a) Vol. **68**, No. 2, p. K211 (1981).
See also *2*, p. *49*; *11*, p. *129*; 21.1.1, p. 375.

T. LEPISTÖ, J. YLI-KAUPPILA, P. KETTUNEN, and P. HAUTOJÄRVI
Voids in Fatigued Copper Single Crystals
(a) Vol. **67**, No. 2, p. K93–K97 (1981).
See also *1.4*, p. *31*; *10.1*, p. *103*.

J. VETTER, J. FRÜHAUF, and H. G. SCHNEIDER
Microstructure of Mn_3Ge_2–Ge and η-Cu_3Si–Si Eutectic Alloys
(a) Vol. **67**, No. 2, p. K99–K100 (1981).
See also *1.4*, p. *32*.

H. K. LACHOWICZ
Remarks on the Extraordinary Hall Effect in Amorphous $Cd_{1-x}Co_x$ Sputtered Films
(a) Vol. **67**, No. 2, p. K131–K135 (1981).
See also *14.1*, p. *203*; *18.2.1*, p. *277*; 21.4, p. 382.

N. V. ERSHOV, A. L. AGEEV, V. V. VASIN, and YU. A. BABANOV
A New Interpretation of EXAFS Spectra in Real Space (II): A Comparison of the Regularization Technique with the Fourier Transformation Method
(b) Vol. **108**, No. 1, p. 103–111 (1981).
See also *20*, p. *299*.

A. I. LANDA, V. E. PANIN, and M. F. ZHOROVKOV
The Pseudopotential Method and Thermodynamics of Eutectic Systems
(b) Vol. **108**, No. 1, p. 113–122 (1981).
See also *1.2*, p. *20*; *2*, p. *50*; 8, p. 88; 13, p. 149; 21, p. 347.

A. ANDREEFF, E. A. GOREMYCHKIN, H. GRIESSMANN, L. P. KAUN, B. LIPPOLD, W. MATZ, O. D. CHISTYAKOV, E. M. SAVITSKII, and P. G. IVANITSKII
The Crystal Field in the Hexagonal Compound $PrCu_5$
(b) Vol. **108**, No. 1, p. 261–267 (1981).
See also 8, p. 88; *13.4*, p. *182*; 18, p. 264; 21.4, p. 383.

M. LÄHDENIEMI, E. OJALA, and M. OKOCHI
XPS Band Emission and Density of States of CoAl
(b) Vol. **108**, No. 1, p. K61–K64 (1981).
See also *13.1*, p. *160*; *17*, p. *258*; 21, p. 348.

A. KORNER and H. P. KARNTHALER
Analysis of the Stacking Faults of Dissociated Glide Dislocations in F.C.C. Metals
(a) Vol. **68**, No. 1, p. 19–28 (1981).
See also *10.1*, p. *103*.

R. PEREZ, K. Z. BOTROS, and S. S. SHEININ
Experimental Evidence of a Breakdown of the Column Approximation in Calculations of Contrast in Electron Microscope Images of Stacking Faults
(a) Vol. **68**, No. 1, p. 91–96 (1981).
See also *10.1*, p. *104*.

M. HASHIMOTO, Y. ISHIDA, R. YAMAMOTO, and M. DOYAMA
Structure and Atomic Vibration of the $\sum = 7$ Tilt Boundary in α-Titanium
(a) Vol. **68**, No. 1, p. 109–116 (1981).
See also *1.4*, p. *32*; 4, p. 61; *10.1*, p. *104*.

R. Ramlua and H. Löffler
Guinier-Preston Zones in an Al-8 at%Zn Alloy
(a) Vol. **68**, No. 2, p. 531–536 (1981).
See also *1.3*, p. *27*; *10.1*, p. *104*; 21, p. 349.

L. Varga, K. Tompa, and T. Schmidt
Thermopower Method for Chemical Composition and Inhomogeneity Measurements in Amorphous Ni–P Samples
(a) Vol. **68**, No. 2, p. 603–606 (1981).
See also *1.3*, p. *27*; 2, p. 51; *15*, p. *250*.

L. Brossard, L. Thomé, A. Traverse, H. Bernas, J. Chaumont, and F. Lalu
Stability of Non-Equilibrium Nickel Hybrides Prepared by Low-Temperature Ion Implantation
(a) Vol. **68**, No. 2, p. 619–627 (1981).
See also *10.1*, p. *104*; 11, p. 132; 14.1, p. 204.

S. V. Sukhvalo, A. I. Demchenko, A. B. Shifrin, and E. M. Zub
Orientation Transitions in Amorphous Gd–Co Films
(a) Vol. **68**, No. 2, p. 637–644 (1981).
See also *1.3*, p. *27*; *2*, p. *51*; 14.1, p. 204; 18, p. 264; 21.4, p. 383.

F. C. Lovey, R. Rapacioli, and M. Chandrasekaran
Precipitates with an Incommensurate Structure in β-Cu–Zn–Al
(a) Vol. **68**, No. 2, p. K105–K108 (1981).
See also *1.3*, p. *27*; 21, p. 349.

A. Zahra and C. Y. Zahra
Comments on "Effects of the Fluctuation of Solute Concentration in Al–10 wt%Zn Alloy on Aging at Low Temperature"
(a) Vol. **68**, No. 2, p. K141–K143 (1981).
See also *1.3*, p. *27*; 8, p. 88; 21, p. 350.

M. Ohta, M. Yamada, T. Kanadani, and A. Sakakibara
Reply to Comments on "Effects of the Fluctuation of Solute Concentration in Al–10wt%Zn Alloy on Aging at Low Temperature"
(a) Vol. **68**, No. 2, p. K145–K148 (1981).
See also *1.3*, p. *27*; *14.1*, p. *204*; 21, p. 350.

Ch. Routsi and J. K. Yakinthos
Magnetic Characteristic of Some TCu_2Si_2 Ternary Alloys (T = Nd, Dy, Ho, Er, Tm, Y, and La)
(a) Vol. **68**, No. 2, p. K153–K157 (1981).
See also *18.2*, p. *275*; *18.4*, p. *285*; 21.4, p. 383.

S. Vaskovskii, J. Schneider, and B. Springmann
Low Field Magnetic Properties of Amorphous $Fe_{40}Ni_{40}P_{14}B_6$ Ribbons
(a) Vol. **68**, No. 2, p. K189–K193 (1981).
See also *2*, p. *52*; *18.2*, p. *275*; 21.1.1, p. 375.

21.1.1 Fe and Fe Alloys

Original Papers and Short Notes

G. Ya. Selyutin
Magnetic Contribution to the Phonon Spectrum of Fe
(b) Vol. **104**, No. 1, p. 341 – 344 (1981).
See also 5, p. 63; *6*, p. *70*; *18*, p. *260*.

Y. D. Yao and S. Arajs
Electrical Resistivity of Fe-Rich Fe–Pd Alloys between 77 and 1250 K
(a) Vol. **64**, No. 1, p. 95 – 104 (1981).
See also 1.2, p. 15; *14.1*, p. *199*.

Y. K. Wu, J. Z. Liang, and K. H. Kuo
Micro Inversion Domains Formed during the Crystallization of the Amorphous Alloy $Fe_{40}Ni_{40}P_{14}B_6$
(a) Vol. **64**, No. 1, p. 113 – 120 (1981).
See also *1.4*, p. *28*; *2*, p. *42*; 21.1, p. 355.

H. J. Blythe, F. Walz, and H. Kronmüller
Magnetic Relaxation in Electron-Irradiated Ordered Ni_3Fe
(a) Vol. **64**, No. 1, p. 169 – 176 (1981).
See also 10.1, p. 98; *11*, p. *123*; *18.2*, p. *269*; 21.1, p. 355.

G. Le Caer and J. M. Dubois
About the Asymmetries in Mössbauer Spectra of Magnetic Amorphous Transition Metal–Metalloid Alloys
(a) Vol. **64**, No. 1, p. 275 – 282 (1981).
See also 2, p. 43; *5*, p. *63*; *18.2*, p. *269*.

J. M. Genin, Ph. Bauer, and M. J. Besnus
Mössbauer Study of the Temperature Dependence of the Magnetization Direction and the Hyperfine Interactions in the Laves Phase Compound $GdFe_2$
(a) Vol. **64**, No. 1, p. 325 – 333 (1981).
See also *5*, p. *63*; 18.3, p. 278.

P. J. Schurer, A. H. Morrish, and M. J. Stavn
Induced Perpendicular Anisotropy and Surface Crystallization in Amorphous $Fe_{78}B_{12}Si_{10}$ Ribbons
(a) Vol. **64**, No. 1, p. 343 – 349 (1981).
See also 2, p. 43; 5, p. 63; *18.2.1*, p. *276*.

A. Zentko, M. Timko, P. Duhaj, and I. Škorvánek
The Effect of Neutron Irradiation on the Electrical Resistivity of Amorphous $Fe_{47}Ni_{25}B_{18}Si_{10}$ Alloys
(a) Vol. **64**, No. 1, p. K19 – K22 (1981).
See also 2, p. 43; *11*, p. *123*; *14.1*, p. *200*.

B. K. Ponomarev and V. G. Thiessen
Temperature Dependence of the Baric Magnetization Coefficient of a Band Ferromagnet
(b) Vol. **104**, No. 2, p. 427 – 434 (1981).
See also *18.2*, p. *269*; 21.1, p. 356.

A. R. McGurn and D. M. Esterling
Liquid Metal Pair-Correlation Functions from Empirical Pair Potentials
(b) Vol. **105**, No. 1, p. 47–53 (1981).
See also *2*, p. *45*; *4*, p. *58*; 21.1, p. 357; 21.6, p. 385.

K. Masuda
Lattice Vibrations around a Vacancy in B.C.C. Transition Metals
(b) Vol. **105**, No. 1, p. 107–113 (1981).
See also *6*, p. *73*; *10.1*, p. *100*; 21, p. 341.

V. I. Malyutin, V. E. Osukhovskii, Yu. D. Vorobiev, A. G. Shishkov, and V. V. Yudin
Structure and Magnetic Properties of Etched NiFeCo Films
(a) Vol. **65**, No. 1, p. 45–52 (1981).
See also *18.2.1*, p. *276*; 21.1, p. 358.

T. Magnin, A. Fourdeux, and J. H. Driver
Cyclic Plastic Deformation of B.C.C. Fe–26 Cr Single Crystals at Room Temperature
(a) Vol. **65**, No. 1, p. 301–314 (1981).
See also *10.1*, p. *100*.

M. Huq
Magnetic Anisotropy in $ErFe_2$ Compound
(a) Vol. **65**, No. 1, p. K29–K32 (1981).
See also *18*, p. *262*; 21.4, p. 381.

K. Uhlmann and F. Storbeck
Relative Determination of L_3 Ionisation Cross Sections of Chromium, Iron, and Nickel
(a) Vol. **65**, No. 1, p. K37–K41 (1981).
See also *17*, p. *257*; 21.1, p. 359.

V. E. Antonov, I. T. Belash, E. G. Ponyatovskii, V. G. Thiessen, and V. I. Shiryaev
Magnetization of Iron Hydride
(a) Vol. **65**, No. 1, p. K43–K48 (1981).
See also *18.2*, p. *271*.

O. F. G. Aczel, G. Haneczok, and J. W. Moroń
The Relaxation Times and the Relaxed Elastic Modulus of Prestressed B.C.C. Metals with Interstitial Impurities
(a) Vol. **65**, No. 1, p. K87–K89 (1981).
See also *12.1*, p. *136*.

K. Hoselitz
The Saturation Magnetization of Some Substituted Fe–Si–B Metallic Glasses
(a) Vol. **65**, No. 1, p. K91–K94 (1981).
See also *18.2*, p. *271*; *18.4*, p. *283*.

P. Valenta, K. Maier, H. Kronmüller, and K. Freitag
Self-Diffusion of Phosphorus in the Amorphous Alloy $Fe_{40}Ni_{40}P_{14}B_6$
(b) Vol. **105**, No. 2, p. 537–542 (1981).
See also 2, p. 46; *9*, p. *90*; 21.1, p. 359.

J. D. Stephenson
An Interpretation of White (Synchrotron) X-Radiation Laue (Topograph) Patterns by Means of the Gnomonic Projection
(a) Vol. **65**, No. 2, p. 601–608 (1981).
See also *1.1*, p. *12*.

B. Rauschenbach, E. Richter, and K. Hohmuth
Spinodal Decomposition in Amorphous Boron-Implanted Iron Films
(a) Vol. **65**, No. 2, p. K103–K105 (1981).
See also *2*, p. *47*; 11, p. 126.

O. Bostanjoglo and Th. Rosin
New Reversal Process of Néel Walls Due to Nanosecond Field Pulses
(a) Vol. **65**, No. 2, p. K117–K119 (1981).
See also *18.2.1*, p. *277*.

K. Terao and M. Shimizu
Phase Transitions in Iron Alloys
(a) Vol. **65**, No. 2, p. K153–K156 (1981).
See also *1.2*, p. *17*; 6, p. 75; 8, p. 86.

P. Valenta, K. Maier, H. Kronmüller, and K. Freitag
Self-Diffusion of Iron in the Amorphous Alloys $Fe_{40}Ni_{40}P_{14}B_6$ and $Fe_{80}B_{20}$
(b) Vol. **106**, No. 1, p. 129–133 (1981).
See also *2*, p. *47*; *9*, p. *90*.

R. Ramji Rao and A. Rajput
Lattice Dynamics and Some Related Thermal Properties of Body-Centred Cubic Metals
(b) Vol. **106**, No. 1, p. 393–406 (1981).
See also *6*, p. *77*; 12.1, p. 136; 21, p. 344; 21.2, p. 376.

C. K. Saw and S. J. Pickart
Aperiodicity in an Amorphous Dense Random Packed Model
(b) Vol. **106**, No. 1, p. K73–K76 (1981).
See also *2*, p. *47*.

G. Dlubek, O. Brümmer, J. Yli-Kauppila, and J. Johansson
Recovery of Quenched-in Vacancies in Fe–Al Alloys Studied by Positron Annihilation
(b) Vol. **106**, No. 1, p. K83–K88 (1981).
See also *10.1*, p. *101*.

D. Melville, K. M. Al-Rawi, and W. I. Khan
Magnetostriction and Magnetisation of $CeFe_2$
(a) Vol. **66**, No. 1, p. 133–137 (1981).
See also *18.2*, p. *272*; 21.4, p. 381.

R. Kuśka and J. W. Moroń
Magnetic Permeability Disaccommodation in α-Fe–C Type Alloys Containing s–i Pairs (I): General Theory
(a) Vol. **66**, No. 1, p. 149–156 (1981).
See also *18.2*, p. *272*.

O. Bostanjoglo and Th. Rosin
Resonance Oscillations of Bloch Lines in Permalloy Films
(a) Vol. **66**, No. 1, p. K5–K10 (1981).
See also *18.2.1*, p. *277*; 21.1, p. 361.

T. Maeda, T. Soumura, and S. Saito
Comparison of the Wall-Pinning Model Theory with Magnetic Damping Experiments
(a) Vol. **66**, No. 1, p. K51–K54 (1981).
See also *18.2*, p. *272*; 21.1, p. 361.

I. G. Salisbury
Comments on Weak-Beam Contrast of Stacking Faults in Transmission Electron Microscopy
(a) Vol. **66**, No. 1, p. K63–K66 (1981).
See also *1.4*, p. *30*; 21.1, p. 361; 22.1.2, p. 416.

R. Kuśka and J. W. Moroń
Magnetic Permeability Disaccommodation in α-Fe–C Type Alloys Containing s–i Pairs (II): Application to α-Fe–V–C Alloys
(a) Vol. **66**, No. 2, p. 431–438 (1981).
See also *18.2*, p. *272*.

J. Filipecki, J. Dworakowski, and B. Całusiński
Investigation of the Metallic Glass $Fe_{40}Ni_{40}B_{20}$ with the Use of Positron Annihilation Methods
(a) Vol. **66**, No. 2, p. 463–466 (1981).
See also *2*, p. *48*; 13, p. 147.

Y. Umakoshi and H. Kronmüller
Observation of Dislocations in Deformed Iron Single Crystals by Means of High-Field Susceptibility Measurements
(a) Vol. **66**, No. 2, p. 509–520 (1981).
See also *10.1*, p. *101*; 18.2, p. 273.

G. N. Gaidukov and J. P. Hirth
The Influence of Thermodynamic Solution Properties on Solute Atmospheres at Dislocations
(a) Vol. **66**, No. 2, p. 601–606 (1981).
See also *1.3*, p. *25*; 8, p. 86; *10.1*, p. *102*.

E. Wieser, A. Handstein, J. Schneider, and K. Závěta
Magnetic Structure of Amorphous $(Fe_{1-x}Mn_x)_{75}P_{15}C_{10}$ Alloys
(a) Vol. **66**, No. 2, p. 607–612 (1981).
See also 2, p. 48; 5, p. 65; *18*, p. *263*; 21.1, p. 362.

A. Handstein, J. Kadlecová, and V. Gregor
Magnetoresistance of Amorphous $(Fe_xNi_{100-x})_{80}B_{10}P_{10}$ Alloys in Magnetic Fields up to 14.5 T
(a) Vol. **66**, No. 2, p. 657–663 (1981).
See also *2*, p. *48*; *14.1*, p. *202*; 18.2, p. 273; 21.1, p. 362.

H. van Swijgenhoven, J. Vanoppen, and L. M. Stals
Vitrification of Recrystallized $Fe_{40}Ni_{38}Mo_4B_{18}$ during 5 keV Ar^+-Sputtering
(a) Vol. **67**, No. 1, p. K5–K9 (1981).
Erratum
(a) Vol. **68**, No. 2, p. K211 (1981).
See also *2*, p. *49*; *11*, p. *129*; 21.1, p. 363.

J. Kamarád, J. Schneider, and Z. Arnold
Pressure Dependence of Curie Temperature of Amorphous $Fe_{90}Zr_{10}$ Alloy
(a) Vol. **67**, No. 1, p. K85–K88 (1981).
See also *2*, p. *49*; *18.2*, p. *273*.

S. Manninen, B. K. Sharma, T. Paakkari, S. Rundqvist, and M. W. Richardson
Electron Momentum Distribution and Charge Transfer Study of FeAl, CoAl, and NiAl
(b) Vol. **107**, No. 2, p. 749–755 (1981).
See also *13.1*, p. *160*; 13.2, p. 164; 21.1, p. 364.

L. Vatskitchev
Planar Hall Effect in Double Layer Magnetic Films with Unidirectional Anisotropy
(a) Vol. **67**, No. 2, p. 591–596 (1981).
See also 14.1, p. 203; *18.2.1*, p. *277*; *18.4*, p. *284*; 21.1, p. 364.

Y. Umakoshi and H. Kronmüller
On Magnetic Domain Patterns and Dislocation Structures in Iron Single Crystals Deformed at 195 and 77 K
(a) Vol. **68**, No. 1, p. 159–172 (1981).
See also *10.1*, p. *104*; *18.2*, p. *274*.

I. Richter and M. Feller-Kniepmeier
Diffusion of Zn in α-Fe Single Crystals
(a) Vol. **68**, No. 1, p. 289–300 (1981).
See also *9*, p. *92*; 18.2, p. 274.

I. A. Tsoukalas
On the Atomic Ordering of Fe-Rich, $Fe_{1-x}Ni_x$ Binary Alloys
(a) Vol. **68**, No. 1, p. K67–K70 (1981).
See also *14.1*, p. *203*; *18.2*, p. *274*; 21.1, p. 366.

M. Szuszkiewicz, B. Rozenfeld, and J. Chojcan
Localization of 3d-Like Electrons in Cr–Fe and V–Fe Alloys
(b) Vol. **108**, No. 2, p. 623–630 (1981).
See also *13*, p. *150*; 21.1, p. 366.

S. Vaskovskii, J. Schneider, and B. Springmann
Low Field Magnetic Properties of Amorphous $Fe_{40}Ni_{40}P_{14}B_6$ Ribbons
(a) Vol. **68**, No. 2, p. K189–K193 (1981).
See also *2*, p. *52*; *18.2*, p. *275*; 21.1, p. 367.

21.2 Alkali Metals

Original Papers and Short Notes

K. N. KHANNA and D. P. KHANDELWAL
Thermodynamic Properties of Liquid Metals and Alloys
(b) Vol. **106**, No. 2, p. 715–721 (1981).
See also *2*, p. *48*; *8*, p. *86*.

L. SKÁLA
The CNDO/BW Study of Lithium Clusters (I): Two-, Three-, and Four-Atomic Clusters
(b) Vol. **107**, No. 1, p. 351–357 (1981).
See also *4*, p. *60*; *13*, p. *147*.

A. P. ZHERNOV
Temperature Dependence of the Righi-Leduc Coefficients for Na and K Metals
(b) Vol. **107**, No. 2, p. 411–421 (1981).
See also 6, p. 79; *8*, p. *87*; *14.1*, p. *202*.

T. SOMA, H. MATSUO, and Y. KOHBU
The Electronic Theory of the Alloy Systems Composed of Alkali Metal Elements (I): Bulk Properties of Cs–K, Cs–Rb, and Rb–K Solid Solutions
(b) Vol. **107**, No. 2, p. 761–768 (1981).
See also *1.3*, p. *26*; 4, p. 60; 12.1, p. 136.

I. GORGAS, P. HERKE, and G. SCHOECK
The Plastic Behaviour of Lithium Single Crystals
(a) Vol. **67**, No. 2, p. 617–623 (1981).
See also *10.1*, p. *103*.

T. SOMA, H. MATSUO, and M. FUNAKI
The Electronic Theory of the Alloy Systems Composed of Alkali Metal Elements (II): Phase Diagrams
(b) Vol. **108**, No. 1, p. 221–227 (1981).
See also *1.2*, p. *20*; 1.3, p. 26; 4, p. 60.

21.3 Alkali-Earth Metals

Review Article

Original Papers and Short Notes

U. De, S. Hoth, and W. Witthuhn
Temperature and Concentration Dependence of the Electric Field Gradient in the Cadmium–Magnesium Alloy System
(b) Vol. **107**, No. 2, p. 569–577 (1981).
See also *1.3*, p. *25*; 21, p. 346.

K. S. Sharma
Phonon Dispersion in Barium
(b) Vol. **108**, No. 2, p. K101–K104 (1981).
See also *4*, p. *61*; *6*, p. *81*.

Y. Umakoshi and M. Yamaguchi
The Stability and Energies of Planar Faults in the DO_{19} Ordered Structure
(a) Vol. **68**, No. 2, p. 457–468 (1981).
See also *1.3*, p. *27*; *10.1*, p. *104*.

21.4 Rare-Earth Metals

Original Papers and Short Notes

T. K. Hatwar, M. N. Ghatikar, B. D. Padalia, S. K. Malik, and R. Vijayaraghavan
X-Ray Absorption Spectroscopic Study of Rare-Earth–Manganese Intermetallic Compounds
(b) Vol. **103**, No. 1, p. 159–166 (1981).
See also *20*, p. *296*; 21.1, p. 351.

A. N. Korshunov, V. M. Fedorov, I. M. Sharshakov, and V. T. Grokhotov
The Effect of Rare-Earth Impurities on the Electronic Transition in Ce
(b) Vol. **103**, No. 1, p. 199–204 (1981).
See also *1.2*, p. *14*; 13, p. 142.

L. Palermo and X. A. da Silva
Induced Magnetism via Conduction Electron Exchange. Application to $PrAl_2$
(b) Vol. **103**, No. 1, p. 419–422 (1981).
See also *13*, p. *142*; *18.2*, p. *267*.

K. N. Khanna
Thermodynamic Properties of Liquid Metals
(b) Vol. **103**, No. 1, p. K15–K19 (1981).
See also *2*, p. *40*; *8*, p. *84*; 21, p. 336; 21.2, p. 376; 21.3, p. 378.

K. Twarowski, H. K. Lachowicz, M. Gutowski, and H. Szymczak
On the Origin of the Perpendicular Anisotropy and Magnetostriction in Amorphous RF Sputtered Gd–Co Films
(a) Vol. **63**, No. 1, p. 103–108 (1981).
See also *2*, p. *40*; *18.2.1*, p. *276*; 18.3, p. 278; 21.1, p. 352.

M. P. Dariel, O. D. McMasters, and K. A. Gschneidner, Jr.
Diffusion of ^{198}Au in Praseodymium Single Crystals
(a) Vol. **63**, No. 1, p. 329–333 (1981).
See also *9*, p. *89*.

R. Żuberek
Columnar Model of Magnetostriction in Amorphous Films
(a) Vol. **63**, No. 1, p. K51–K54 (1981).
See also *2*, p. *41*; *18.2.1*, p. *276*; 18.3, p. 278; 21.1, p. 352.

P. V. Ashrit and M. A. Angadi
Electrical Properties of Thin Ytterbium Films
(a) Vol. **63**, No. 1, p. K77–K82 (1981).
See also *14.1*, p. *199*.

O. V. Farberovich
Band Structure and Electron–Electron Interaction in Samarium Monosulfide
(b) Vol. **104**, No. 1, p. 365–374 (1981).
See also *13.1*, p. *154*; 17, p. 256; 20.1, p. 305; 22, p. 392.

H. Oesterreicher and F. T. Parker
Erratum to: Aspects of Intrinsic Coercivity via Gd Analogs
(a) Vol. **64**, No. 1, p. 419 (1981).
See also *18.2*, p. *269*; 21.1, p. 356.

W. Zahn
Inductance of Small Window Type Tunnel Junctions
(a) Vol. **64**, No. 1, p. K43–K47 (1981).
See also *14.2*, p. *206*; 21, p. 339.

B. V. Karpenko
On the Magnetism in Narrow-Band Conductors
(b) Vol. **104**, No. 2, p. 407–413 (1981).
See also *13.1*, p. *154*; *18.1*, p. *265*.

F. Brouers and O. L. T. de Menezes
Electron–Phonon Interaction in Mixed Valence Systems
(b) Vol. **104**, No. 2, p. 541–546 (1981).
See also *6*, p. *71*; *13*, p. *144*.

P. C. M. Gubbens, A. M. van der Kraan, and K. H. J. Buschow
^{161}Dy Mössbauer Effect and Magnetic Properties of Some Amorphous and Crystalline Dy–Co Alloys
(a) Vol. **64**, No. 2, p. 657–663 (1981).
See also 2, p. 44; *5*, p. *63*; *18*, p. *261*; 21.1, p. 357.

J. Jarosz
The Application of Kramers-Kronig Relations to the Analysis of the Resonance Lines of Metallic Samples in ESR Spectroscopy
(b) Vol. **105**, No. 1, p. 155–160 (1981).
See also *19*, p. *289*; 21, p. 341.

D. Fruchart, F. Vaillant, E. Roudaut, A. Nemoz, and X. G. Tessema
Neutron Diffraction Study of $CeRu_2D_5$
(a) Vol. **65**, No. 1, p. K19–K24 (1981).
See also *1.3*, p. *24*.

M. Huq
Magnetic Anisotropy in $ErFe_2$ Compound
(a) Vol. **65**, No. 1, p. K29–K32 (1981).
See also *18*, p. *262*; 21.1.1, p. 371.

D. Melville, K. M. Al-Rawi, and W. I. Khan
Magnetostriction and Magnetisation of $CeFe_2$
(a) Vol. **66**, No. 1, p. 133–137 (1981).
See also *18.2*, p. *272*; 21.1.1, p. 372.

G. Herbst and H. Kronmüller
Magnetic After-Effect of Hydrogen in Gadolinium
(a) Vol. **66**, No. 1, p. 255–260 (1981).
See also *10.1*, p. *101*; *18.2*, p. *272*.

A. Andreeff, E. A. Goremychkin, H. Griessmann, L. P. Kaun, B. Lippold, W. Matz, O. D. Chistyakov, E. M. Savitskii, and P. G. Ivanitskii
The Crystal Field in the Hexagonal Compound $PrCu_5$
(b) Vol. **108**, No. 1, p. 261–267 (1981).
See also 8, p. 88; *13.4*, p. *182*; 18, p. 264; 21.1, p. 365.

L. G. Khvostantsev and N. A. Nikolaev
Polymorphism of Cerium under Hydrostatic Pressure up to 8.0 GPa
(a) Vol. **68**, No. 1, p. 193–197 (1981).
See also *1.2*, p. *20*.

K. P. Hoffmann, H. Drulis, and B. Staliński
Electron Paramagnetic Resonance of Er^{3+} Ions in Lutetium Hydrides
(a) Vol. **68**, No. 2, p. 439–446 (1981).
See also *19*, p. *295*.

S. V. Sukhvalo, A. I. Demchenko, A. B. Shifrin, and E. M. Zub
Orientation Transitions in Amorphous Gd–Co Films
(a) Vol. **68**, No. 2, p. 637–644 (1981).
See also *1.3*, p. *27*; *2*, p. *51*; 14.1, p. 204; 18, p. 264; 21.1, p. 367.

Ch. Routsi and J. K. Yakinthos
Magnetic Characteristic of Some TCu_2Si_2 Ternary Alloys (T = Nd, Dy, Ho, Er, Tm, Y, and La)
(a) Vol. **68**, No. 2, p. K153–K157 (1981).
See also *18.2*, p. *275*; *18.4*, p. *285*; 21.1, p. 367.

21.6 Noble Metals

Original Papers and Short Notes

A. R. McGurn and D. M. Esterling
Liquid Metal Pair-Correlation Functions from Empirical Pair Potentials
(b) Vol. **105**, No. 1, p. 47–53 (1981).
See also *2*, p. *45*; *4*, p. *58*; 21.1, p. 357; 21.1.1, p. 371.

S. Tsugane, K. Morigaki, and C. Nagashima
Anomalous Magnetoresistance in Amorphous Si–Au System
(b) Vol. **105**, No. 1, p. K5–K8 (1981).
See also *2*, p. *45*; *14.3.1*, p. *220*; 22.1.2, p. 413.

G. van Tendeloo and S. Amelinckx
Electron Diffraction and High Resolution Electron Microscopy Study of Ordering in the Gold–Manganese System (I): Lozenge Shaped Island Structures Derived from the DO_{22} Structure $Au_{3+x}Mn$
(a) Vol. **65**, No. 1, p. 73–86 (1981).
See also *1.1*, p. *11*; 1.3, p. 23; *1.4*, p. *29*; 21.1, p. 358.

D. G. Kanhere, R. M. Singru, and R. Harthoorn
Electron Momentum Distribution and Compton Profiles of Ag
(b) Vol. **105**, No. 2, p. 715–722 (1981).
See also *13.1*, p. *156*; 13.2, p. 163; *20*, p. *298*.

S. M. M. Rahman
A Model Calculation on the Stability of the Glassy $Au_{0.82}Si_{0.18}$
(b) Vol. **105**, No. 2, p. K115–K118 (1981).
See also *2*, p. *46*; *4*, p. *59*; 22.1.2, p. 414.

G. van Tendeloo and S. Amelinckx
Electron Diffraction and High Resolution Electron Microscopy Study of Ordering in the Gold–Manganese System (II): Square Island Structures Derived from the Au_4Mn Structure $(Au_{4-x}Mn_1)$
(a) Vol. **65**, No. 2, p. 431–446 (1981).
See also *1.1*, p. *11*; 1.3, p. 24; *1.4*, p. *30*; 21.1, p. 359.

V. V. Nemoshkalenko, A. V. Zhalko-Titarenko, V. N. Antonov, Vl. N. Antonov, and E. Mrosan
Anisotropy of Cyclotron Masses in Gold
(b) Vol. **106**, No. 2, p. K117–K121 (1981).
See also *13.2*, p. *163*; *19*, p. *292*.

J. Sołtys
X-Ray Diffraction Research of the Order–Disorder Transitions in the Ternary Heusler Alloys B_2MnAl (B = Cu, Ni, Co, Pd, Pt)
(a) Vol. **66**, No. 2, p. 485–491 (1981).
See also *1.2*, p. *18*; *1.3*, p. *24*; 21, p. 344; 21.1, p. 361.

G. Frommeyer and H. G. Brion
Hall Effect in Directionally Solidified and Deformed Eutectic Silver–Copper Alloys with Fibre Structure
(a) Vol. **66**, No. 2, p. 559–564 (1981).
See also *14.1*, p. *202*; 21.1, p. 362.

21.7 Semi-Metals (Semiconducting Alloys See 22.7)

Original Papers and Short Notes

P. Středa and P. Vašek
Thermopower in Strong Magnetic Fields and the Mott Rule
(b) Vol. **103**, No. 2, p. K137–K140 (1981).
See also **14.1**, p. 199; *15*, p. *247*.

J. Sosnovski, S. Bednarski, W. Bührer, A. Czachor, and E. Maliszewski
Phonon Dispersion Relations in the $Bi_{0.95}Sb_{0.05}$ Alloy
(b) Vol. **104**, No. 1, p. 97–102 (1981).
See also *6*, p. *69*.

B. Pardo, C. Pariset, and D. Renard
Study of Moire Fringes in Electron Diffraction by the Many-Beam Dynamic Theory and Application to Sb–Bi Double Layers
(a) Vol. **64**, No. 1, p. 283–295 (1981).
See also *1.4*, p. *29*

L. G. Khvostantsev and V. A. Sidorov
High-Pressure Polymorphism of Antimony. Thermoelectrical Properties and Electrical Resistance Studies
(a) Vol. **64**, No. 1, p. 379–384 (1981).
See also 1.2, p. 16; *14.1*, p. *199*; *15*, p. *247*.

R. Herrmann, B. Fellmuth, R. Weissflog, and V. G. Golubev
Erratum to: Effect of the Anisotropic of the Fermi Surface on the Propagation of Helicon Waves Parallel to the Trigonal Axis
(b) Vol. **104**, No. 2, p. K169 (1981).
See also *13.2*, p. *163*; 14.3, p. 212; *19*, p. *288*.

D. Kashchiev
Growth of Crystallites in Deposition from Vapours
(a) Vol. **64**, No. 2, p. 715–721 (1981).
See also *3*, p. *54*.

W. Kraak, R. Kuhl, F. Thom, H. Haefner, and P. Schmidt
Size Effects and Temperature Dependence of the Electrical Resistance in Bismuth Single Crystals of High Perfection
(b) Vol. **105**, No. 2, p. 597–604 (1981).
See also *14.1*, p. *201*.

D. V. Gitsu, F. M. Muntyanu, and M. I. Onu
Quantum Oscillations of Thermomagnetic Power and Magnetoresistance in Antimony-Bismuth Alloys
(b) Vol. **105**, No. 2, p. K95–K98 (1981).
See also *14.1*, p. *201*; *15*, p. *248*.

O. Bostanjoglo
New Phase Transitions in Arsenic Films
(a) Vol. **65**, No. 2, p. 595–600 (1981).
See also *1.2*, p. *17*; *1.5*, p. *35*; 2, p. 46.

L. B. Dubovskii
Density of States in Multilayer Sandwiches
(b) Vol. **106**, No. 1, p. 315–325 (1981).
See also *13*, p. *146*; 14.1, p. 201; 14.2, p. 206; 21, p. 344.

T. Soma, H. Matsuo, and T. Katoh
The Heat of Mixing in Liquid Si–Ge, Si–Sn, and Ge–Sn Mixtures
(b) Vol. **106**, No. 1, p. K5–K7 (1981).
See also *1.2*, p. *17*; *1.3*, p. *24*; 22.1.1, p. 405; 22.1.2, p. 415.

H. Müller
Simulation of Spherulitic Crystallization in Thin Amorphous Antimony Films
(a) Vol. **66**, No. 1, p. 199–205 (1981).
See also *1.5*, p. *36*; *3*, p. *54*.

B. Fellmuth, H. Krüger, R. Rudolph, and R. Herrmann
Composition Dependence of the Anisotropy of the Cyclotron Masses in n-Type Semiconducting $Bi_{1-x}Sb_x$
(b) Vol. **106**, No. 2, p. 561–567 (1981).
See also *13.1*, p. *158*; *13.2*, p. *163*.

W. Braune, U. Straubel, N. Kubicki, J. Lebech, and K. Saermark
Relaxation Time and Effective Masses for Charge Carriers in Semimetallic BiSb Alloys versus Alloy Composition
(b) Vol. **106**, No. 2, p. 587–592 (1981).
See also *13.5.2*, p. *194*; *13.5.4*, p. *197*; 19, p. 291.

M. N. Ghatikar, T. K. Hatwar, B. D. Padalia, E. V. Sampathkumaran, L. C. Gupta, and R. Vijayaraghavan
Study of X-Ray Absorption Edges of Rare Earths and Gallium in RAl_2Ga_2 Intermetallics
(b) Vol. **106**, No. 2, p. K89–K93 (1981).
See also *20*, p. *298*; 21.4, p. 382.

M. Wołcyrz, R. Kubiak, and S. Maciejewski
X-Ray Investigation of Thermal Expansion and Atomic Thermal Vibrations of Tin, Indium, and Their Alloys
(b) Vol. **107**, No. 1, p. 245–253 (1981).
See also 1.3, p. 25; *6*, p. *78*; *8*, p. *87*.

W. Braune, N. Kubicki, U. Staubel, and R. Herrmann
Electronic Peculiarities in $Bi_{1-x}Sb_x$ Alloys Close to the Semiconductor–Semimetal Transition at $x \approx 0.22$
(b) Vol. **107**, No. 1, p. K67–K70 (1981).
See also *13.1*, p. *159*; *19*, p. *292*.

V. G. Chudinov, V. P. Gogolin, B. N. Goshichitskii, and V. I. Protasov
Simulation of Collision Cascades in Intermetallic Nb_3Sn Compounds
(a) Vol. **67**, No. 1, p. 61–67 (1981).
See also *11*, p. *128*; 21, p. 346.

Yu. A. Pospelov and G. S. Grachev
About the Character of Temperature Variations of Bi-Group Semimetal Electron Spectra
(b) Vol. **107**, No. 2, p. K155–K158 (1981).
See also *13.1*, p. *160*.

A. H. Abou El Ela, S. Mahmoud, and M. A. Mahmoud
Electrical Conduction of Thin Bismuth Films
(a) Vol. **67**, No. 2, p. K155–K160 (1981).
See also *14.1*, p. *203*.

K. Saermark, J. Lebech, W. Braune, and N. Kubicki
Non-Local Calculations of the Dispersion of Electromagnetic Waves in Bismuth. Dispersion Relation for the Voigt Configuration: $\boldsymbol{q} \parallel \boldsymbol{n} \parallel \boldsymbol{C}_1$, $\boldsymbol{q} \perp \boldsymbol{B}$
(b) Vol. **108**, No. 1, p. 275–280 (1981).
See also *13*, p. *149*; *20*, p. *299*.

A. Baldereschi, K. Maschke, A. Milchev, R. Pickenhain, and K. Unger
Thomas-Fermi Approximation for the Valence Electron Densities in Cubic Semiconductors and Insulators
(b) Vol. **108**, No. 2, p. 511–520 (1981).
See also *13.1*, p. *161*; 14.3, p. 218; 22.1.1, p. 407; 22.1.2, p. 422; 22.2.1, p. 431; 22.4.2, p. 453; 22.5.2, p. 469.

K. Kranjc, D. Kunstelj, and V. Marinković
Parallel Moiré Effect in Electron Micrographs of Crystal Sandwiches (II): Comparison of Theory and Experiment
(a) Vol. **68**, No. 2, p. 363–372 (1981).
See also *1.4*, p. *32*; 1.5, p. 37; 21, p. 349; 21.1, p. 366; 22.4.3, p. 455.

D. Colaïtis, D. van Dyck, and S. Amelinckx
Electron Microscopic Study of the System mPbS–$n$$Bi_2S_3$
(a) Vol. **68**, No. 2, p. 419–438 (1981).
See also *1.3*, p. *27*; 1.4, p. 32.

V. Damodara Das and N. Jayaprakash
Initial Lattice Distortion Energy Spectra of Bismuth Thin Films and the Effect of Deposition Rate
(a) Vol. **68**, No. 2, p. K159–K164 (1981).
See also 10.1, p. 105; *13.4*, p. *185*; *14.1*, p. *204*.

22. Semiconductors and Ionic Crystals

Review Articles

A. B. Roitsin
Generalized Spin-Hamiltonian and Low-Symmetry Effects in Paramagnetic Resonance
(b) Vol. **104**, No. 1, p. 11–35 (1981).
See also 13.4, p. 168; *19*, p. *286*.

K. M. van Vliet and H. Mehta
Theory of Transport Noise in Semiconductors
(b) Vol. **106**, No. 1, p. 11–30 (1981).
See also *14.3*, p. *209*.

Original Papers and Short Notes

V. Čápek
Role of Magnetization in the Electron Quasiparticle Energy in the s–d Model of Ferromagnetic Semiconductors
(b) Vol. **103**, No. 1, p. 107–113 (1981).
See also *13.1*, p. *151*; *18.2*, p. *267*.

V. Szőcs, V. Čápek, and B. Velický
Simple Two-Band s–d Model for Ferromagnetic Semiconductors with Hybridization (II): Further Development and Optical Properties
(b) Vol. **103**, No. 1, p. 139–149 (1981).
See also *13.1*, p. *151*; 18.2, p. 267.

G. Bader and Y. Lépine
A Two-Particle Model for the Ground State Energy of the Polaron
(b) Vol. **103**, No. 1, p. 177–184 (1981).
See also *13.5.3*, p. *196*.

J. Sinkkonen
AC Properties of a Random Barrier Network
(b) Vol. **103**, No. 1, p. 231–237 (1981).
See also *14.3.4*, p. *232*.

R. J. Jerrard, H. Ueba, and S. G. Davison
Virtual Surface States of Thin Films
(b) Vol. **103**, No. 1, p. 353–359 (1981).
See also *13.3*, p. *165*.

A. A. Gogolin
Mobility of a Small Polaron at Low Temperatures
(b) Vol. **103**, No. 1, p. 397–409 (1981).
See also *13.5.3*, p. *196*; *14.3*, p. *209*.

Z. W. Gortel, J. Szymański, and L. Świerkowski
Acceptor States in Zero-Gap Semiconductors in Magnetic Field. Theory for Short-Range Defect Potential
(b) Vol. **103**, No. 1, p. 429–439 (1981).
See also 13.1, p. 151; *13.4*, p. *169*; 18, p. 260.

E. V. Goncharova, V. V. Kaminskii, V. A. Kapustin, and I. A. Smirnov
Defects and Isomorphic Semiconductor–Metal Phase Transition in Samarium Monosulphide
(b) Vol. **103**, No. 1, p. K9–K11 (1981).
See also *1.2*, p. *14*; 13.4, p. 169.

T. P. Humphreys and G. P. Srivastava
Electronic Charge Density for Two Isoelectronic Series: Ge–GaAs–ZnSe and α-Sn–InSb–CdTe
(b) Vol. **103**, No. 1, p. K85–K90 (1981).
See also 4, p. 57; *13.1*, p. *152*.

B. J. Mulder and S. van Heusden
Mechanism of the Darkening of Soda–Lime Glass by a Low-Pressure Mercury Vapour Discharge
(a) Vol. **63**, No. 1, p. 137–141 (1981).
See also *2*, p. *40*; 9, p. 89. 20.1, p. 302.

J. Morawiec and E. Mizera
X-Ray and TEM Investigations of PbTe Crystal
(a) Vol. **63**, No. 1, p. K1–K2 (1981).
See also *1.1*, p. *11*; 10.2, p. 106.

Sh. A. Gulamov, Yu. A. Katulevskii, and R. A. Muminov
Peculiarities of an Electromotive Force Arising in a Semiconductor in a Non-Homogeneous Magnetic Field (I)
(b) Vol. **103**, No. 2, p. 567–572 (1981).
See also *14.3*, p. *210*; 18, p. 260.

S. Ghosh, A. N. Basu, and S. Sengupta
Extended Deformable Shell Model Calculation of the Lattice Statics and Dynamics of LiH and LiD Crystals
(b) Vol. **103**, No. 2, p. 595–603 (1981).
See also *6*, p. *67*; *12.2*, p. *138*.

T. Chattopadhyay and H. G. von Schnering
Infrared Absorption Studies on Crystalline α-P_4S_3
(b) Vol. **103**, No. 2, p. 637–642 (1981).
See also *6*, p. *68*; *20.1*, p. *303*.

R. Pässler
Temperature Dependences of the Nonradiative Multiphonon Carrier Capture and Ejection Properties of Deep Traps in Semiconductors (II): Interpretation and Extrapolation of Capture Data
(b) Vol. **103**, No. 2, p. 673–686 (1981).
See also *13.4*, p. *170*.

M. Wautelet
Jahn-Teller-Type Phase Transition in Solids during Laser Irradiation
(b) Vol. **103** No. 2, p. 703–707 (1981).
See also *1.2*, p. *14*; *13.4*, p. *170*.

N. G. Bebenin
The Critical Dynamics of Hot Electrons in a Ferromagnetic Semiconductor at $T = T_c$
(b) Vol. **104**, No. 1, p. K65–K70 (1981).
See also *13*, p. *143*; *14.3.4*, p. *233*; 18.2, p. 269.

V. K. Arora
Effect of Spin-Splitting on the Magnetoresistance in Parabolic Semiconductors
(b) Vol. **104**, No. 1, p. K71–K73 (1981).
See also *13.1*, p. *154*; *14.3*, p. *210*; 18, p. 260.

D. Schalch, A. Scharmann, and H. Weber
Electrical Conduction in Reactively Sputtered Silicon Nitride Films at Medium Field Strengths
(a) Vol. **64**, No. 1, p. 159–167 (1981).
See also *2*, p. *42*; 13.4, p. 172; *14.3.3*, p. *226*; 20.1, p. 306.

B. Esser and I. P. Zvyagin
On the Time Dependence of the Capture of Non-Equilibrium Hopping Charge Carriers
(b) Vol. **104**, No. 2, p. 523–529 (1981).
See also *2*, p. *43*; *13.4*, p. *173*.

I. Gregora, B. Velický, and M. Závětová
Refractive Index of Crystalline and Amorphous GeS
(b) Vol. **104**, No. 2, p. K95–K100 (1981).
See also 2, p. 44; *20.1*, p. *307*.

V. M. Koleshko and N. Yu. Trifonov
Fine Impurity Band Structure of One-Dimensional Disordered Crystal
(b) Vol. **104**, No. 2, p. K105–K108 (1981).
See also *13.1*, p. *155*; *13.4*, p. *173*.

M. Stordeur and H. T. Langhammer
The Lorenz Number for Many-Valley Semiconductors on Special Transport Conditions
(b) Vol. **104**, No. 2, p. K121–K124 (1981).
See also *8*, p. *85*; *14.3*, p. *211*.

R. Claus
Directional Dispersion of Polar Phonon Frequencies Derived by a Simple Electrostatic Argument
(b) Vol. **104**, No. 2, p. K163–K164 (1981).
See also *6*, p. *72*.

E. M. Raspopova, M. Závětová, J. Krištofik, and B. T. Kolomiets
The Influence of Defects Generated during Densification on the Properties of the As–Se System
(a) Vol. **64**, No. 2, p. 457–460 (1981).
See also *2*, p. *44*; 10.2, p. 109; 14.4, p. 238; *20.1*, p. *308*.

F. Stöckmann
An Exact Evaluation of Steady-State Space-Charge-Limited Currents for Arbitrary Trap Distributions
(a) Vol. **64**, No. 2, p. 475–483 (1981).
See also *13.4*, p. *174*; *14.3.4*, p. *233*.

M. KITAO, H. AKAO, T. ISHIKAWA, and S. YAMADA
Influence of Copper Addition on Electrical and Optical Properties of Amorphous As_2Se_3
(a) Vol. **64**, No. 2, p. 493–498 (1981).
See also 2, p. 44; 13.4, p. 174; *14.3*, p. *212*; *20.1*, p. *308*.

V. P. ROMANOV and YU. A. CHAPLYGIN
A Theoretical Model for the Stationary Distribution of Mobile Ions in a Double-Layer Insulator
(a) Vol. **64**, No. 2, p. 525–531 (1981).
See also *14.3.3*, p. *226*; *14.4*, p. *239*.

D. LUKÁŠIK and P. MACKO
Creation of Thermal Spikes in the GeS_{11} Glassy System by Fast Neutron Irradiation
(a) Vol. **64**, No. 2, p. 677–682 (1981).
See also 2, p. 44; *11*, p. *124*.

G. VAN TENDELOO, J. VAN LANDUYT, and S. AMELINCKX
Real Space Observation of Charge Density Waves in the Incommensurate Room Temperature Phase of TaS_2
(a) Vol. **64**, No. 2, p. K105–K106 (1981).
See also *1.3*, p. *23*; *1.4*, p. *29*.

S. N. KOZLOV and N. L. LEVSHIN
On the Distribution Function of Activation Barriers Surmounted by Charge Carriers when Captured to Slow Surface States of a Semiconductor
(a) Vol. **64**, No. 2, p. K169–K172 (1981).
See also *13.3*, p. *165*; *14.3.2*, p. *223*.

Y. ISHIHARA and I. NAKADA
Electrical Conduction of GeSe at Low Temperatures
(b) Vol. **105**, No. 1, p. 285–290 (1981).
See also *13.4*, p. *175*; *14.3*, p. *212*.

A. ANDERSON, P. GASH, and F. LÜTY
Lattice Dynamics of KCN and NaCN in the Anti-Ferroelectric Phase
(b) Vol. **105**, No. 1, p. 315–325 (1981).
See also *6*, p. *74*; 14.4.1, p. 243; *20.1*, p. *309*.

V. G. BARU and E. V. GREKOV
Van der Walls Theory to the Phase Transition in Quasi-Two-Dimensional Electron Gas in Magnetic Semiconductors
(b) Vol. **105**, No. 1, p. 369–376 (1981).
See also *13*, p. *145*; 14.3.3, p. 227; 18.2, p. 270.

SH. A. GULAMOV, YU. A. KATULEVSKII, and R. A. MUMINOV
Peculiarities of an Electromotive Force Arising in a Semiconductor in a Non-Homogeneous Magnetic Field (II)
(b) Vol. **105**, No. 1, p. 405–408 (1981).
See also *14.3*, p. *213*; 18, p. 261.

A. I. Bobrysheva, V. T. Zyukov, and S. A. Moskalenko
The Excited States of the Surface Biexciton
(b) Vol. **105**, No. 1, p. K45–K49 (1981).
See also *13.3*, p. *165*; *13.5.1*, p. *190*.

L. V. Chernysh, V. A. Kovarskii, and E. P. Sinyavskii
Laser Induced Auger Recombination in Doped Semiconductors
(b) Vol. **105**, No. 1, p. K51–K53 (1981).
See also 6, p. 74; *13.4*, p. *175*; 16, p. 253.

V. M. Efimov
Study of Thin Silicon Nitride Films by Isothermal Depolarization on MNOS Structures
(a) Vol. **65**, No. 1, p. 177–187 (1981).
See also *13.4*, p. *176*; *14.3.1*, p. *221*; 14.3.3, p. 227.

D. J. Robbins, P. T. Landsberg, and E. Schöll
Threshold Switching as a Generation–Recombination Induced Non-Equilibrium Phase Transition (II)
(a) Vol. **65**, No. 1, p. 353–364 (1981).
See also 2, p. 46; 13.4, p. 176; *14.3.4*, p. *234*.

N. M. Ravindra, V. K. Srivastava, and B. Mishra
Temperature Dependent Electrical Conductivity of Natural Galena Crystals
(a) Vol. **65**, No. 1, p. K59–K61 (1981).
See also 13.1, p. 156; *14.3*, p. *213*.

Z. S. Gribnikov, V. M. Ivastchenko, and V. V. Mitin
Nonlocality of Carrier Multiplication in Semiconductor Depletion Layers
(b) Vol. **105**, No. 2, p. 451–459 (1981).
See also 13.4, p. 176; *14.3.3*, p. *227*.

J. Szczyrbowski
The Exponential Shape of the Optical Absorption Edge Tail
(b) Vol. **105**, No. 2, p. 515–524 (1981).
See also *13.1*, p. *156*; *20.1*, p. *310*.

H. Stolz, R. Zimmermann, and G. Röpke
Correlated Hartree-Fock Theory of the Electron–Hole Plasma Containing Exciton Bound States
(b) Vol. **105**, No. 2, p. 585–595 (1981).
See also *13.5*, p. *186*; *13.5.1*, p. *190*.

S. N. Klymin and E. P. Pokatilov
Non-Linear Optical Properties of the Degenerate Band Charge Carrier Gas
(b) Vol. **105**, No. 2, p. 641–648 (1981).
See also *20*, p. *297*.

V. K. Arora
Size-Dependent Electric Conductivity in Semiconducting Thin Wires
(b) Vol. **105**, No. 2, p. 707–713 (1981).
See also *14.3*, p. *214*.

T. HEJWOWSKI and M. SUBOTOWICZ
The Mn^{2+} Ion Spectra in $Pb_xSn_{1-x}Te$ Alloys
(b) Vol. **106**, No. 1, p. 373–377 (1981).
See also **14.3**, p. 215; *19*, p. *291*.

Y. ISHIHARA and I. NAKADA
Negative Magnetoresistance of GeSe in a Weak Magnetic Field
(b) Vol. **106**, No. 1, p. K27–K31 (1981).
See also *14.3*, p. *215*.

I. A. TALE
Trap Spectroscopy by the Fractional Glow Technique
(a) Vol. **66**, No. 1, p. 65–75 (1981).
See also *13.4*, p. *178*; *20.3*, p. *330*.

E. M. TRUKHANOV and S. I. STENIN
Initial Stage of Plastic Deformation of Substrates in Heterostructures
(a) Vol. **66**, No. 1, p. 125–131 (1981).
See also 1.5, p. 35; *10.2*, p. *112*; 22.1.1, p. 405; 22.1.2, p. 415; 22.6, p. 478.

M. T. KOSTYSHIN and YU. V. USHENIN
He–Ne Laser Radiation-Stimulated Diffusion of Ag into Vitreous As_2S_3
(a) Vol. **66**, No. 1, p. K47–K50 (1981).
See also *9*, p. *91*.

A. M. BEREZHKOVSKII, A. A. OVCHINNIKOV, and R. A. SURIS
A New Method for Detecting the Wannier-Stark Ladder in a Semiconductor in a Strong Electric Field
(b) Vol. **106**, No. 2, p. 461–466 (1981).
See also 13.1, p. 158; *14.3.4*, p. *235*; *20.1*, p. *314*.

V. D. ISKRA
Phonon Assisted Interband Optical Transitions in Semiconductors Substaining a Coulomb Random Field
(b) Vol. **106**, No. 2, p. 627–633 (1981).
See also 6, p. 77; *13.1*, p. *159*; *20.1*, p. *315*.

M. J. GELTEN and L. A. BOSCH
A New Method to Determine the Coupled Mode Parameters of a Plasmon–Multiphonon System
(b) Vol. **106**, No. 2, p. 635–645 (1981).
See also *6*, p. *77*; *13.5.2*, p. *195*; 20.1, p. 315.

M. K. AGARWAL, J. V. PATEL, and N. G. PATEL
Stacking Fault Energy Measurements in WSe_2 Single Crystals Using Weak-Beam Techniques
(a) Vol. **66**, No. 2, p. 425–429 (1981).
See also *1.4*, p. *30*; *10.2*, p. *114*.

N. SINGH, G. L. MARWAHA, and V. K. MATHUR
Luminescence Centres and Charge Compensation in CaS Phosphors
(a) Vol. **66**, No. 2, p. 761–765 (1981).
See also **13.4**, p. 179; *20.3*, p. *331*.

S. Ghosh, N. Apte, and V. K. Agarwal
Excitation of the Hybrid Mode Due to Modulation of an Intense Laser Beam in a Transversely Magnetised Semiconductor Plasma
(b) Vol. **107**, No. 1, p. 75–81 (1981).
See also *7*, p. *83*; 18, p. 263; *20*, p. *298*.

J. van Royen, F. Beleznay, and J. T. Devreese
On the Equivalence of Different Approximations in Linear Electron Transport Theory
(b) Vol. **107**, No. 1, p. 335–345 (1981).
See also 13.5.3, p. 196; *14.3*, p. *216*.

G. L. Bleris, J. G. Antonopoulos, Th. Karakostas, and P. Delavignette
Grain Boundary Analysis in TEM (V): A Criterion for the Identification of a CSL and a Characterization of an Approximate CSL
(a) Vol. **67**, No. 1, p. 249–257 (1981).
See also *1.1*, p. *12*; *1.4*, p. *31*.

G. A. Medvedkin and Yu. V. Rud
The Parameters of Polarization Photosensitivity of Isotropic Semiconductors
(a) Vol. **67**, No. 1, p. 333–337 (1981).
See also *16*, p. *255*.

M. Leon, N. Terao, and F. Rueda
On the Phase Transition of Chalcocite Cu_2S Thin Films
(a) Vol. **67**, No. 1, p. K11–K14 (1981).
See also *1.2*, p. *19*; *1.5*, p. *36*.

B. B. Baizakov, R. A. Moominov, S. H. Shamirzaiev, and A. Ya. Yafasov
Bias Current in Semiconductors during Overcharging of Impurity Centres
(a) Vol. **67**, No. 1, p. K59–K62 (1981).
See also *13.4*, p. *181*; *14.3*, p. *217*.

B. Olejníková, Ľ. Hrivnák, and M. Kedro
The Model Potential for Charged Centres in Semiconductors
(b) Vol. **107**, No. 2, p. 451–460 (1981).
See also *13.4*, p. *181*.

D. K. Sinha and M. Gupta
Nonlinear Parametric Interaction in a Piezosemiconducting Medium
(b) Vol. **107**, No. 2, p. 469–478 (1981).
See also 6, p. 79; *7*, p. *83*.

Doan Nhat Quang, B. Esser, and R. Keiper
On the Theory of Resonance Raman Scattering in Heavily Doped Degenerate Semiconductors
(b) Vol. **107**, No. 2, p. 535–542 (1981).
See also *6*, p. *79*; *20.1*, p. *317*.

I. K. Bandrovskaya, V. Ya. Prokhorenko, V. G. Romanov, and B. I. Sokolovskii
Optical and Thermoelectric Properties of Arsenic Selenide in Glassy and Liquid States
(b) Vol. **107**, No. 2, p. K95–K98 (1981).
See also 2, p. 50; *14.3*, p. *217*; 15, p. 249; *20.1*, p. *318*.

Y. AOYAGI, Y. SEGAWA, S. NAMBA, T. SUHARA, H. NISHIHARA, and H. GAMO
Dynamic Behavior of the Photodarkening Process in As_2S_3 Chalcogenide Glass
(a) Vol. **67**, No. 2, p. 669–676 (1981).
See also *2*, p. *50*; *10.2*, p. *117*; 20, p. 299.

B. R. NAG and A. N. CHAKRAVARTI
Generalized Einstein Relation – A Suggested Method for Experimental Verification
(a) Vol. **67**, No. 2, p. K113–K117 (1981).
See also *14.3*, p. *217*.

N. N. KOREN and T. E. BORISENKO
Space-Charge-Limited Currents in GeS Single-Crystal Layers
(a) Vol. **67**, No. 2, p. K147–K150 (1981).
See also *14.3.4*, p. *235*.

E. KLIER
A Simple Method of Evaluation of TSC Measurements
(a) Vol. **67**, No. 2, p. K151–K154 (1981).
See also *13.4*, p. *182*; *14.3*, p. *217*.

M. N. ALIEV
Theory of the Nuclear Magnetic Resonance Lineshape in Magnetically Diluted Crystals
(b) Vol. **108**, No. 1, p. 59–64 (1981).
See also *19*, p. *293*.

B. OLEJNÍKOVÁ
The Influence of the Electric Field on the Electron Capture Coefficient of Screened Coulomb Centres
(b) Vol. **108**, No. 1, p. 79–85 (1981).
See also *13.4*, p. *182*; *14.3.4*, p. *235*.

Z. V. POPOVIĆ and H. J. STOLZ
Infrared and Raman Spectra of Germanium Dichalcogenides (II): $GeSe_2$
(b) Vol. **108**, No. 1, p. 153–163 (1981).
See also *6*, p. *80*; *20.1*, p. *319*.

B. ESSER
On the Lifetime of Excited States in One-Dimensional Disordered Systems
(b) Vol. **108**, No. 1, p. 195–204 (1981).
See also *2*, p. *50*; *13.4*, p. *182*; 20.3, p. 333.

T. CHATTOPADHYAY and H. G. VON SCHNERING
Temperature Dependence of the Infrared Absorption Spectrum of P_4S_3
(b) Vol. **108**, No. 1, p. 241–244 (1981).
See also *6*, p. *80*; *20.1*, p. *319*.

A. L. DAWAR, S. K. PARADKAR, P. KUMAR, O. P. TANEJA, and P. C. MATHUR
Electrical Transport Properties of Ag-Doped p-Type PbTe Epitaxial Thin Films
(a) Vol. **68**, No. 1, p. 227–232 (1981).
See also *14.3.1*, p. *221*.

B. R. NAG, A. N. CHAKRAVARTI, and P. K. BASU
Generalized Einstein Relation for the Diffusity–Mobility Ratio in Multi-Band Degenerate Semiconductors
(a) Vol. **68**, No. 1, p. K75–K80 (1981).
See also 13.1, p. 161; *14.3*, p. *218*.

V. P. GUPTA, A. GUPTA, and V. K. SRIVASTAVA
Electronic Properties of Gallium Chalcogenides (GaS, GaSe, and GaTe)
(b) Vol. **108**, No. 2, p. 323–327 (1981).
See also *4*, p. *61*; 14.4, p. 241.

A. N. GUSATINSKII, M. A. BLOKHIN, P. SHUMICHEN, M. A. BUNIN, and S. S. MIKHAILOVA
X-Ray L_1Spectra and XPS Investigation of the Electronic Structure of Indium Compounds
(b) Vol. **108**, No. 2, p. 405–410 (1981).
See also *13.1*, p. *161*; *20*, p. *299*; 21, p. 349; 22.2, p. 426; 22.3, p. 442.

W. GEHLHOFF and K. H. SEGSA
Changes of EPR Spectra by Inhomogeneous Plane Stress Distributions in a Two-Layer System
(b) Vol. **108**, No. 2, p. 421–433 (1981).
See also *12.2*, p. *141*; *19*, p. *294*.

W. NOLTING and A. M. OLÉS
Simple Spectral Density Approach to the Problem of Mixed Valency
(b) Vol. **108**, No. 2, p. 459–466 (1981).
See also *13*, p. *150*.

U. BERG, T. CHASSÉ, and O. BRÜMMER
Investigation of the XPS Valence Band Structure from Sn Chalcogenides
(b) Vol. **108**, No. 2, p. 507–510 (1981).
See also *13.1*, p. *161*; 17, p. 258.

V. ČÁPEK and R. W. MUNN
From Band to Hopping Diffusion of Excitons and Charge Carriers in Molecular Crystals (I): Principles
(b) Vol. **108**, No. 2, p. 521–530 (1981).
See also *13.5.1*, p. *192*.

I. A. CHAIKOVSKII, G. M. SHMELEV, and N. A. ENAKI
Light Absorption in Semiconductors with Superlattice in Quantizing Magnetic Field
(b) Vol. **108**, No. 2, p. 559–566 (1981).
See also *20.1*, p. *320*.

A. M. GHODGAONKAR and K. RAMANI
Evaluation of Bulk Modulus, Pressure Derivative of Bulk Modulus, and Cohesive Energy by Modified Morse-Kratzer Potential
(b) Vol. **108**, No. 2, p. K133–K138 (1981).
See also *4*, p. *61*; *12.2*, p. *141*; 22.5.1, p. 461; 22.5.3, p. 472.

J. Gazsó

Determination of the Density of Localized States from the Field Effect in Wide Gap Semiconductors

(a) Vol. **68**, No. 2, p. 675–688 (1981).

See also 2, p. 51; *13.4*, p. *184*; 14.3.1, p. 222; *14.3.4*, p. *236*.

22.1 Elements

Original Papers and Short Notes

E. Kühbt and R. Wedell
Theoretical Description of Planar Channeling Process at Small Depths
(b) Vol. **107**, No. 2, p. 665–675 (1981).
See also *11*, p. *129*; 22.1.2, p. 420.

H. E. Roosendaal, B. Schmiedeskamp, H. H. Hubbes, and H. O. Lutz
Planar Channeling Trajectories in Diamond and Silicon
(b) Vol. **107**, No. 2, p. 677–687 (1981).
See also *11*, p. *129*; 22.1.2, p. 420.

A. A. Bednyakov, Yu. V. Bulgakov, V. S. Nikolaev, and V. L. Chernov
Energy Straggling of Hydrogen and Helium Ions in Al, C, and Polystyrene at Energies of Tens and Hundreds keV per Nucleon
(a) Vol. **68**, No. 1, p. 187–192 (1981).
See also *11*, p. *131*; 21, p. 348; 22.9, p. 512.

A. C. Sharma and S. Auluck
Small Wave Vector ***q*** and Frequency *ω* Dependent Longitudinal Dielectric Function for a Model Semiconductor
(b) Vol. **108**, No. 2, p. 587–600 (1981).
See also *13.1*, p. *161*; *14.4*, p. *241*; 22.2, p. 426; 22.4, p. 446.

V. K. Bashenov, A. G. Gontar, and A. G. Petukhov
Excited States of a Shallow Acceptor in Semiconducting Diamond from Photoconductivity Spectra
(b) Vol. **108**, No. 2, p. K139–K142 (1981).
See also *13.4*, p. *184*; *16*, p. *255*.

O. Bostanjoglo and G. Schlotzhauer
Impulse Stimulated Crystallization of Sb Films Investigated by Time Resolved TEM
(a) Vol. **68**, No. 2, p. 555–560 (1981).
See also 1.2, p. 21; 2, p. 51; *3*, p. *56*.

22.1.1 Germanium

Original Papers and Short Notes

S. R. Bakhchieva, M. G. Kekua, and P. P. Seregin
A Study of Tin Impurity Atoms in Silicon, Germanium, and Silicon–Germanium Solid Solutions by Means of Mössbauer Spectroscopy
(a) Vol. **63**, No. 1, p. 23 – 30 (1981).
See also *5*, p. *62*; 10.2, p. 106; 22.1.2, p. 408.

A. Cavallini, P. Gondi, and A. Castaldini
Shallow Dislocation States in Ge. Thermal Treatment Effects
(a) Vol. **63**, No. 1, p. 143 – 149 (1981).
See also *10.2*, p. *106*; *13.4*, p. *169*; 14.3, p. 209.

S. Kończak and M. Nowak
The Estimation of Semiconductor Parameters Using Least Squares in Photomagnetoelectric Investigations
(a) Vol. **63**, No. 1, p. 305 – 311 (1981).
See also *16*, p. *251*; 22.2.3, p. 435.

R. Ts. Gabrielyan and V. G. Kohn
The Forbidden-Reflection Method for Absorption Coefficient Measurement in Many-Wave X-Ray Diffraction in Single Crystals
(a) Vol. **63**, No. 1, p. 345 – 353 (1981).
See also *1*, p. *9*.

A. I. Kolyubakin and S. A. Shevchenko
On the Spectrum of Dislocation States in Germanium
(a) Vol. **63**, No. 2, p. 677 – 685 (1981).
See also *10.2*, p. *107*; *13.4*, p. *171*; 14.3, p. 210.

M. Misra and N. P. Sinha
Photoresponse and Barrier Height of Al–aGe–nSi–Mn Structures
(a) Vol. **63**, No. 2, p. K207 – K210 (1981).
See also *14.3.3*, p. *226*; *16*, p. *252*; 22.1.2, p. 410.

W. Ebeling, W. Richert, W. D. Kraeft, and W. Stolzmann
Padé Approximations for the Thermodynamic Functions of Weakly Interacting Coulombic Quantum Systems
(b) Vol. **104**, No. 1, p. 193 – 202 (1981).
See also *13*, p. *143*; 13.5, p. 186.

T. Soma
Pressure Derivatives of Elastic Stiffness Constants under Pressure of Si and Ge
(b) Vol. **104**, No. 1, p. 293 – 298 (1981).
See also *12.2*, p. *139*; 22.1.2, p. 410.

M. I. Tarasik, V. D. Tkachev, V. U. Yavid, and A. M. Yanchenko
Capture Mechanism of Holes by Radiation Defects with $E_c - 0.20$ eV Level in Sb-Doped Ge
(b) Vol. **104**, No. 2, p. K117 – K120 (1981).
See also *11*, p. *123*; *13.4*, p. *173*.

A. F. Khokhlov, A. I. Mashin, and A. M. Satanin
The Dependence of Magnetic Ordering Temperature in Amorphous Semiconductors on Paramagnetic Centre Concentration
(b) Vol. **105**, No. 1, p. 129–136 (1981).
See also 2, p. 45; 11, p. 124; *18.1*, p. *266*; *18.4*, p. *282*; 19, p. 289; 22.1.2, p. 413.

S. Cristoloveanu
Magnetic Field and Surface Influences on Double Injection Phenomena in Semiconductors (II): The Magnetodiode Effect Theory for the Diffusion Regimes, Magnetodiode Experiments and Applications
(a) Vol. **65**, No. 1, p. 281–292 (1981).
See also *14.3.3*, p. *227*; 14.3.4, p. 234; *18*, p. *262*; 22.1.2, p. 414.

F. R. Kessler and J. W. Mangelsdorf
Free Carrier Compression and Galvanomagnetic Luminescence of Germanium at High Lorentz Fields
(b) Vol. **105**, No. 2, p. 525–535 (1981).
See also 14.3, p. 214; *20.3*, p. *328*.

B. M. Ashkinadze
Role of Free Electrons and Holes in the Exciton Condensation Process in Semiconductors
(b) Vol. **105**, No. 2, p. 633–640 (1981).
See also *13.5.1*, p. *190*; 20.3, p. 328.

S.-L. Chang
Excitations and Phases of Two-Beam Dynamical Bragg Reflection of X-Rays for Infinitely Thick Crystals
(a) Vol. **65**, No. 2, p. 553–562 (1981).
See also *1.1*, p. *11*.

T. Soma, H. Matsuo, and T. Katoh
The Heat of Mixing in Liquid Si–Ge, Si–Sn, and Ge–Sn Mixtures
(b) Vol. **106**, No. 1, p. K5–K7 (1981).
See also *1.2*, p. *17*; *1.3*, p. *24*; 21.7, p. 388; 22.1.2, p. 415.

U. Pietsch
Bond Charge Determination from X-Ray Data for Germanium
(b) Vol. **106**, No. 1, p. K9–K11 (1981).
See also *1.1*, p. *12*; *4*, p. *59*.

V. F. Kisilev, S. N. Kozlov, and N. L. Levshin
On the Mechanism of Dissipation of Energy Released in the Capture of Charge Carriers to Adsorptive Slow States of a Semiconductor
(a) Vol. **66**, No. 1, p. 93–101 (1981).
See also *13.3*, p. *166*; 14.3.2, p. 224.

E. M. Trukhanov and S. I. Stenin
Initial Stage of Plastic Deformation of Substrates in Heterostructures
(a) Vol. **66**, No. 1, p. 125–131 (1981).
See also 1.5, p. 35; *10.2*, p. *112*; 22, p. 397; 22.1.2, p. 415; 22.6, p. 478.

K. Masuda
Calculation of Atomic Relaxation near the (111) 1 × 1 Surface of Covalent Semiconductors: Tight-Binding Green's Function Approach
(b) Vol. **107**, No. 2, p. 529–533 (1981).
See also *1.6*, p. *39*; *13.3*, p. *166*; 22.1, p. 402; 22.1.2, p. 419.

R. Zimmermann, M. Rösler, and V. M. Asnin
Direct Absorption inside the Electron–Hole Liquid in Ge: Warping Effects and Final State Interactions
(b) Vol. **107**, No. 2, p. 579–590 (1981).
See also *13.5.1*, p. *192*; 13.5.2, p. 195; *20.1*, p. *317*.

F. Bechstedt, R. Enderlein, and R. Wischnewski
Binding Energies and Chemical Shifts of Least Bound Core Electron Excitations in Cubic A_NB_{8-N} Semiconductors
(b) Vol. **107**, No. 2, p. 637–651 (1981).
See also *4*, p. *60*; *13*, p. *148*; 17, p. 258; 22.1, p. 402; 22.1.2, p. 419; 22.2, p. 426; 22.4, p. 446.

V. L. Vinetskii, I. R. Entinzon, and G. A. Kholodar
Probability of Defect Formation in an Elastic Impact of Atoms in Crystals
(a) Vol. **67**, No. 2, p. 477–483 (1981).
See also *10.2*, p. *117*; *11*, p. *130*; 22.1.2, p. 420.

D. A. Aronov, R. Mamatkulov, and V. V. Rubinov
On the Theory of Carrier Exclusion in "Pure" Intrinsic Semiconductors
(a) Vol. **68**, No. 1, p. 61–68 (1981).
See also *14.3.4*, p. *236*.

Bhanwar Singh and K. L. Chopra
A Novel Bi-Layer System for High Resolution Optical Imaging
(a) Vol. **68**, No. 1, p. K5–K9 (1981).
See also *2*, p. *50*; *20.1*, p. *319*; 22.1.3, p. 425.

A. Baldereschi, K. Maschke, A. Milchev, R. Pickenhain, and K. Unger
Thomas-Fermi Approximation for the Valence Electron Densities in Cubic Semiconductors and Insulators
(b) Vol. **108**, No. 2, p. 511–520 (1981).
See also *13.1*, p. *161*; 14.3, p. 218; 21.7, p. 389; 22.1.2, p. 422; 22.2.1, p. 431; 22.4.2, p. 453; 22.5.2, p. 469.

S. N. Karyagin
On the Nature of Centers Responsible for the Spin-Dependent Recombination in Si and Ge
(a) Vol. **68**, No. 2, p. K113–K116 (1981).
See also *10.2*, p. *120*; *19*, p. *295*; 22.1.2, p. 424.

22.1.2 Silicon

Review Articles

Original Papers and Short Notes

A. Szekeres, K. Kirov, and S. Alexandrova
Plasma Photoresist Stripping in a Planar Reactor
(a) Vol. **63**, No. 1, p. 371–374 (1981).
See also *14.3.2*, p. *223*.

M. S. Abrahams, J. L. Hutchison, and G. R. Booker
Direct Observation of a Silicon–Sapphire Hetero-Epitaxial Interface by High Resolution Transmission Electron Microscopy
(a) Vol. **63**, No. 1, p. K3–K6 (1981).
See also *1.4*, p. *28*; 3, p. 53; 22.6, p. 474.

A. N. Chakravarti, A. K. Chowdhury, and K. P. Ghatak
Oscillatory Electronic Heat Capacity in Inversion Layers on ⟨100⟩ Oriented p-Type Si in a Quantizing Magnetic Field
(a) Vol. **63**, No. 1, p. K97–K101 (1981).
See also *8*, p. *84*; *14.3.2*, p. *223*; 14.3.3, p. 225.

V. A. Grazhulis, V. V. Kveder, and Yu. A. Osipyan
Investigation of the Dislocation Spin System in Silicon as Model of One-Dimensional Spin Chains
(b) Vol. **103**, No. 2, p. 519–528 (1981).
See also *13.4*, p. *170*; *19*, p. *287*.

P. John, I. M. Odeh, M. J. K. Thomas, M. J. Tricker, and J. I. B. Wilson
Silicon–Hydrogen Stretching Modes in Amorphous Silicon. A Semi-Empirical Calculation of the Infrared Band Widths
(b) Vol. **103**, No. 2, p. K141–K146 (1981).
See also 2, p. 41; *6*, p. *68*; *20.1*, p. *304*.

J. Tatarkiewicz
On Some Proton Radiation Effects in Silicon
(a) Vol. **63**, No. 2, p. 423–429 (1981).
See also *11*, p. *121*; *20.1*, p. *304*.

V. I. Khrupa, E. N. Kislovskii, and L. I. Datsenko
Energy Absorption at the X-Ray Interference Transmission in an Elastically Bent Thin Crystal
(a) Vol. **63**, No. 2, p. 439–443 (1981).
See also *12.2*, p. *139*.

J. Gong, K. M. van Vliet, A. D. Sutherland, and E. R. Chenette
Noise Measurements on Photo Avalanche Diodes
(a) Vol. **63**, No. 2, p. 445–460 (1981).
See also *14.3.3*, p. *225*; 16, p. 251.

P. H. Tsien, S. C. Tsou, M. Takai, D. Röschenthaler, M. Ramin, H. Ryssel, I. Ruge, and K. Wittmaack
Annealing of Boron-Implanted Silicon Using a CW CO_2 Laser
(a) Vol. **63**, No. 2, p. 547–555 (1981).
See also *1.5*, p. *34*; *3*, p. *53*; 11, p. 121; 14.3.1, p. 220.

A. Le Bloa, P. N. Favennec, and Y. Colin
DLTS Method Using a Single Temperature Scanning
(a) Vol. **64**, No. 1, p. 85–93 (1981).
Erratum
(a) Vol. **66**, No. 2, p. 807 (1981).
See also *13.4*, p. *172*; 14.3, p. 210; 22.2.1, p. 428.

C. Wagner, A. El-Sadek, and H.-J. Mechelke
The Annealing Characteristics of Arsenic-Implanted Silicon Investigated at Low Temperatures
(a) Vol. **64**, No. 1, p. 143–150 (1981).
See also *11*, p. *122*; *14.3*, p. *211*.

K. Schmalz and P. Gaworzewski
On the Donor Activity of Oxygen in Silicon at Temperatures from 500 to 800 °C
(a) Vol. **64**, No. 1, p. 151–158 (1981).
See also *13.4*, p. *172*; 14.3, p. 211.

G. F. Golovanova, V. F. Kiselev, S. N. Kozlov, A. S. Petrov, and E. A. Silaev
On the Effect of Halogens on Proton Processes in the Si–SiO_2 Structures
(a) Vol. **64**, No. 1, p. 177–186 (1981).
See also *13.3*, p. *165*; 19, p. 288; 22.6, p. 475.

H. Lemke
Doping Levels of Iron in Silicon
(a) Vol. **64**, No. 1, p. 215–224 (1981).
See also *13.4*, p. *172*; 14.3, p. 211.

V. G. Kohn and M. V. Kovalchuk
On the Theory of the External Photoeffect Accompanying X-Ray Diffraction in an Ideal Crystal with Disturbed Surface Layer
(a) Vol. **64**, No. 1, p. 359–366 (1981).
See also *1.6*, p. *38*; *17*, p. *256*.

J. P. Hauck, M. M. Beguwala, C. L. Hayes, and R. F. Palys
Laser Annealing of Silicon Wafers at 10.6 μm
(a) Vol. **64**, No. 1, p. 407–412 (1981).
See also *11*, p. *123*; 14.3, p. 211.

Pa Tu and He Yizhen
Erratum to: Motion of Dislocations from an Indentation Rosette on Silicon Crystals
(a) Vol. **64**, No. 1, p. 419 (1981).
See also *10.2*, p. *108*.

S. V. Joshi and M. C. Joshi
Investigation of the E_c – 0.55 eV Level of Iron in Silicon
(a) Vol. **64**, No. 1, p. K7–K11 (1981).
See also *13.4*, p. *172*.

P. Saxena, K. Bahl, and P. C. Mathur
1/*f* Noise in Si Avalanche Diodes at X-Band
(a) Vol. **64**, No. 1, p. K49–K53 (1981).
See also *14.3.3*, p. *226*.

K. H. Herrmann and T. X. Hoai
Saturated Open Circuit Photovoltage — a New Method for Measuring the Carrier Lifetime at High Excitation Level
(a) Vol. **64**, No. 1, p. K59—K62 (1981).
See also *14.3.3*, p. *226*; *16*, p. *252*.

N. A. Drozdov, A. A. Patrin, and V. D. Tkachev
Modification of the Dislocation Luminescence Spectrum by Oxygen Atmospheres in Silicon
(a) Vol. **64**, No. 1, p. K63—K65 (1981).
See also *10.2*, p. *108*; *20.3*, p. *326*.

R. Klabes, R. Grötzschel, M. Voelskow, S. Panzer, and H. Bartsch
High-Speed Electron Beam Annealing of Arsenic and Gallium Implanted Silicon
(a) Vol. **64**, No. 1, p. K73—K75 (1981).
See also *11*, p. *123*.

J. Doerschel and F.-G. Kirscht
Differences in Plastic Deformation Behaviour of CZ and FZ-Grown Silicon Crystals
(a) Vol. **64**, No. 1, p. K85—K88 (1981).
See also *10.2*, p. *108*.

P. John, I. M. Odeh, M. J. K. Thomas, M. J. Tricker, and J. I. B. Wilson
Studies of the Oscillator Strengths of Infrared Vibrational Modes in Glow-Discharge Hydrogenated Amorphous Silicon
(b) Vol. **104**, No. 2, p. 607—612 (1981).
See also *2*, p. *43*; 6, p. 72; *20.1*, p. *307*.

P. Kh. Besirganyan, A. K. Kocharyan, and K. G. Truni
Diffraction of a Narrow X-Ray Beam on the Double-Block System with a Wedge-Shaped Crystal
(a) Vol. **64**, No. 2, p. 431—434 (1981).
See also *1.1*, p. *11*.

G. C. Jain, B. C. Chakravarty, and A. Prasad
Effect of Interfacial Stress at the Si/SiO_2 Interface on the Diffusion of Ga in Si through SiO_2
(a) Vol. **64**, No. 2, p. 485—491 (1981).
See also *9*, p. *90*; 22.6, p. 476.

B. Lukasch
Correction of the Limited Energy Resolution in RBS Spectra
(a) Vol. **64**, No. 2, p. 533—538 (1981).
See also *11*, p. *124*.

H. Lemke
Properties of Manganese and Vanadium Doping Levels in Silicon
(a) Vol. **64**, No. 2, p. 549—556 (1981).
See also *13.4*, p. *174*.

R. S. Korde and M. S. Tyagi
Breakdown Voltage Reduction of High Voltage PIN Diodes by a Process Induced Linear Defect
(a) Vol. **64**, No. 2, p. K89 – K91 (1981).
See also *14.3.3*, p. *227*.

J. Marczewski
Lwo-Temperature Annealing of Surface States at the Si–SiO_2 Interfaces
(a) Vol. **64**, No. 2, p. K107 – K110 (1981).
See also *13.3*, p. *165*; *14.3.2*, p. *223*.

A. F. Khokhlov, A. I. Mashin, and A. M. Satanin
The Dependence of Magnetic Ordering Temperature in Amorphous Semiconductors on Paramagnetic Centre Concentration
(b) Vol. **105**, No. 1, p. 129 – 136 (1981).
See also 2, p. 45; 11, p. 124; *18.1*, p. *266*; *18.4*, p. *282*; 19, p. 289; 22.1.1, p. 405.

A. F. Burenkov, F. F. Komarov, and M. M. Temkin
Influence of the Interatomic Potential and the Elastic Scattering Cross-Section Approximation on the Spatial Distribution of Implanted Ions
(b) Vol. **105**, No. 1, p. 201 – 212 (1981).
See also *11*, p. *125*; 22.1, p. 402.

N. I. Zimin, R. Wedell, and O. Greschner
Theory of Planar Channelling of Relativistic Positrons
(b) Vol. **105**, No. 1, p. 257 – 264 (1981).
See also *11*, p. *125*.

H. Dersch, J. Stuke, and J. Beichler
Electron Spin Resonance of Doped Glow-Discharge Amorphous Silicon
(b) Vol. **105**, No. 1, p. 265 – 274 (1981).
See also 2, p. 45; *13.4*, p. *175*; *19*, p. *289*.

S. Tsugane, K. Morigaki, and C. Nagashima
Anomalous Magnetoresistance in Amorphous Si–Au System
(b) Vol. **105**, No. 1, p. K5 – K8 (1981).
See also *2*, p. *45*; *14.3.1*, p. *220*; 21.6, p. 385.

M. A. Kumakhov and V. A. Muralev
Spatial Distribution of Channeled Ions
(a) Vol. **65**, No. 1, p. 107 – 117 (1981).
See also *11*, p. *125*.

W. Wesch, E. Glaser, G. Götz, H. Karge, and R. Prager
Correlation between Structural Defects and Optical Properties in Ion-Implanted Silicon
(a) Vol. **65**, No. 1, p. 225 – 232 (1981).
See also *11*, p. *125*; *20.1*, p. *310*.

B. Jenichen and R. Köhler
Simulated X-Ray Contrast of Striations for Plane Waves in the Laue Case
(a) Vol. **65**, No. 1, p. 245 – 252 (1981).
See also *1.4*, p. *29*; 10.2, p. 110.

A. B. Danilin, A. V. Dvurechenskii, I. A. Ryazantsev, P. A. Timofeev, and V. D. Verner
Silicon Amorphization by Ion Beam with Radiation Heating
(a) Vol. **65**, No. 2, p. 453–461 (1981).
See also *2*, p. *46*; *11*, p. *126*.

B. Jenichen and R. Köhler
Experimental and Simulated X-Ray Contrast of Striations for Nearly Plane Waves in the Laue Case
(a) Vol. **65**, No. 2, p. 535–543 (1981).
See also *1.4*, p. *30*; *10.2*, p. *111*.

M. A. Kumakhov and V. A. Muralev
Spatial Distributions of Channeled Ions. Comparison with Experiments
(a) Vol. **65**, No. 2, p. 623–636 (1981).
See also *11*, p. *126*.

C. Donolato
An Analytical Model of SEM and STEM Charge Collection Images of Dislocations in Thin Semiconductor Layers (I): Minority Carrier Generation, Diffusion, and Collection
(a) Vol. **65**, No. 2, p. 649–658 (1981).
See also *10.2*, p. *111*; *14.3.2*, p. *223*; 14.3.4, p. 234.

G. Götz, H. D. Geiler, M. Wagner, K. H. Heinig, and H. Woittenneck
Laser Annealing of Thin Buried Amorphous Layers in Silicon
(a) Vol. **65**, No. 2, p. 677–682 (1981).
See also *2*, p. *47*; *11*, p. *126*.

H. Van Cong
Heavy Doping Effects in Bipolar Silicon Transistors and p–n Junction Silicon Solar Cells
(a) Vol. **65**, No. 2, p. 683–693 (1981).
See also 13.1, p. 157; *14.3.3*, p. *227*.

W. Gehlhoff and K. H. Segsa
Line Broadening and Line Displacement of EPR Transitions Produced by Inhomogeneous Plane Stresses in a Two-Layer System
(b) Vol. **106**, No. 1, p. 171–182 (1981).
See also *12.2*, p. *140*; *19*, p. *291*; 22.6, p. 477.

T. Soma, H. Matsuo, and T. Katoh
The Heat of Mixing in Liquid Si–Ge, Si–Sn, and Ge–Sn Mixtures
(b) Vol. **106**, No. 1, p. K5–K7 (1981).
See also *1.2*, p. *17*; *1.3*, p. *24*; 21.7, p. 388; 22.1.1, p. 405.

E. M. Trukhanov and S. I. Stenin
Initial Stage of Plastic Deformation of Substrates in Heterostructures
(a) Vol. **66**, No. 1, p. 125–131 (1981).
See also 1.5, p. 35; *10.2*, p. *112*; 22, p. 397; 22.1.1, p. 405; 22.6, p. 478.

B. A. Komarov and V. I. Sopryakov
Capacitance Method Applied in Studying Defects in Heavily Doped Semiconductors
(a) Vol. **66**, No. 1, p. 139–147 (1981).
See also *13.4*, p. *178*; 14.3.3, p. 228; 14.3.4, p. 235.

S. S. Dyunaidov, N. A. Urmanov, and M. V. Gafurova
The Cadmium Levels in Silicon
(a) Vol. **66**, No. 1, p. K79–K81 (1981).
See also *13.4*, p. *178*; 14.3, p. 215.

A. Altshuler, S. N. Bezryadin, Yu. Kh. Vekilov, and V. D. Verner
On the Pseudopotential Calculation from First Principles
(b) Vol. **106**, No. 2, p. 447–459 (1981).
See also *13*, p. *146*; 21, p. 344; 21.1, p. 361; 21.2, p. 376; 22.1, p. 402; 22.1.1, p. 406.

C. Donolato
An Analytical Model of SEM and STEM Charge Collection Images of Dislocations in Thin Semiconductor Layers (II): EBIC Images of Dislocations
(a) Vol. **66**, No. 2, p. 445–454 (1981).
See also 1.5, p. 36; *10.2*, p. *114*; *14.3.3*, p. *228*.

L. N. Aleksandrov and M. I. Zotov
Dynamic Characteristics of Structural Defects in Semiconductors by Internal Friction Data
(a) Vol. **66**, No. 2, p. 467–475 (1981).
See also *10.2*, p. *115*; *12.2*, p. *140*; 22.1.1, p. 406; 22.2.1, p. 430; 22.4.1, p. 449.

H. Lemke
Properties of Repulsive Centres in Silicon
(a) Vol. **66**, No. 2, p. 541–549 (1981).
See also *13.4*, p. *179*.

E. F. Krimmel, H. Oppolzer, H. Runge, and W. Wondrak
Kinetics of Scanned Electron Beam Annealing of High-Energy As Ion Implanted Silicon
(a) Vol. **66**, No. 2, p. 565–571 (1981).
See also *3*, p. *55*; *11*, p. *127*; 14.3.3, p. 228.

M. Kittler and W. Seifert
On the Sensitivity of the EBIC Technique as Applied to Defect Investigations in Silicon
(a) Vol. **66**, No. 2, p. 573–583 (1981).
See also 1.4, p. 30; *10.2*, p. *115*.

P. Saxena and P. C. Mathur
Intrinsic Noise in Avalanche Diodes (II): Experiment
(a) Vol. **66**, No. 2, p. 737–742 (1981).
See also *14.3.3*, p. *229*; 22.2.1, p. 430.

B. A. Komarov and V. I. Sopryakov
Study of the Field Dependence for the Parameters of the A-Centre in Si by Capacitance Transient Spectroscopy
(a) Vol. **66**, No. 2, p. 783–791 (1981).
See also *13.4*, p. *180*; *14.3.3*, p. *229*; 14.3.4, p. 235.

J. Tatarkiewicz and K. Wieteska
IR Spectroscopy and X-Ray Topography Study of Annealing of Proton Bombarded Silicon
(a) Vol. **66**, No. 2, p. K101–K103 (1981).
See also *10.2*, p. *115*; *11*, p. *128*; 20.1, p. 316.

E. KÜHRT, K. LENKEIT, and F. TÄUBNER
Measurements of the Stopping Power of 40 to 300 keV Protons in Silicon
(a) Vol. **66**, No. 2, p. K131–K133 (1981).
See also *11*, p. *128*.

J. SZUBER
On the Correlation of Paramagnetic Properties and Electronic Structure of (111) Si Surfaces Baked at Low Temperature in High Vacuum
(a) Vol. **66**, No. 2, p. K141–K144 (1981).
See also *13.3*, p. *166*; 17 p. 257.

K. HEILIG, H. FLIETNER, J. REINEKE, and LÊ T. T. TUYÊN
Surface State Distributions at Silicon Surfaces, Influenced by Hydrogen Fluoride
(a) Vol. **66**, No. 2, p. K165–K168 (1981).
See also *13.3*, p. *166*.

M. V. KOVALCHUK, N. HERTEL, M. K. MELKONYAN, R. M. IMAMOV, and P. A. ALEKSANDROV
Fluorescence Yield Curves under X-Ray Diffraction Conditions
(a) Vol. **66**, No. 2, p. K173–K177 (1981).
See also *17*, p. *257*; *20.3*, p. *331*; 22.1.1, p. 406.

P. F. ROBUSTO and R. BRAUNSTEIN
Optical Properties of the Surface Plasmon of n^+ Silicon
(b) Vol. **107**, No. 1, p. 127–138 (1981).
See also *13.5.2*, p. *195*; *20.1*, p. *316*.

H. OVERHOF and W. BEYER
Density of State in Amorphous Si:H
(b) Vol. **107**, No. 1, p. 207–213 (1981).
See also *13.2*, p. *163*; 14.3, p. 216; 15, p. 249.

H. DERSCH, J. STUKE, and J. BEICHLER
Temperature Dependence of ESR Spectra of Doped a-Si:H
(b) Vol. **107**, No. 1, p. 307–317 (1981).
See also 2, p. 49; *13.4*, p. *180*; 14.3, p. 216; *19*, p. *292*.

S. M. DAVIDSON, R. M. INNES, and S. M. LINDSAY
The Sublinear Behavior of Electron-Beam and Photo-Induced Currents in a p–n Junction as a Test of Injection Levels
(b) Vol. **107**, No. 1, p. K9–K12 (1981).
See also *14.3.3*, p. *229*; 14.3.4, p. 235.

U. PIETSCH and K. UNGER
Asphericity and Temperature Dependence of the Bond Charge
(b) Vol. **107**, No. 1, p. K19–K23 (1981).
See also *4*, p. *60*; 13, p. 147.

V. V. SIDORIK and A. N. ZHEVNO
Jahn-Teller Distortions of A- and E-Centers in Silicon
(b) Vol. **107**, No. 1, p. K57–K61 (1981).
See also *13.4*, p. *180*.

I. ZÓLOMY
The Effect of Non-Zero Switch-Off Voltage upon the Transient of MISS Diodes
(a) Vol. **67**, No. 1, p. 69–74 (1981).
See also *14.3.3*, p. *229*; 14.3.4, p. 235.

B. Gruska and G. Götz
Depth Profiling of Extended Defects in Silicon by Rutherford Backscattering Measurements
(a) Vol. **67**, No. 1, p. 129–139 (1981).
See also *11*, p. *128*.

J. Leroueille
Influence of Carbon on Oxygen Behavior in Silicon
(a) Vol. **67**, No. 1, p. 177–181 (1981).
See also *13.4*, p. *180*.

S. K. Krawczyk, A. Jakubowski, and M. Żurawska
Effect of Temperature on Electrical and Photoelectrical Properties of Silicon MIS Solar Cells
(a) Vol. **67**, No. 1, p. 313–323 (1981).
See also *14.3.3*, p. *229*; *16*, p. *254*.

V. D. Akhmetov, V. V. Bolotov, and L. S. Smirnov
Optical Properties of Defects Produced at Annealing of Neutron-Induced Clusters in Silicon
(a) Vol. **67**, No. 1, p. K33–K37 (1981).
See also 11, p. 129; *13.4*, p. *181*; *20.1*, p. *317*.

B. Pohorytes
Hydrogen Passivation of Defects in Deformed Silicon
(a) Vol. **67**, No. 1, p. K75–K80 (1981).
See also *10.2*, p. *116*.

L. N. Safronov, L. N. Aleksandrov, and R. N. Lovyagin
Luminescent Analysis of an Impurity in the Near-Surface Layer of Thermally Evaporated Silicon
(b) Vol. **107**, No. 2, p. 461–468 (1981).
See also *13.3*, p. *166*; 13.4, p. 181; *20.3*, p. *332*.

A. M. Taratin and S. A. Vorobiev
Planar Channeling of Relativistic Protons in Bent Crystal
(b) Vol. **107**, No. 2, p. 521–528 (1981).
See also *11*, p. *129*; 13, p. 148.

K. Masuda
Calculation of Atomic Relaxation near the (111) 1 × 1 Surface of Covalent Semiconductors: Tight-Binding Green's Function Approach
(b) Vol. **107**, No. 2, p. 529–533 (1981).
See also *1.6*, p. *39*; *13.3*, p. *166*; 22.1, p. 402; 22.1.1, p. 407.

F. Bechstedt, R. Enderlein, and R. Wischnewski
Binding Energies and Chemical Shifts of Least Bound Core Electron Excitations in Cubic A_NB_{8-N} Semiconductors
(b) Vol. **107**, No. 2, p. 637–651 (1981).
See also *4*, p. *60*; *13*, p. *148*; 17, p. 258; 22.1, p. 402; 22.1.1, p. 407; 22.2, p. 426; 22.4, p. 446.

E. Kühet and R. Wedell
Theoretical Description of Planar Channeling Process at Small Depths
(b) Vol. **107**, No. 2, p. 665–675 (1981).
See also *11*, p. *129*; 22.1, p. 403.

H. E. Roosendaal, B. Schmiedeskamp, H. H. Hubbes, and H. O. Lutz
Planar Channeling Trajectories in Diamond and Silicon
(b) Vol. **107**, No. 2, p. 677–687 (1981).
See also *11*, p. *129*; 22.1, p. 403.

A. D. Pogrebnyak, V. A. Kuzminikh, V. I. Raszov, and A. S. Rudnev
Positron Annihilation and Positron Profiles in Si, Irradiated by Super Dense Electron Pulses
(b) Vol. **107**, No. 2, p. K79–K82 (1981).
See also *11*, p. *129*.

S. K. Maksimov
Determination of the Sign of the Displacement Vector of Defects from Electron Microscopy Fringe Contrast Images Obtained under Low Anomalous Absorption
(a) Vol. **67**, No. 2, p. 387–394 (1981).
See also 1.4, p. 31; *10.2*, p. *116*.

A. P. Dolgolenko and I. I. Fishchuk
A-Centres Build-Up Kinetics in the Conductive Matrix of Pulled n-Type Silicon with Calculation of Their Recharges at Defect Clusters
(a) Vol. **67**, No. 2, p. 407–411 (1981).
See also *11*, p. *129*.

D. Wieluńska, L. Wieluński, and A. Turos
The Effect of Dislocations on the Planar Dechanneling (I): Analytical Model
(a) Vol. **67**, No. 2, p. 413–419 (1981).
See also *10.2*, p. *117*; *11*, p. *129*.

N. M. Olekhnovich and A. I. Olekhnovich
Dynamic Effects of Diffuse X-Ray Scattering Near Bragg Reflections
(a) Vol. **67**, No. 2, p. 427–433 (1981).
See also *1.1*, p. *12*.

V. L. Vinetskii, I. R. Entinzon, and G. A. Kholodar
Probability of Defect Formation in an Elastic Impact of Atoms in Crystals
(a) Vol. **67**, No. 2, p. 477–483 (1981).
See also *10.2*, p. *117*; *11*, p. *130*; 22.1.1, p. 407.

P. Gaworzewski and G. Ritter
On the Out-Diffusion of Oxygen from Silicon
(a) Vol. **67**, No. 2, p. 511–516 (1981).
See also *9*, p. *92*.

F. Bechstedt and K. Hübner
Mean Free Path of Photoelectrons in Silicon and Silicon Oxides
(a) Vol. **67**, No. 2, p. 517–526 (1981).
See also *13*, p. *148*; *17*, p. *258*; 22.6, p. 479.

V. J. Rao, W. A. Anderson, G. Rajeswaran, and M. Thayer
Grain Boundary Effects in Poly-Si as Determined by Low Frequency Capacitance and Conductance Data
(a) Vol. **67**, No. 2, p. 709–715 (1981).
See also 13.3, p. 167; *14.3.3*, p. *230*.

H. Runge, H. Oppolzer, and H. Mader
Dependence of the Lateral Distribution of Implanted Ions on Tilt Angle
(a) Vol. **67**, No. 2, p. K111–K112 (1981).
See also *11*, p. *130*.

J. Piekoszewski, M. Gryziński, J. Langner, and Z. Werner
Pulse Ion Implantation – New Single Step Doping Technique
(a) Vol. **67**, No. 2, p. K163–K166 (1981).
See also *11*, p. *130*; 14.3.3, p. 230.

S. Marklund
Energy Levels of Intrinsic and Extrinsic Stacking Faults in Silicon
(b) Vol. **108**, No. 1, p. 97–102 (1981).
See also *10.2*, p. *117*; *13.4*, p. *182*.

A. V. Tulupov
Some Characteristic Features of the Kumakhov Radiation in an X-Ray Range
(b) Vol. **108**, No. 1, p. K11–K14 (1981).
See also *11*, p. *130*; 13, p. 149; *20*, p. *299*.

O. G. Kostareva, D. E. Popov, and S. A. Vorobiev
Effect of Preferential Population of Quantum States in Angular Distributions of 4.9 MeV Electrons under Planar Channeling Conditions in Silicon
(b) Vol. **108**, No. 1, p. K47–K51 (1981).
See also *11*, p. *130*.

V. I. Fineberg, R. V. Konakova, L. V. Shcherbina, and N. S. Boltovets
Negative Resistance and Current Filament in Avalanche Diodes
(a) Vol. **68**, No. 1, p. 39–43 (1981).
See also *14.3.3*, p. *230*; *14.3.4*, p. *236*.

D. Wieluńska, L. Wieluński, and A. Turos
The Effect of Dislocations on the Planar Dechanneling (II): Transparency Effect
(a) Vol. **68**, No. 1, p. 45–51 (1981).
See also *10.2*, p. *118*; *11*, p. *130*.

I. E. Bondarenko, V. G. Eremenko, B. Ya. Farber, V. I. Nikitenko, and E. B. Yakimov
On the Real Structure of Monocrystalline Silicon near Dislocation Slip Planes
(a) Vol. **68**, No. 1, p. 53–60 (1981).
See also *10.2*, p. *118*.

A. F. Akkerman and S. A. Akkerman
Range Distribution Characteristics of H^+ and He^{++} Ions with Energies (5 to 2) $\times$ 10^3 keV Implanted into Silicon
(a) Vol. **68**, No. 1, p. 69–75 (1981).
See also *11*, p. *131*.

L. I. Datsenko, V. I. Khrupa, and E. N. Kislovskii
Borrmann Effect in Elastically Bent Silicon Crystals with Structural Defects
(a) Vol. 68, No. 2, p. 399–404 (1981).
See also 10.2, p. 119; *12.2*, p. *141*; *20*, p. *300*.

J. Finster and D. Schulze
Studies of the Si/SiO_2 Interface by Angular Dependent X-Ray Photoelectron Spectroscopy
(a) Vol. 68, No. 2, p. 505–517 (1981).
See also *1.5*, p. *37*; *17*, p. *258*; 22.6, p. 480.

F. F. Komarov, V. D. Kuryazov, V. S. Solovev, and S. Yu. Shiryaev
Investigation of Defect Formation in Boron Ion-Implanted Silicon during Annealing
(a) Vol. 68, No. 2, p. 519–529 (1981).
See also *10.2*, p. *119*; *11*, p. *131*.

C. A. B. Ball, F. D. Auret, and H. C. Snyman
Transmission Electron Microscopy and Auger Electron Spectroscopy Studies of Boron Implanted Silicon Single Crystals
(a) Vol. 68, No. 2, p. 545–553 (1981).
See also 9, p. 92; 10.2, p. 120; *11*, p. *131*.

N. S. Minaev and A. V. Mudryi
Thermally-Induced Defects in Silicon Containing Oxygen and Carbon
(a) Vol. 68, No. 2, p. 561–565 (1981).
See also *10.2*, p. *120*; *13.4*, p. *184*; 20.3, p. 335.

M. Sato, K. Sumino, and K. Hiraga
Impurity Effect in Stacking Fault Energy of Silicon Crystals Studied by High Resolution Electron Microscopy
(a) Vol. 68, No. 2, p. 567–577 (1981).
See also *10.2*, p. *120*.

H. G. Bach, W. R. Fahrner, and D. Bräunig
Equivalent Circuit for Ion Implanted Counterdoped Layers as Determined by MOS Admittance and Crosstalk Measurements
(a) Vol. 68, No. 2, p. 589–601 (1981).
See also 11, p. 131; *14.3.2*, p. *224*; *14.3.3*, p. *230*.

K. F. Heidemann
The Propagation of Light Waves through Oxygen Irradiation Induced Depth Profiles of the Complex Refractive Index in Silicon
(a) Vol. 68, No. 2, p. 607–617 (1981).
See also 11, p. 132; *20*, p. *300*.

H. van Cong, S. Charar, and S. Brunet
Minority-Hole Diffusion Coefficient in an n-Type Heavily Doped Semiconductor Region of Silicon Devices
(a) Vol. 68, No. 2, p. 697–702 (1981).
See also *14.3.3*, p. *231*.

S. N. Karyagin

On the Nature of Centers Responsible for the Spin-Dependent Recombination in Si and Ge
(a) Vol. **68**, No. 2, p. K113–K116 (1981).
See also *10.2*, p. *120*; *19*, p. *295*; 22.1.1, p. 407.

Ch. Schnittler, G. Teichmann, and W. Kaufmann

Energetic Calculations Concerning the A-Defect Structure and Formation in Silicon
(a) Vol. **68**, No. 2, p. K135–K139 (1981).
See also *1.4*, p. *33*; *10.2*, p. *120*.

M. S. Gilenko, P. M. Karageorgy-Alkalaev, and A. Yu. Leiderman

On the Magnetosensitivity of Semiconductor Diode Structures with Strong Carrier Accumulation
(a) Vol. **68**, No. 2, p. K165–K169 (1981).
See also *14.3.3*, p. *231*; 14.3.4, p. 236.

J. Kodeš, J. Samek, and T. Šimko

Lifetime Measurements on Electron Irradiated Silicon
(a) Vol. **68**, No. 2, p. K177–K179 (1981).
See also *11*, p. *132*; *14.3*, p. *219*.

22.1.3 Group VI Elements

Original Papers and Short Notes

E. López-Cruz and J. L. Herrera P.
Dark Decay of Positive Surface Potential in Orthorhombic Sulfur
(a) Vol. **63**, No. 1, p. K35–K38 (1981).
See also *14.3.4*, p. *232*.

G. Fleury, A. Hamou, C. Viger, and C. Vautier
Conductivity and Crystallization of Amorphous Selenium
(a) Vol. **64**, No. 1, p. 311–316 (1981).
See also *2*, p. *43*; *14.3.1*, p. *220*.

V. Balynas, Z. Dobrovolskis, W. Hoerstel, and A. Krotkus
The Effect of Hydrostatic Pressure on the Anomalous Sign Reversal of the Hall Coefficient in Tellurium
(a) Vol. **64**, No. 2, p. K115–K117 (1981).
See also 12.2, p. 139; 13.1, p. 155; *14.3*, p. *212*.

J.-C. Bernède
Switchings in Selenium Thin Films. Influence of Electrodes
(a) Vol. **65**, No. 1, p. 97–102 (1981).
See also *14.3.1*, p. *220*; *14.3.4*, p. *234*.

V. Z. Zhokhov, A. A. Stepurenko, V. I. Pustovoit, and M. M. Mazur
Some Features of Acoustoelectronic Interaction in p-Te in Magnetic Field
(a) Vol. **65**, No. 2, p. 491–496 (1981).
See also *7*, p. *82*; *14.3.4*, p. *234*; 18, p. 262.

L. V. Rao, B. S. Naidu, and P. J. Reddy
Electrical Conductivity and Thermopower Measurements on Silver Doped Tellurium Films
(a) Vol. **65**, No. 2, p. K135–K138 (1981).
See also 1.5, p. 35; *14.3.1*, p. *221*; *15*, p. *248*.

J. Sinkkonen
Photoconductivity of an Inhomogeneous Semiconductor
(a) Vol. **67**, No. 2, p. 555–564 (1981).
See also *13.4*, p. *182*; *16*, p. *255*; 22.4.1, p. 450.

Bhanwar Singh and K. L. Chopra
A Novel Bi-Layer System for High Resolution Optical Imaging
(a) Vol. **68**, No. 1, p. K5–K9 (1981).
See also *2*, p. *50*; *20.1*, p. *319*; 22.1.1, p. 407.

K. Harnisch, S. M. Neubacher, and O. Ennemoser
Optical Loss in Selenium Thin Films as a Measure of Crystallinity
(a) Vol. **68**, No. 1, p. K87–K91 (1981).
See also *1.5*, p. *37*; 3, p. 56; 20, p. 299.

22.2 III–V Compounds

Original Papers and Short Notes

V. A. Tyagai†, O. V. Snitko, A. M. Evstigneev, and A. N. Krasiko
Electroreflectance of InN Semimetallic Thin Films
(b) Vol. **103**, No. 2, p. 589–594 (1981).
See also *14.3.1*, p. *220*; *20.1*, p. *303*.

P. Bogusławski
The Theory of the Electron Spin–Lattice Relaxation in Zincblende Semiconductors
(b) Vol. **104**, No. 1, p. 89–95 (1981).
See also *6*, p. *69*; *13.1*, p. *153*; 22.4, p. 445.

E. Neubauer
Surface Erosion during Annealing of $M^{III}X^{V}:F^{IIb}$ Systems and Its Reduction
(a) Vol. **64**, No. 1, p. 233–237 (1981).
See also *1.2*, p. *16*; *9*, p. *90*.

G. Antonioli, D. Bianchi, and P. Franzosi
Intrinsic Urbach Rule and Electron–Phonon Interaction in GaAs and Related III–V Compounds
(b) Vol. **106**, No. 1, p. 79–84 (1981).
See also *6*, p. *75*; *20.1*, p. *312*; 22.2.1, p. 429.

H. G. Brühl
Determination of the Bulk Charge Density of Elemental Semiconductors and III–V Compounds by X-Ray Diffraction Methods (I): Fundamentals
(b) Vol. **107**, No. 1, p. 289–295 (1981).
See also *1*, p. *10*; *13*, p. *147*; 22.1, p. 402.

F. Bechstedt, R. Enderlein, and R. Wischnewski
Binding Energies and Chemical Shifts of Least Bound Core Electron Excitations in Cubic A_NB_{8-N} Semiconductors
(b) Vol. **107**, No. 2, p. 637–651 (1981).
See also *4*, p. *60*; *13*, p. *148*; 17, p. 258; 22.1, p. 402; 22.1.1, p. 407; 22.1.2, p. 419; 22.4, p. 446.

A. N. Gusatinskii, M. A. Blokhin, P. Shumichen, M. A. Bunin, and S. S. Mikhailova
X-Ray L_1Spectra and XPS Investigation of the Electronic Structure of Indium Compounds
(b) Vol. **108**, No. 2, p. 405–410 (1981).
See also *13.1*, p. *161*; *20*, p. 299; 21, p. 349; 22, p. 400; 22.3, p. 442.

A. C. Sharma and S. Auluck
Small Wave Vector ***q*** and Frequency *ω* Dependent Longitudinal Dielectric Function for a Model Semiconductor
(b) Vol. **108**, No. 2, p. 587–600 (1981).
See also *13.1*, p. *161*; *14.4*, p. *241*; 22.1, p. 403; 22.4, p. 446.

G. E. Archangelskii, M. V. Fock, S. I. Pačesová, and L. Jastrabík
EPR Study of Oxygen Centers in AlN:O
(b) Vol. **108**, No. 2, p. K117–K122 (1981).
See also *19*, p. *294*.

22.2.1 Arsenides

Original Papers and Short Notes

U. PIETSCH
X-Ray Bond Charge in GaAs and InSb
(b) Vol. **103**, No. 1, p. 93–100 (1981).
See also *4*, p. *57*; 22.2.3, p. 435.

N. S. ORLOVA
X-Ray Thermal Diffuse Scattering Measurements of the [110] Phonon Dispersion Curves of Indium Arsenide
(b) Vol. **103**, No. 1, p. 115–121 (1981).
See also *6*, p. *66*; 20, p. 296.

A. M. DYACHENKO, G. V. BEHRENSHTEIN, and I. E. MARONCHUK
The Internal Stresses in Semiconductors Induced by Microinclusions
(a) Vol. **63**, No. 1, p. K27–K30 (1981).
See also *1.4*, p. *28*; *10.2*, p. *106*; 22.2.4, p. 437.

T. SOMA, Y. SAITO, and H. MATSUO
Phonon Dispersion Curves near the Covalent–Metallic Transition Pressure of GaAs, GaSb, and InSb
(b) Vol. **103**, No. 2, p. K173–K177 (1981).
See also *6*, p. *69*; 8, p. 84; 22.2.3, p. 435.

V. M. LOMAKO and P. YA. STAROSTIN
Simulation of Electrical Characteristics and Properties of Defect Clusters in Irradiated Semiconductors (I): Approximation of Electric Charge Density
(a) Vol. **63**, No. 2, p. 585–593 (1981).
See also *13.4*, p. *171*; 14.3.4, p. 233.

S. SARKAR
Second Harmonic Dependence of Domain Excess Voltage in GaAs
(a) Vol. **63**, No. 2, p. K171–K173 (1981).
See also *14.3.4*, p. *233*.

TH. HARBICH and G. MAHLER
The Electron–Hole Plasma in a Strong Electromagnetic Field. Optical Properties in a Simplified Model
(b) Vol. **104**, No. 1, p. 185–191 (1981).
See also *13.1*, p. *153*; *13.5*, p. *186*; 20.1, p. 305.

U. PIETSCH and K. UNGER
The Temperature Dependence and Dynamical Properties of the Bond Charge in GaAs and InSb
(b) Vol. **104**, No. 1, p. 253–260 (1981).
See also *4*, p. *58*; 6, p. 69; *13*, p. *143*; 22.2.3, p. 435.

A. Le Bloa, P. N. Favennec, and Y. Colin
DLTS Method Using a Single Temperature Scanning
(a) Vol. **64**, No. 1, p. 85–93 (1981).
Erratum
(a) Vol. **66**, No. 2, p. 807 (1981).
See also *13.4*, p. *172*; 14.3, p. 210; 22.1.2, p. 411.

F. Voillot, J. Barrau, M. Brousseau, and J. C. Brabant
The Fine Structure of the 0.84 eV No-Phonon Luminescence in GaAs:Cr
(a) Vol. **64**, No. 1, p. K39–K41 (1981).
See also *20.3*, p. *326*.

D. A. Aronov and V. Zaitova
The Influence of Surface Recombination on the Photoelectromagnetic (PEM) Effect in Semiconductors with Deep Traps
(b) Vol. **104**, No. 2, p. K159–K162 (1981).
See also 13.4, p. 174; *16*, p. 252.

J. H. Basson and H. Booyens
The Effect of Substrate Relaxation on the Introduction of Misfit Dislocations in Misfitting Epitaxial Layers
(a) Vol. **64**, No. 2, p. 777–786 (1981).
See also 1.5, p. 35; *10.1*, p. *99*; *12.1*, p. *135*; 22.2.2, p. 433; 22.2.4, p. 437.

G. Weimann, W. Schlapp, and H. Burkhard
MBE Growth of GaAs and GaAlAs for the Fabrication of DH Lasers
(a) Vol. **64**, No. 2, p. K99–K103 (1981).
See also *3*, p. *54*; *20.2*, p. *322*; 22.2.4, p. 437.

K. D. Glinchuk, K. Lukat, and A. V. Prokhorovich
Non-Trivial Excitation Dependence of the Kinetics of the Extrinsic Luminescence Decay in GaAs
(a) Vol. **64**, No. 2, p. K139–K143 (1981).
See also *13.4*, p. *174*; *20.3*, p. *326*.

S. J. Pearton
Magnetic Field Dependence of Defect State in GaAs
(b) Vol. **105**, No. 1, p. K19–K22 (1981).
See also *13.4*, p. *175*; 14.3, p. 213; 18, p. 261.

V. M. Lomako and P. Ya. Starostin
Simulation of Electrical Characteristics and Properties of Defect Clusters in Irradiated Semiconductors (II): Self-Consistent Solution
(a) Vol. **65**, No. 1, p. 39–43 (1981).
See also *13.4*, p. *175*; 14.3.4, p. 234.

J. Garrido, J. L. Castaño, and J. Piqueras
Photoluminescence in Gallium Arsenide Irradiated with Thermal Neutrons
(a) Vol. **65**, No. 1, p. 103–106 (1981).
See also 11, p. 125; *14.3*, p. *213*; *20.3*, p. *327*.

V. L. Alperovich, A. F. Kravchenko, N. A. Pakhanov, and A. S. Terekhov
Effect of Electron Heating by Surface Electric Fields on Oscillating Photoresponse Spectra of Gallium Arsenide–Metal Structures
(a) Vol. **65**, No. 1, p. 259–263 (1981).
See also *14.3.2*, p. *223*; *16*, p. *253*.

G. Jungk
On the Experimental Determination of Many-Particle Effects at Interband Transitions
(b) Vol. **105**, No. 2, p. 551–556 (1981).
See also *13.1*, p. *156*; 13.5.1, p. 190; *20.1*, p. *311*; 22.2.2, p. 433.

H. Neumann, Yu. I. Tsipivka, and K. Unger
Influence of Dielectric Screening and Band Nonparabolicity on Ionized Impurity Scattering in Doped Semiconductors
(b) Vol. **105**, No. 2, p. 577–583 (1981).
See also 13.1, p. 156; *14.3*, p. *214*; 14.4, p. 239; 22.2.3, p. 436.

B. Tuck and A. J. N. Houghton
Shallow Diffusions of Zinc in GaAs at 700 °C
(a) Vol. **65**, No. 2, p. 643–647 (1981).
See also *9*, p. *90*.

G. D. Bagratishvili, R. B. Dzhanelidze, D. A. Jishiashvili, L. V. Piskanovskii, A. N. Zyuganov, V. N. Mikhelashvili, and P. S. Smertenko
Mechanism of Charge Flow through the M–Ge_3N_4–GaAs Structure
(a) Vol. **65**, No. 2, p. 701–707 (1981).
See also 13.3, p. 166; *14.3.4*, p. *234*.

G. Antonioli, D. Bianchi, and P. Franzosi
Intrinsic Urbach Rule and Electron–Phonon Interaction in GaAs and Related III–V Compounds
(b) Vol. **106**, No. 1, p. 79–84 (1981).
See also *6*, p. *75*; *20.1*, p. *312*; 22.2, p. 426.

E. A. Sozontov, M. V. Kruglov, and B. G. Zakharov
Photoemission in Bragg Diffraction of X-Rays by Bicrystals
(a) Vol. **66**, No. 1, p. 303–310 (1981).
See also *1.1*, p. *12*; *17*, p. *257*; 22.1.1, p. 406; 22.2.4, p. 438.

M. Kusaka, M. Hirai, and S. Okazaki
Correlations between the Change in Capacitance Caused by Piezoelectric Effect and the Surface States in III–V Compound MOS Diodes
(a) Vol. **66**, No. 1, p. 317–325 (1981).
See also *14.3.2*, p. *224*; 14.3.3, p. 228.

P. Saxena and P. C. Mathur
Intrinsic Noise in Avalanche Diodes (I): Theory
(a) Vol. **66**, No. 1, p. 383–398 (1981).
See also *14.3.3*, p. *228*.

I. Singh and G. P. Srivastava
Semiempirical Pseudopotential Calculations for the Electronic Structure of Zincblende Semiconductors
(b) Vol. **108**, No. 2, p. 467–473 (1981).
See also *13.1*, p. *161*; 22.2.2, p. 434; 22.4.3, p. 455.

A. Baldereschi, K. Maschke, A. Milchev, R. Pickenhain, and K. Unger
Thomas-Fermi Approximation for the Valence Electron Densities in Cubic Semiconductors and Insulators
(b) Vol. **108**, No. 2, p. 511–520 (1981).
See also *13.1*, p. *161*; 14.3, p. 218; 21.7, p. 389; 22.1.1, p. 407; 22.1.2, p. 422; 22.4.2, p. 453; 22.5.2, p. 469.

A. Matulionis, J. Požela, and E. Starikov
Hot Electron Velocity Correlation and Diffusion in a Many Layered Heterostructure
(a) Vol. **68**, No. 2, p. K149–K152 (1981).
See also *13*, p. *150*; *14.3.4*, p. *236*; 22.2.4, p. 439.

22.2.2 Phosphides

Original Papers and Short Notes

J. H. BASSON and H. BOOYENS
The Effect of Substrate Relaxation on the Introduction of Misfit Dislocations in Misfitting Epitaxial Layers
(a) Vol. **64**, No. 2, p. 777–786 (1981).
See also 1.5, p. 35; *10.1*, p. *99*; *12.1*, p. *135*; 22.2.1, p. 428; 22.2.4, p. 437.

K. WANDEL, G. WURBS, R. BINDEMANN, A. KIRSCH, and R. SCHWABE
Minority Carrier Lifetime in VPE-GaP:N, Te at Various Excitation Levels
(a) Vol. **65**, No. 1, p. 233–243 (1981).
See also *13.4*, p. *176*; *20.3*, p. *327*.

V. D. AKHMETOV, I. S. BZINKOVSKAYA, V. V. BOLOTOV, and R. N. LOVYAGIN
Depth Distribution of Growth Defects in GaP Films on Sapphire Synthesized in Vacuum
(a) Vol. **65**, No. 1, p. K9–K12 (1981).
See also 3, p. 54; *10.2*, p. *110*; *20.1*, p. *310*.

G. JUNGK
On the Experimental Determination of Many-Particle Effects at Interband Transitions
(b) Vol. **105**, No. 2, p. 551–556 (1981).
See also *13.1*, p. *156*; 13.5.1, p. 190; *20.1*, p. *311*; 22.2.1, p. 429.

U. PIETSCH
Bond Charge Reconstruction for GaP from X-Ray Powder Data
(b) Vol. **105**, No. 2, p. K135–K137 (1981).
See also *1.1*, p. *11*; *4*, p. *59*.

H. NICKEL and E. KUPHAL
Deep Level Spectroscopy and Schottky Barrier Characteristics of LPE n- and p-InP
(a) Vol. **65**, No. 2, p. 583–588 (1981).
See also *13.4*, p. *177*; *14.3.4*, p. *234*.

G. OELGART and G. ARNOLD
Impurity Profile on GaP Diodes in the Neighbourhood of a p–n Junction
(a) Vol. **65**, No. 2, p. 609–616 (1981).
See also *10.2*, p. *111*; *14.3.3*, p. *227*.

N. ERBEN, P. PAUFLER, K. LÖSCHKE, S. DÄBRITZ, and K. KLEINSTÜCK
Kossel Line Profiles near Dislocations in GaP
(a) Vol. **65**, No. 2, p. K175–K178 (1981).
See also *10.2*, p. *112*.

G. OELGART, J. FIDDICKE, and R. REULKE
Investigation of Minority-Carrier Diffusion Lengths by Means of the Scanning Electron Microprobe (SEM)
(a) Vol. **66**, No. 1, p. 283–292 (1981).
See also 11, p. 127; *14.3.3*, p. *228*; 22.1.2, p. 416.

H. KLOSE and M. RIETH
Influence of Al Postimplantation on p–n Junctions in Gallium Phosphide
(a) Vol. **66**, No. 2, p. K153–K156 (1981).
See also *11*, p. *128*; 14.3.3, p. 229; 20.3, p. 331.

I. Singh and G. P. Srivastava

Semiempirical Pseudopotential Calculations for the Electronic Structure of Zincblende Semiconductors

(b) Vol. **108**, No. 2, p. 467–473 (1981).

See also *13.1*, p. *161*; 22.2.1, p. 431; 22.4.3, p. 455.

H.-G. Brühl and U. Pietsch

Investigation of the Bond Charge in GaP with X-Ray Data

(a) Vol. **68**, No. 2, p. 689–695 (1981).

See also *4*, p. *61*; *13*, p. *150*.

H. Berek and P. Kirsten

Observation of Recombination-Enhanced Defect Reactions (REDR) at Low Temperature by Means of Cathodoluminescence and Photoluminescence in GaP

(a) Vol. **68**, No. 2, p. K203–K206 (1981).

See also 11, p. 132; *20.3*, p. *335*.

22.2.3 Antimonides

Original Papers and Short Notes

U. Pietsch
X-Ray Bond Charge in GaAs and InSb
(b) Vol. **103**, No. 1, p. 93–100 (1981).
See also *4*, p. *57*; 22.2.1, p. 427.

S. Kończak and M. Nowak
The Estimation of Semiconductor Parameters Using Least Squares in Photomagnetoelectric Investigations
(a) Vol. **63**, No. 1, p. 305–311 (1981).
See also *16*, p. *251*; 22.1.1, p. 404.

I. Gorczyca
Scattering on Short-Range Potentials in InSb. A Pseudopotential Calculation
(b) Vol. **103**, No. 2, p. 529–533 (1981).
See also *13.4*, p. *170*; 14.3, p. 210; 15, p. 247.

S. Guha and P. Sen
Parametric Decay of Alfvén Waves in Piezoelectric Semiconductors
(b) Vol. **103**, No. 2, p. K131–K136 (1981).
See also *13*, p. *143*; *20*, p. *296*.

T. Soma, Y. Saito, and H. Matsuo
Phonon Dispersion Curves near the Covalent–Metallic Transition Pressure of GaAs, GaSb and InSb
(b) Vol. **103**, No. 2, p. K173–K177 (1981).
See also *6*, p. *69*; 8, p. 84; 22.2.1, p. 427.

U. Pietsch and K. Unger
The Temperature Dependence and Dynamical Properties of the Bond Charge in GaAs and InSb
(b) Vol. **104**, No. 1, p. 253–260 (1981).
See also *4*, p. *58*; 6, p. 69; *13*, p. *143*; 22.2.1, p. 427.

J. Wilke, V. T. Bublik, A. G. Braginskaya, A. N. Popkov, G. P. Kolchina, V. B. Osvenskii, S. P. Grishina and O. G. Stolyarov
Study of the Homogeneous Region in Undoped Gallium Antimonide
(a) Vol. **64**, No. 1, p. K31–K33 (1981).
See also *1.3*, p. *23*.

S. Porowski, L. Konczewcz, J. Kowalski, R. L. Aulombard, and J. L. Robert
High Magnetic Field Observation of the Resonance Donor States of S in InSb
(b) Vol. **104**, No. 2, p. 657–666 (1981).
See also *13.4*, p. *173*; *14.3*, p. *211*.

L. M. Brescansin and A. Fazzio
Theoretical Study of Single Vacancies in GaSb
(b) Vol. **105**, No. 1, p. 339–345 (1981).
See also 10.2, p. 110; *13.4*, p. *175*.

V. K. Malyutenko, K. Yu. Guga, and Yu. M. Malozovskii
Magnetoconcentration Effect at Auger Recombination of Current Carriers
(a) Vol. **65**, No. 1, p. 131–140 (1981).
See also *14.3*, p. *213*; 18, p. 261.

H. Neumann, Yu. I. Tsipivka, and K. Unger
Influence of Dielectric Screening and Band Nonparabolicity on Ionized Impurity Scattering in Doped Semiconductors
(b) Vol. **105**, No. 2, p. 577–583 (1981).
See also 13.1, p. 156; *14.3*, p. *214*; 14.4, p. 239; 22.2.1, p. 429.

S. W. Koch, S. Schmitt-Rink, and H. Haug
Theory of Optical Nonlinearities in InSb
(b) Vol. **106**, No. 1, p. 135–140 (1981).
See also 13.1, p. 157; *20.1*, p. *312*.

E. Litwin-Staszewska, W. Szymańska, and R. Piotrzkowski
The Electron Mobility and Thermoelectric Power in InSb at Atmospheric and Hydrostatic Pressures
(b) Vol. **106**, No. 2, p. 551–559 (1981).
See also *14.3*, p. *215*; *15*, p. *249*.

I. Gorczyca
Calculations of Deformation-Potential Constants in InSb
(b) Vol. **108**, No. 1, p. 205–209 (1981).
See also *6*, p. *80*; *13*, p. *149*.

22.2.4 Mixed Crystals

Original Papers and Short Notes

C. Ance and A. Amamou
Photoemission Spectra of the Ternary $Ga_{1-x}Al_xSb$ Alloy
(b) Vol. **103**, No. 1, p. 441–449 (1981).
See also *13.1*, p. *152*; 17, p. 256.

A. M. Dyachenko, G. V. Berenshtein, and I. E. Maronchuk
The Internal Stresses in Semiconductors Induced by Microinclusions
(a) Vol. **63**, No. 1, p. K27–K30 (1981).
See also *1.4*, p. *28*; *10.2*, p. *106*; 22.2.1, p. 427.

R. V. Konakova, Yu. A. Tkhorik, I. L. Zaitsevskii, P. Kordoš, M. Morvic, and J. Červenák
Avalanche Breakdown Peculiarities of Al_xGa_{1-x} As IMPATT Diodes
(a) Vol. **63**, No. 2, p. K163–K166 (1981).
See also *14.3.3*, p. *226*; *14.3.4*, p. *233*.

K. Sierański and J. Szatkowski
Theory of the Free-Carrier Absorption in Zincblende Narrow-Gap Semiconductors
(b) Vol. **104**, No. 1, p. 57–67 (1981).
See also 10.2, p. 107; *20.1*, p. *304*; 22.4.4, p. 456.

J. H. Basson and H. Booyens
The Effect of Substrate Relaxation on the Introduction of Misfit Dislocations in Misfitting Epitaxial Layers
(a) Vol. **64**, No. 2, p. 777–786 (1981).
See also 1.5, p. 35; *10.1*, p. *99*; *12.1*, p. *135*; 22.2.1, p. 428; 22.2.2, p. 433.

G. Weimann, W. Schlapp, and H. Burkhard
MBE Growth of GaAs and GaAlAs for the Fabrication of DH Lasers
(a) Vol. **64**, No. 2, p. K99–K103 (1981).
See also *3*, p. *54*; *20.2*, p. *322*; 22.2.1, p. 428.

A. K. Saxena
Non-Γ Deep Levels and the Conduction Band Structure of $Ga_{1-x}Al_xAs$ Alloys
(b) Vol. **105**, No. 2, p. 777–787 (1981).
See also *13.1*, p. *156*; *13.4*, p. *177*; 14.3, p. 214.

V. F. Kovalenko, G. P. Peka, and L. G. Shepel
Photoluminescent Methods for Determination of Graded Band-Gap Semiconductor Parameters (I): Analysis of Methods
(a) Vol. **65**, No. 2, p. 529–534 (1981).
See also 13.1, p. 157; 13.3, p. 165; *20.3*, p. *328*.

P. Swarup, R. K. Jain, S. V. Verma, Sh. Charan, and D. M. Tandle
Preparation and Properties of GaAsSb and GaAlAsSb Epi-Layers
(a) Vol. **65**, No. 2, p. K183–K187 (1981).
See also *3*, p. *54*; *20.1*, p. 312; 20.3, p. 329.

E. A. SOZONTOV, M. V. KRUGLOV, and B. G. ZAKHAROV
Photoemission in Bragg Diffraction of X-Rays by Bicrystals
(a) Vol. **66**, No. 1, p. 303–310 (1981).
See also *1.1*, p. *12*; *17*, p. *257*; 22.1.1, p. 406; 22.2.1, p. 429.

V. F. KOVALENKO, G. P. PEKA, and L. G. SHEPEL
Photoluminescent Methods for Determination of Graded Band-Gap Semiconductor Parameters (II): Experimental Investigation of $Al_xGa_{1-x}As$
(a) Vol. **66**, No. 1, p. 407–414 (1981).
See also 13.1, p. 158; 13.3, p. 166; *20.3*, p. *330*.

W. RÜHLE and P. BROSSON
Single Longitudinal Mode Operation Due to Excessive Gain Supression ? A Comparative Study of Laser Structures with and without Optical Confinement
(a) Vol. **66**, No. 1, p. K31–K35 (1981).
See also *20.2*, p. *322*.

W. RÜHLE, P. BROSSON, and J. E. RIPPER
Microwave-Induced Oscillations in Channeled Substrate Planar and V-Groove GaAlAs Lasers
(a) Vol. **66**, No. 2, p. K87–K91 (1981).
See also *20.2*, p. *322*.

P. KLEINERT and E. JAHNE
A Theoretical Study of Raman Scattering and Infrared Absorption Spectra of $Ga_{1-x}In_xP$ Mixed Crystals
(b) Vol. **107**, No. 1, p. 177–183 (1981).
See also *6*, p. *78*; *20.1*, p. *316*.

U. PIETSCH
Determination of Static Bond Charge Properties in $Ga_{1-x}In_xAs$ and $GaAs_{1-y}P_y$ Solid Solutions
(b) Vol. **107**, No. 1, p. 185–194 (1981).
See also *4*, p. *60*.

D. HENNIG and R. STREHLOW
Disorder-Induced Broadening for Free Direct Excitons in II–VI and III–V Semiconducting Mixed Crystals
(b) Vol. **107**, No. 1, p. 283–288 (1981).
See also *13.5.1*, p. *192*; *20.1*, p. *316*; 22.4.4, p. 457.

S. MODESTI, A. FROVA, J. L. STAEHLI, M. GUZZI, and M. CAPIZZI
Electron–Hole Plasma Expansion in Direct-Gap $GaAs_{1-x}P_x$
(b) Vol. **108**, No. 1, p. 281–288 (1981).
See also *13.5*, p. *187*; 20.3, p. 333.

S. K. SUTRADHAR and D. CHATTOPADHYAY
Transverse Magnetoresistance in n-Type $InAs_{0.65}P_{0.35}$
(b) Vol. **108**, No. 1, p. K15–K18 (1981).
See also *14.3*, p. *218*.

W. Kowalsky, A. Schlachetzki, and F. Fiedler
Near-Band-Gap Absorption of InGaAsP at 1.3 μm Wavelength
(a) Vol. **68**, No. 1, p. 153–158 (1981).
See also *20.1*, p. *319*.

U. Pietsch
Bond Charge Determination in Ternary Chalcopyrite Semiconductors by Means of X-Ray Structure Amplitude Investigations
(a) Vol. **68**, No. 1, p. 321–327 (1981).
See also *4*, p. *61*.

K. Starosta
RF Sputtering of $Al_xIn_{1-x}N$ Thin Films
(a) Vol. **68**, No. 1, p. K55–K57 (1981).
See also *1.5*, p. *37*.

S. K. Maksimov and E. N. Nagdaev
Electron Microscopy Images of Composition-Modulated Crystals (I): Theory
(a) Vol. **68**, No. 2, p. 645–652 (1981).
See also *1.4*, p. *32*; 10.2, p. 120.

B. Balland, J. L. Paviot, B. de Cremoux, and P. Hirtz
Sn–X Centers in GaAlAs Thin Layer Emitters
(a) Vol. **68**, No. 2, p. 661–673 (1981).
See also 10.2, p. 120; *13.4*, p. *184*; 14.3.1, p. 222; *14.3.3*, p. *231*.

A. Matulionis, J. Požela, and E. Starikov
Hot Electron Velocity Correlation and Diffusion in a Many Layered Heterostructure
(a) Vol. **68**, No. 2, p. K149–K152 (1981).
See also *13*, p. *150*; *14.3.4*, p. *236*; 22.2.1, p. 431.

22.3 IV-IV, II-V, and III-VI Compounds

Review Article

P. MÜLLER
Electric Relaxation Currents in High-Resistivity Solids
(a) Vol. **67**, No. 1, p. 11–60 (1981).
See also 14.3.3, p. 225; 14.3.4, p. 232; *14.4*, p. *237*; 22.5.1, p. 461; 22.5.2, p. 462.

Original Papers and Short Notes

C. CARLONE, S. JANDL, and H. R. SHANKS
Optical Phonons and Crystalline Symmetry of InSe
(b) Vol. **103**, No. 1, p. 123–130 (1981).
See also 1.3, p. 22; *6*, p. *66*; *20.1*, p. *301*.

G. B. DUBROVSKII and V. I. BELYAVSKII
Selection Rules and Absorption Lineshapes for Electron Transitions between Minizones in a Superlattice
(b) Vol. **103**, No. 1, p. 131–137 (1981).
See also *13.1*, p. *151*; *20.1*, p. *301*.

A. N. CHAKRAVARTI, K. P. GHATAK, K. K. GHOSH, S. GOHSH, and A. DHAR
Effect of Carrier Degeneracy on the Screening Length in n-Cd_3As_2
(b) Vol. **103**, No. 1, p. K55–K60 (1981).
See also 13.1, p. 152; *14.3*, p. *209*.

T. IKARI, S. SHIGETOMI, Y. KOGA, and S. SHIGETOMI
Photoluminescence of Zn Doped InSe Single Crystals
(b) Vol. **103**, No. 1, p. K81–K83 (1981).
See also *20.3*, p. *323*.

G. L. BELENKII, E. YU. SALAEV, R. A. SULEIMANOV, and E. I. MIRZOEV
The Influence of Pressure on the Electronic Spectrum of Layered Crystals
(a) Vol. **63**, No. 1, p. 97–102 (1981).
See also *13.1*, p. *152*; 20.1, p. 302; 20.3, p. 324.

A. U. SHELEG, E. PELJO, and P. SUORTTI
Structure and Thermal Motion Parameters of Tetragonal ZnP_2 between 300 and 740 K
(a) Vol. **63**, No. 2, p. 751–758 (1981).
See also *1.1*, p. *11*; 6, p. 69.

V. RIEDE, H. NEUMANN, F. LÉVY, and H. SOBOTTA
Two-Phonon Absorption Spectra in ε-GaSe
(b) Vol. **104**, No. 1, p. 277–282 (1981).
See also *6*, p. *70*; *20.1*, p. *305*.

J. V. S. S. NARAYANA MURTY and R. RAMJI RAO
Lattice Dynamics of ZrC Using Keating's Approach
(b) Vol. **104**, No. 2, p. 691–696 (1981).
See also 4, p. 58; *6*, p. *72*.

K. Kambas, J. Spyridelis, and M. Balkanski
Far Infrared and Raman Optical Study of α-and β-In_2S_3 Compounds
(b) Vol. **105**, No. 1, p. 291–296 (1981).
See also *6*, p. *73*; *20.1*, p. *309*.

T. Soma, H. Matsuo, and S. Kagaya
Phase Diagrams of the Si–Sn and Ge–Sn Systems
(b) Vol. **105**, No. 1, p. 311–314 (1981).
See also *1.2*, p. *16*; *1.3*, p. *23*.

Yu. P. Gnatenko and P. A. Skubenko
Exciton Absorption of GaSe Crystals in the Indirect Transition Region
(b) Vol. **105**, No. 1, p. K9–K12 (1981).
See also *13.5.1*, p. *189*; *20.1*, p. *309*.

J. Misiewicz and J. A. Gaj
Optical Anisotropy in the Zn_3P_2 Band Gap Region
(b) Vol. **105**, No. 1, p. K23–K25 (1981).
See also 16, p. 252; *20.1*, p. *309*.

E. F. Bagirzade, B. G. Tagiev, G. M. Mamedov, I. M. Askerov, and F. N. Sultanov
Interference Oscillations of Photocurrent in GaSe:Sn and $GeSe_{0.95}Te_{0.05}$:Sn Single Crystals
(a) Vol. **65**, No. 1, p. K55–K58 (1981).
See also 14.3.4, p. 234; *16*, p. *253*; *20.1*, p. *310*.

K. Nauka and J. Misiewicz
Minority-Carrier Diffusion Length in Zn_3P_2
(a) Vol. **65**, No. 1, p. K95–K97 (1981).
See also *14.3.3*, p. *227*; 16, p. 253.

A. N. Chakravarti, K. P. Ghatak, A. Dhar, and S. Ghosh
Effect of Magnetic Quantization on the Diffusivity–Mobility Ratio in n-Cd_3As_2
(b) Vol. **105**, No. 2, p. 691–697 (1981).
See also *14.3*, p. *214*; 18, p. 262.

V. V. Artamonov, M. Ya. Valakh, and M. P. Lisitsa
Interlayer Interaction in Phonon Spectra of ε-GaSe Crystals
(b) Vol. **105**, No. 2, p. K103–K106 (1981).
See also *6*, p. *75*; *20.1*, p. *311*.

V. S. Ballandovich, Yu. M. Tairov, and G. N. Violina
Acceptor Boron in α-SiC (6H): Investigation by the Photocapacitance Method
(a) Vol. **65**, No. 2, p. 709–716 (1981).
See also *13.4*, p. *177*; *16*, p. *253*; 20.3, p. 329.

Z. Januškevičius, N. Korech, A. Sakalas, and I. Tychina
Electrical Properties of CdP_2 Single Crystals
(a) Vol. **65**, No. 2, p. K149–K151 (1981).
See also 13.4, p. 177; *14.3*, p. *215*.

B. OLEŚ, H. J. STOLZ, and G. H. VON SCHNERING
Raman Spectra of $CaAs_3$, $EuAs_3$, $SrAs_3$, $BaAs_3$, and α-EuP_3 and the Mixed Phases $Eu_{1-x}Sr_xAs_3$, and $Eu(P_{1-x}As_x)_3$
(b) Vol. **106**, No. 1, p. 157–170 (1981).
See also *6*, p. *76*; *20.1*, p. *312*; 22.3.1, p. **444**.

N. M. GASANLY, B. M. DZHAVADOV, A. S. RAGIMOV, V. I. TAGIROV, and R. E. GUSEINOV
Long-Wave Optical Phonons in In_6S_7 Layer Crystals
(b) Vol. **106**, No. 1, p. K47–K51 (1981).
See also *6*, p. *77*; *20.1*, p. *313*.

F. M. GASHIMZADE and G. S. ORUDZHEV
Effect of Pressure and Temperature on the Band Structure of TlSe
(b) Vol. **106**, No. 1, p. K67–K71 (1981).
See also *13.1*, p. *158*.

T. OGAWA
Morphologies of CVD ZrC Crystals
(a) Vol. **66**, No. 1, p. 175–181 (1981).
See also *1*, p. *10*; *3*, p. *54*; **4**, p. **59**.

YU. P. GNATENKO, Z. D. KOVALYUK, and P. A. SKUBENKO
Influence of Ni Impurity on the Absorption Spectrum of Layered GaSe Crystals
(b) Vol. **106**, No. 2, p. 621–626 (1981).
See also *13.4*, p. *179*; *20.1*, p. *315*.

S. SEN and D. N. BOSE
Preparation and Electrical Properties of Ga_2Te_3 Thin Films
(a) Vol. **66**, No. 2, p. K117–K119 (1981).
See also *14.3.1*, p. *221*.

J. HUMLIČEK and E. SCHMIDT
Precise Determination of Gap Energies from Modulation Spectra
(b) Vol. **107**, No. 2, p. K105–K108 (1981).
See also *13.1*, p. *160*; 20.1, p. 318.

S. R. SINGH, G. SINGH, and G. VAN TENDELOO
Structure Determination of a 40 H Polytype of SiC and Its Intergrowth Structures Using Lattice Imaging Technique
(a) Vol. **67**, No. 2, p. 625–632 (1981).
See also *1.1*, p. *12*.

A. N. GUSATINSKII, M. A. BLOKHIN, P. SHUMICHEN, M. A. BUNIN, and S. S. MIKHAILOVA
X-Ray L_1-Spectra and XPS Investigation of the Electronic Structure of Indium Compounds
(b) Vol. **108**, No. 2, p. 405–410 (1981).
See also *13.1*, p. *161*; *20*, p. *299*; 21, p. 349; 22, p. 400; 22.2, p. 426.

H. Wipf, M. V. Klein, and W. S. Williams
Vacancy-Induced and Two-Phonon Raman Scattering in ZrC_x, NbC_x, HfC_x, and TaC_x
(b) Vol. **108**, No. 2, p. 489–500 (1981).
See also *6*, p. *81*; *20.1*, p. *320*.

A. N. Chakravarti, K. P. Ghatak, K. K. Ghosh, S. Ghosh, and H. M. Mukherjee
Effect of Size Quantization on the Einstein Relation in Ultrathin Films of n-Cd_3As_2
(b) Vol. **108**, No. 2, p. 609–615 (1981).
See also *9*, p. *92*; *14.3.1*, p. *222*.

22.3.1 Mixed Crystals

Original Papers and Short Notes

S. G. ABDULLAEVA, G. L. BELENKII, M. O. GODZHAEV, and N. T. MAMEDOV
Excitons in $TlGaSe_2$
(b) Vol. **103**, No. 1, p. K61 – K63 (1981).
See also *13.5.1*, p. *188*; 20.1, p. 302.

V. ANGELINI, M. CASALBONI, M. GRANDOLFO, and P. VECCHIA
Exciton Thermoabsorption of $GaSe_{1-x}S_x$ Solid Solutions
(b) Vol. **105**, No. 1, p. 63 – 68 (1981).
See also *13.5.1*, p. *189*; *20.1*, p. *309*.

B. OLEŚ, H. J. STOLZ, and H. G. VON SCHNERING
Raman Spectra of $CaAs_3$, $EuAs_3$, $SrAs_3$, $BaAs_3$, and α-EuP_3 and the Mixed Phases $Eu_{1-x}Sr_xAs_3$ and $Eu(P_{1-x}As_x)_3$
(b) Vol. **106**, No. 1, p. 157 – 170 (1981).
See also *6*, p. *76*; *20.1*, p. *312*; 22.3, p. 442.

22.4 II–VI Compounds

Original Papers and Short Notes

T. Skettrup
Urbach's Rule and Phase Fluctuations of the Transmitted Light
(b) Vol. **103**, No. 2, p. 613–621 (1981).
See also 6, p. 67; 13.5.1, p. 188; *20.1*, p. *303*.

N. Rihon
Band Bending in Field Emission Spectroscopy from the Zinc Oxide (0001) Polar Face
(a) Vol. **63**, No. 2, p. 617–624 (1981).
See also *17.1*, p. *259*.

P. Bogusławski
The Theory of the Electron Spin–Lattice Relaxation in Zincblende Semiconductors
(b) Vol. **104**, No. 1, p. 89–95 (1981).
See also *6*, p. *69*; *13.1*, p. *153*; 22.2, p. 426.

S. Takata, T. Minami, H. Nanto, and T. Kawamura
Temperature Dependence of Electro- and Photoluminescence in ZnO Single Crystals
(a) Vol. **65**, No. 1, p. K83–K86 (1981).
See also *20.3*, p. *327*.

G. Neumann
On the Defect Structure of Zinc-Doped Zinc Oxide
(b) Vol. **105**, No. 2, p. 605–612 (1981).
See also *10.2*, p. *111*; *19*, p. *290*.

H. Nanto, T. Minami, and S. Takata
Photoluminescence in Sputtered ZnO Thin Films
(a) Vol. **65**, No. 2, p. K131–K134 (1981).
See also *20.3*, p. *329*.

E. Ziegler, A. Heinrich, H. Oppermann, and G. Stöver
Electrical Properties and Non-Stoichiometry in ZnO Single Crystals
(a) Vol. **66**, No. 2, p. 635–648 (1981).
See also 3, p. 55; *14.3*, p. *216*.

R. Kuhnert, R. Helbig, and K. Hümmer
Photoluminescence of ZnO Single Crystals. Influence of *k*-Linear Interaction and Non-Analytic Long-Range Exchange on the Emission Spectrum of the Free A-Exciton
(b) Vol. **107**, No. 1, p. 83–92 (1981).
See also *13.5.1*, p. *191*; *20.3*, p. *331*.

G. Blattner, C. Klingshirn, R. Helbig, and R. Meinl
The Influence of a Magnetic Field on the Ground and Excited States of Bound Exciton Complexes in ZnO
(b) Vol. **107**, No. 1, p. 105–115 (1981).
See also *13.5.1*, p. *191*; 20.1, p. 316; 20.3, p. 332.

F. Bechstedt, R. Enderlein, and R. Wischnewski
Binding Energies and Chemical Shifts of Least Bound Core Electron Excitations in Cubic A_NB_{8-N} Semiconductors
(b) Vol. **107**, No. 2, p. 637–651 (1981).
See also *4*, p. *60*; *13*, p. *148*; 17, p. 258; 22.1, p. 402; 22.1.1, p. 407; 22.1.2, p. 419; 22.2, p. 426.

W. Moldenhauer, K.-H. Bäther, W. Brückner, D. Hinz, and D. Bühling
Degradation Phenomena of ZnO Varistors
(a) Vol. **67**, No. 2, p. 533–542 (1981).
See also *14.3.4*, p. *235*.

A. C. Sharma and S. Auluck
Small Wave Vector $\boldsymbol{q}$ and Frequency ω Dependent Longitudinal Dielectric Function for a Model Semiconductor
(b) Vol. **108**, No. 2, p. 587–600 (1981).
See also *13.1*, p. *161*; *14.4*, p. *241*; 22.1, p. 403; 22.2, p. 426.

22.4.1 Sulphides

Review Article

K. W. Böer
The CdS/Cu_2S Solar Cell (II): The Junction and Junction–Emitter Relation
(a) Vol. **66**, No. 1, p. 11–43 (1981).
See also 13.4, p. 168; *14.3.3*, p. *225*; 14.3.4, p. 232; *16*, p. *251*.

Original Papers and Short Notes

D. Goede and R. Müller
Calorimetric Measurements of Luminescence Quantum Efficiency in Crystals
(b) Vol. **103**, No. 1, p. K75–K79 (1981).
See also *20.3*, p. *323*.

I. D. Joshi and D. Sharma
Preparation and Properties of CdS Sintered Layers
(a) Vol. **63**, No. 1, p. K23–K26 (1981).
See also *16*, p. *251*.

S. W. Koch
On the Dynamics of the Plasma Phase Transition in Highly Excited Direct Gap Semiconductors
(b) Vol. **103**, No. 2, p. 687–695 (1981).
See also *13.5*, p. *186*.

P. L. Kukk, H. A. Aarna, and M. P. Voogne
High-Temperature Point Defects in CdS:Cu
(a) Vol. **63**, No. 2, p. 389–399 (1981).
See also *10.2*, p. *107*; *14.3*, p. *210*.

J. Godlewski
On the Interpretation of Unipolar Current Flow in Insulators with Exponential Trap Energy Distribution
(a) Vol. **63**, No. 2, p. K199–K202 (1981).
See also *14.3.4*, p. *233*; 16, p. 252.

P. Dub
Reflection of Light on a Spatially Dispersive Crystal
(b) Vol. **104**, No. 1, p. 109–117 (1981).
See also *13.5.1*, p. *189*; *20.1*, p. *305*.

G. Blasse
Some Remarks on the Ce^{3+} and Bi^{3+} Luminescence in Alkaline-Earth Chalcogenides
(b) Vol. **104**, No. 1, p. K57–K59 (1981).
See also *13.4*, p. *172*; *20.3*, p. *325*; 22.6, p. 475.

L. N. Tripathi, V. K. Dubey, B. R. Chaubey, and C. P. Mishra
Dielectric Dependence of EL Brightness of Powdered Phosphor and Particle Size Distribution
(a) Vol. **64**, No. 1, p. 297–302 (1981).
See also *20.3*, p. *325*.

N. B. Lukyanchikova, T. M. Pavelko, G. S. Pekar, N. N. Tkachenko, and M. K. Sheinkman
Mechanism of Electroluminescence Excitation in Forward-Biased MS and MIS Light-Emitting Diodes Based on Wide-Band-Gap II–VI Compounds
(a) Vol. **64**, No. 2, p. 697–706 (1981).
See also *14.3.3*, p. *226*; *20.3*, p. *326*.

G. A. Soler, J.-P. Aicardi, and J.-P. Leyris
Red Photoluminescence of α-HgS Single Crystals
(b) Vol. **105**, No. 1, p. 249–255 (1981).
See also 16, p. 252; *20.3*, p. *326*.

S. Asano and N. Yamashita
Luminescence and Electron–Phonon Interaction in the MgS:Bi^{3+} Phosphor (I): Experiments
(b) Vol. **105**, No. 1, p. 305–310 (1981).
See also *6*, p. *74*; 13.4, p. 175; *20.3*, p. *327*; 22.6, p. 476.

N. E. Korsunskaya, I. V. Markevich, T. V. Torchinskaya, and M. K. Sheinkman
Erratum to: Photosensitivity Degradation Mechanism in CdS:Cu Single Crystals
(a) Vol. **65**, No. 1, p. 403 (1981).
See also *13.4*, p. *176*; *16*, p. *253*.

I. B. Ermolovich and V. V. Gorbunov
On Specific Features of the Defect Formation Mechanism in Cadmium Chalcogenides under Thermal Neutron Irradiation
(a) Vol. **65**, No. 1, p. K13–K18 (1981).
See also *11*, p. *125*; 22.4.2, p. 452; 22.4.3, p. 454.

S. Asano
Luminescence and Electron–Phonon Interaction in the MgS:Bi^{3+} Phosphor (II): Theory of $^3T_{1u}$ and $^3A_{1u}$ Mixing
(b) Vol. **105**, No. 2, p. 613–621 (1981).
See also *6*, p. *74*; 13.4, p. 177; *20.3*, p. *328*.

T. C. Leslie and J. W. Allen
Thermal Quenching of Photoluminescence in ZnS:Mn and ZnSe:Mn
(a) Vol. **65**, No. 2, p. 545–551 (1981).
See also 13.4, p. 177; *20.3*, p. *328*; 22.4.2, p. 452.

I. B. Ermolovich, V. V. Gorbunov, and G. I. Matvievskaya
Properties of the R_n and IR-1 Luminescence Bands in Cadmium Sulfide Irradiated with Thermal Neutrons
(b) Vol. **106**, No. 1, p. 45–53 (1981).
See also *11*, p. *126*; 13.4, p. 178; 16, p. 254; *20.3*, p. *329*.

V. K. Jain and J. Shanker
Analysis of Interionic Potentials in II–VI Chalcogenide Crystals with NaCl Structure
(b) Vol. **106**, No. 1, p. 287–293 (1981).
See also *4*, p. *59*; 22.4.2. p. 453; 22.4.3, p. 454; 22.6, p. 477 .

E. Yu. Brailovskii, L. A. Matveeva, G. N. Semenova, Yu. A. Tkhorik, and L. S. Khazan
Radiation-Stimulated Relaxation of Internal Stresses in Heteroepitaxial Systems
(a) Vol. **66**, No. 1, p. K59–K62 (1981).
See also *10.2*, p. *113*; *11*, p. *127*; 22.1.1, p. 406; 22.2.1, p. 430.

A. Król, W. Nazarewicz, L. Gluziński, and M. J. Kozielski
Infrared Studies of Defect in ZnS Crystals Double-Doped with Li and Al, Ga, and In
(b) Vol. **106**, No. 2, p. 489–498 (1981).
See also *6*, p. *77*; 10.2, p. 114; *20.1*, p. *314*.

A. Stahl
Model of the Excitonic Surface Layer in Normal Incidence Reflection
(b) Vol. **106**, No. 2, p. 575–580 (1981).
See also *13.5.1*, p. *191*; *20.1*, p. *314*; 22.2.1, p. 430.

S. Chatterjee, S. Ghosh, and A. N. Basu
Lattice Sums for Zinc Sulphide Structure Crystals
(b) Vol. **106**, No. 2, p. 693–698 (1981).
See also *1*, p. *10*; 4, p. 59.

L. N. Aleksandrov and M. I. Zotov
Dynamic Characteristics of Structural Defects in Semiconductors by Internal Friction Data
(a) Vol. **66**, No. 2, p. 467–475 (1981).
See also *10.2*, p. *115*; *12.2*, p. *140*; 22.1.1, p. 406; 22.1.2, p. 417; 22.2.1, p. 430.

N. B. Lukyanchikova, T. M. Pavelko, and G. S. Pekar
Physical Processes in Forward-Biased ZnS MIS Light-Emitting Diodes with a "Thick" Interfacial Layer
(a) Vol. **66**, No. 2, p. 749–759 (1981).
See also *14.3.3*, p. *229*; *20.3*, p. *331*.

K. Kempf, G. Schmieder, G. Kurtze, and C. Klingshirn
Excitation Induced Renormalization Effects of the Excitonic Polariton Dispersion in CdS
(b) Vol. **107**, No. 1, p. 297–306 (1981).
See also *13.5.1*, p. *192*; *20.1*, p. *316*.

J. Puls, H. Redlin, and J. Voigt
Evidence of Quadrupole Transitions to Excited States of the I_2-Bound Exciton in CdS
(b) Vol. **107**, No. 1, p. K71–K74 (1981).
See also *13.5.1*, p. *192*; *20.3*, p. *332*.

L. Romero and J. Campos
Luminescence Decay of ZnS:Ag in the 10^{-7} to 1 s Range by Means of Laser Excitation
(a) Vol. **67**, No. 1, p. 259–262 (1981).
See also *20.3*, p. *332*.

J. Dziesiaty, D. Backs, and D. Mantwitz
Photosensitive EPR Spectra of Pb-Doped ZnS Single Crystals
(b) Vol. **107**, No. 2, p. 553–557 (1981).
See also *13.4*, p. *181*; *19*, p. *293*.

J. A. Majewski
Electronic Structure of Transition Metal Impurities in Zinc Sulphide
(b) Vol. **108**, No. 2, p. 663–672 (1981).
See also *13.4*, p. *184*.

G. Döding and R. Labusch
Anisotropic Conductivity of CdS after Plastic Deformation and Conduction along Dislocations (II): Measurements on a Microscopic Scale
(a) Vol. 68, No. 2, p. 469–476 (1981).
See also *10.2*, p. *119*; *14.3*, p. *219*.

22.4.2 Selenides

Original Papers and Short Notes

J. Collet, M. Pugnet, A. Cornet, M. Brousseau, B. S. Razbirin, and G. V. Mikhailov
Picosecond Spectroscopy of CdSe at High Excitation
(b) Vol. **103**, No. 1, p. 367–379 (1981).
See also *13.5.1*, p. *188*; *20.3*, p. *323*.

G. N. Ivanova, D. D. Nedeoglo, A. V. Simashkevich, and I. N. Timchenko
Negative Magnetoresistance in Zinc Selenide
(b) Vol. **103**, No. 2, p. 643–652 (1981).
See also **13.4**, p. 170; *14.3*, p. *210*; 18, p. 260.

G. J. Russell, B. Vincent, and J. Woods
Precipitation Effects in Indium-Doped Zinc Selenide
(a) Vol. **63**, No. 2, p. 573–584 (1981).
See also *1.4*, p. *28*.

H. Yoshida, H. Saito, and S. Shionoya
Picosecond Relaxation Processes of High Density Electron–Hole Plasma in CdSe
(b) Vol. **104**, No. 1, p. 331–340 (1981).
See also **13.5**, p. 186; *20.3*, p. *325*.

Yu. G. Semenov and B. D. Shanina
Exchange Interaction between Paramagnetic Centers and Valence Band Electrons in Semiconductors with Cubic Lattices
(b) Vol. **104**, No. 2, p. 631–639 (1981).
See also *13.1*, p. *154*; *13.4*, p. *173*; 18.1, p. 265; 22.4.3, p. 454.

I. N. Timchenko and D. D. Nedeoglo
Hopping Conduction and Energy Spectrum of Localized States in Low-Doped Intermediate-Compensated n-ZnSe
(b) Vol. **105**, No. 1, p. 55–61 (1981).
See also *13.4*, p. *174*; *14.3*, p. *212*.

I. B. Ermolovich and V. V. Gorbunov
On Specific Features of the Defect Formation Mechanism in Cadmium Chalcogenides under Thermal Neutron Irradiation
(a) Vol. **65**, No. 1, p. K13–K18 (1981).
See also *11*, p. *125*; 22.4.1, p. 448; 22.4.3, p. 454.

J. R. Meyer and F. J. Bartoli
Ionized Impurity Scattering in n-Type ZnSe from Partial Wave Phase Shifts
(b) Vol. **105**, No. 2, p. K77–K80 (1981).
See also *13.4*, p. *177*; *14.3*, p. *214*.

T. C. Leslie and J. W. Allen
Thermal Quenching of Photoluminescence in ZnS:Mn and ZnSe:Mn
(a) Vol. **65**, No. 2, p. 545–551 (1981).
See also **13.4**, p. 177; *20.3*, p. *328*; 22.4.1, p. 448.

S. Uthanna and P. J. Reddy
Electrical Properties of CdSe Thin Films
(a) Vol. **65**, No. 2, p. K113–K115 (1981).
See also *1.5*, p. *35*; *14.3.1*, p. *221*; 20.1, p. 312.

V. K. Jain and J. Shanker
Analysis of Interionic Potentials in II–VI Chalcogenide Crystals with NaCl Structure
(b) Vol. **106**, No. 1, p. 287–293 (1981).
See also *4*, p. *59*; 22.4.1, p. 448; 22.4.3, p. 454; 22.6, p. 477.

Tran Chot
A Modified Forward *I–U* Plot for Schottky Diodes with High Series Resistance
(a) Vol. **66**, No. 1, p. K43–K45 (1981).
See also *14.3.3*, p. *228*; *14.3.4*, p. *235*; 22.1.2, p. 416.

A. M. Aliev, M. A. Nizametdinova, and V. Ya. Steinshreiber
Lattice Dynamics and Specific Heat of Thallium Monoselenide
(b) Vol. **107**, No. 2, p. K151–K154 (1981).
See also *6*, p. *80*; *8*, p. *88*.

Y. Colin, B. Fortin, and F. Raoult
Resistance Variation of a Semiconducting Thin Film during a Thermal Desorption
(a) Vol. **67**, No. 2, p. 485–495 (1981).
See also *14.3.1*, p. *221*.

H. L. Oczkowski
Thermoluminescence in Conductive n-Type ZnSe Crystals
(a) Vol. **68**, No. 1, p. 199–207 (1981).
See also *13.4*, p. *183*; *20.3*, p. *333*.

A. Baldereschi, K. Maschke, A. Milchev, R. Pickenhain, and K. Unger
Thomas-Fermi Approximation for the Valence Electron Densities in Cubic Semiconductors and Insulators
(b) Vol. **108**, No. 2, p. 511–520 (1981).
See also *13.1*, p. *161*; 14.3, p. 218; 21.7, p. 389; 22.1.1, p. 407; 22.1.2, p. 422; 22.2.1, p. 431; 22.5.2, p. 469.

S. Asano and N. Yamashita
Effect of a Magnetic Field on the Luminescence of the Pb^{2+} Ion in CaO, CaS, CaSe, and MgS Phosphors
(b) Vol. **108**, No. 2, p. 549–558 (1981).
See also *6*, p. *81*; 13.4, p. 184; 18, p. 264; *20.3*, p. *334*; 22.4.1, p. 450; 22.6, p. 480.

22.4.3 Tellurides

Original Papers and Short Notes

P. Höschl, P. Moravec, V. Prosser, A. Sakalas, and R. Jašinskaite
Electrical Characterization of ZnTe Crystals Grown from In Solvent
(a) Vol. **66**, No. 1, p. K41–K42 (1981).
See also *13.4*, p. *178*; *14.3*, p. *215*.

T. Takebe, H. Ono, J. Saraie, and T. Tanaka
Photocapacitance Study of Deep Levels in n-CdTe Single Crystals
(a) Vol. **66**, No. 2, p. 725–736 (1981).
See also *13.4*, p. *179*; 16, p. 254.

V. Leute, H. M. Schmidtke, W. Stratmann, and W. Winking
Interdiffusion in Doped and Undoped Semiconductor Alloys $Hg_wCd_{1-w}Te$
(a) Vol. **67**, No. 1, p. 183–192 (1981).
See also *9*, p. *91*; *10.2*, p. *116*; 22.4.4, p. 457.

S. Öberg
Electronic State Calculations of Dislocations in Cadmium Telluride
(b) Vol. **108**, No. 2, p. 357–362 (1981).
See also *13.4*, p. *183*.

I. Singh and G. P. Srivastava
Semiempirical Pseudopotential Calculations for the Electronic Structure of Zincblende Semiconductors
(b) Vol. **108**, No. 2, p. 467–473 (1981).
See also *13.1*, p. *161*; 22.2.1, p. 431; 22.2.2, p. 434.

K. Kranjc, D. Kunstelj, and V. Marinković
Parallel Moiré Effect in Electron Micrographs of Crystal Sandwiches (II): Comparison of Theory and Experiment
(a) Vol. **68**, No. 2, p. 363–372 (1981).
See also *1.4*, p. *32*; 1.5, p. 37; 21, p. 349; 21.1, p. 366; 21.7, p. 389.

22.4.4 Mixed Crystals

Original Papers and Short Notes

D. B. Goer, P. S. Kossobutskii, and S. G. Shevel
Non-Classical Excitonic Reflectivity Spectra of Mixed Single Crystals of the II–VI Group
(b) Vol. **103**, No. 1, p. K47 – K50 (1981).
See also *13.5.1*, p. *188*; *20.1*, p. *302*.

A. M. Rastogi and S. L. Mor
Studies on the Electroluminophor $(Ca_xSr_{1-x})S:Cu, Nd$
(a) Vol. **63**, No. 1, p. 75 – 83 (1981).
See also *20.3*, p. *323*.

V. V. Sobolev, O. G. Maksimova, and S. G. Kroitoru
Reflectivity Spectra and Band Structure of the ZnTe–CdTe System
(b) Vol. **103**, No. 2, p. 499 – 509 (1981).
See also *13.1*, p. *152*; *20.1*, p. *303*.

K. Sierański and J. Szatkowski
Theory of the Free-Carrier Absorption in Zincblende Narrow-Gap Semiconductors
(b) Vol. **104**, No. 1, p. 57 – 67 (1981).
See also 10.2, p. 107; *20.1*, p. *304*; 22.2.4, p. 437.

M. Ya. Valakh, A. P. Litvinchuk, G. S. Pekar, and G. N. Polysskii
Multiple-Phonon Resonant Raman Scattering in $Zn_{1-x}Cd_xSe$ Crystals
(b) Vol. **104**, No. 2, p. 743 – 750 (1981).
See also *6*, p. *72*; 13.5.1, p. 189; *20.1*, p. *307*.

A. Olzewski, W. Wojdowski, and W. Nazarewicz
Infrared Spectra of $Mn_xZn_{1-x}Te$ Mixed Crystals
(b) Vol. **104**, No. 2, p. K155 – K157 (1981).
See also *6*, p. *72*; *20.1*, p. *308*.

U. Dębska, M. Dietl, G. Grabecki, E. Janik, E. Kierzek-Pecold, and M. Klimkiewicz
Energy Band Gaps of $Hg_xCd_yMn_zTe$ ($x + y + z = 1$) Epitaxial Layers
(a) Vol. **64**, No. 2, p. 707 – 713 (1981).
See also 3, p. 54; 13.1, p. 155; *20.1*, p. *308*.

J. J. Dubowski
Influence of Electron–Electron Scattering on Lorentz Number in Narrow-Gap $Cd_xHg_{1-x}Te$
(b) Vol. **105**, No. 1, p. 399 – 404 (1981).
See also *14.3*, p. *212*; *15*, p. *248*.

M. Dobrowolska, W. Dobrowolski, R. R. Gałązka, and A. Mycielski
Interband Magnetooptical Studies on the Semimagnetic Semiconductor $Hg_{1-x}Mn_xSe$
(b) Vol. **105**, No. 2, p. 477 – 484 (1981).
See also 13.1, p. 156; *18*, p. *262*; *20*, p. *297*.

O. Goede, W. Heimbrodt, and R. Müller
$CdS_{1-x}Te_x$ as Persistence-Type Semiconductor Mixed Crystals
(b) Vol. **105**, No. 2, p. 543–550 (1981).
See also *13.5.1*, p. *190*; *20.1*, p. *311*; 20.3, p. 328.

S. Permogorov, A. Reznitskii, V. Travnikov, S. Verbin, G. O. Müller, P. Flögel, and M. Nikiforova
Emission of Localized Excitons in Mixed $CdS_{1-x}Se_x$ Crystals
(b) Vol. **106**, No. 1, p. K57–K60 (1981).
See also *13.5.1*, p. *191*; *20.3*, p. *330*.

J. Calas, J. Allègre, and C. Fau
Recombination Processes in 3 to 5 μm HgCdTe
(b) Vol. **107**, No. 1, p. 275–282 (1981).
See also *13.4*, p. *180*; *20.3*, p. *332*.

D. Hennig and R. Strehlow
Disorder-Induced Broadening for Free Direct Excitons in II–VI and III–V Semiconducting Mixed Crystals
(b) Vol. **107**, No. 1, p. 283–288 (1981).
See also *13.5.1*, p. *192*; *20.1*, p. *316*; 22.2.4, p. 438.

H. Heidrich
Magnetic Susceptibility and Photomagnetic Measurements on ZnS:MnS under Low-Field Conditions with a SQUID Magnetometer
(a) Vol. **67**, No. 1, p. 163–170 (1981).
See also *18*, p. *263*; *20*, p. *298*.

V. Leute, H. M. Schmidtke, W. Stratmann, and W. Winking
Interdiffusion in Doped and Undoped Semiconductor Alloys $Hg_wCd_{1-w}Te$
(a) Vol. **67**, No. 1, p. 183–192 (1981).
See also *9*, p. *91*; *10.2*, p. *116*; 22.4.3, p. 455.

A. V. Voitsekhovskii and Yu. V. Lilenko
Recombination Effects in Narrow-Gap Semiconductors p-$Hg_{1-x}Cd_xTe$
(a) Vol. **67**, No. 2, p. 381–386 (1981).
See also 13.4, p. 181; *16*, p. *255*.

22.5 Halides

Original Papers and Short Notes

M. A. Blokhin, E. G. Orlova, and I. G. Schweitzer
The Influence of the Environment of the Absorbing Atom on the Photoelectric Yield Spectra of Beryllium Compounds in the Soft X-Ray Region
(b) Vol. **103**, No. 1, p. 63--69 (1981).
See also *4*, p. *57*; *17*, p. *256*; 22.6, p. 474; 22.8, p. 484; 22.9, p. 507.

Nguyen Ba An, Nguyen Van Hieu, Nguyen Toan Thang, and Nguyen Ai Viet
Interband Electronic Raman Scattering of Polarized Light in CuCl
(b) Vol. **103**, No. 1, p. 205–210 (1981).
See also *13.1*, p. *151*; 13.5.1, p. 188; *20*, p. *296*.

J. Biellmann, B. Meyer, and C. Schwab
Anisotropy of Exciton Polaritons and Polariton–Phonon Interaction in 2H-PbI_2 Crystals (I): Structure of the Fundamental Excitonic Emission
(b) Vol. **103**, No. 1, p. 361–365 (1981).
See also *13.5.1*, p. *188*; 20.1, p. 302; *20.3*, p. *323*.

J. Danko, D. Pacheco, and B. di Bartolo
Thermal Behavior of the Position and Width of the 10488 Å Line of Nd^{3+} in $RbMnF_3$
(a) Vol. **63**, No. 1, p. K31–K34 (1981).
See also *20.3*, p. *324*.

J. Biellmann, V. Gubanov, B. Meyer, and C. Schwab
Anisotropy of Exciton Polaritons and Polariton–Phonon Interaction in 2H–PbI_2 Crystals (II): Anisotropy of the Photoluminescence of 2H–PbI_2
(b) Vol. **103**, No. 2, p. 697–701 (1981).
See also 6, p. 68; *13.5.1*, p. *189*; *20.3*, p. *324*.

D. P. Yadav, H. N. Acharya, and K. V. Rao
Effect of Quenching and X-Ray Irradiation on the Dielectric Properties of PbI_2 Single Crystals
(a) Vol. **64**, No. 1, p. 413–417 (1981).
See also 11, p. 123; *14.4*, p. *238*.

G. Gevers, J. Pichon, and A. S. Barrière
Optical Absorption Study of Nickel Fluoride Thin Films
(b) Vol. **104**, No. 2, p. 641–647 (1981).
See also 13.4, p. 173; *20.1*, p. *307*.

A. R. Hassan and R. Raouf
Polarization Dependence of Three-Photon Absorption in Solids
(b) Vol. **104**, No. 2, p. 703–709 (1981).
See also 13.1, p. 155; *20.1*, p. *307*.

V. N. Borzunov, A. E. Kadyshevich, and A. P. Rusakov
Self-Consistent Band Structure Calculation of CuCl
(b) Vol. **104**, No. 2, p. K113–K116 (1981).
See also *13.1*, p. *155*.

V. S. Mironov, G. V. Ionova, V. S. Lubimov, I. N. Groznov, and S. P. Ionov
Cooperative Jahn-Teller Effect and Spin–Orbit Coupling in d-Systems
(b) Vol. **105**, No. 1, p. 333–338 (1981).
See also *13.4*, p. *175*; 18.2, p. 270; 18.4, p. 282.

J. Shanker, J. P. Singh, and G. D. Jain
Evaluation of Third-Order and Fourth-Order Elastic Constants of Cesium Halides and Thallium Halides
(b) Vol. **105**, No. 1, p. 385–390 (1981).
See also *4*, p. *58*; *12.2*, p. *140*; 22.5.2, p. 464.

J. M. Hvam and A. Bivas
Erratum to: Optical Gain and Induced Absorption in High-Density Exciton System in CuCl
(b) Vol. **105**, No. 1, p. 439 (1981).
See also *13.5.1*, p. *189*; *20.1*, p. *309*.

T. Mohammed-Brahim
Defect Levels in HgI_2 and Illumination Effect
(a) Vol. **65**, No. 1, p. K1–K6 (1981).
See also *13.4*, p. *176*; *14.3*, p. *213*.

B. Trieman and N. Kristianpoller
Defects Induced in $RbMgF_3$ Crystals by X- and UV-Irradiation
(b) Vol. **105**, No. 2, p. 739–746 (1981).
See also *11*, p. *126*; *20.3*, p. *328*.

K. M. Häusler
Infrared Spectra of Magnetic Excitations in CoF_2
(b) Vol. **105**, No. 2, p. K81–K83 (1981).
See also *13.5.4*, p. *197*; *20.1*, p. *311*.

N. Piślewski and L. T. H. Ferris
Investigation of the Molecular Motions in $Cd(NH_3)_6Cl_2$ by NMR Spectroscopy
(b) Vol. **106**, No. 1, p. 123–128 (1981).
See also 1.2, p. 17; *19*, p. *290*.

T. Sauder, B. Stébé, and M. Certier
Acceptor Bound Exciton States in High-Purity CuCl
(b) Vol. **106**, No. 1, p. 279–286 (1981).
See also 13.4, p. 178; *13.5.1*, p. *190*; *20.1*, p. *313*; 20.3, p. 329.

B. Ghosh and R. K. Mukherjee
Electronic Absorption Spectrum of Rb_2MnCl_4 Single Crystals
(b) Vol. **106**, No. 2, p. 699–706 (1981).
See also *20.1*, p. *315*.

Nguyen Van Hieu
Resonant Photon–Photon Scattering and Approximate Spherical Symmetry in CuCl
(b) Vol. **107**, No. 1, p. 69–73 (1981).
See also *13.5.1*, p. *191*.

22.5.1 Silver Halides

Review Article

P. Müller
Electric Relaxation Currents in High-Resistivity Solids
(a) Vol. **67**, No. 1, p. 11–60 (1981).
See also 14.3.3, p. 225; 14.3.4, p. 232; *14.4*, p. *237*; 22.3, p. 440; 22.5.2, p. 462.

Original Papers and Short Notes

R. K. Singh and P. Khare
Study of Cohesion and Allied Properties of Silver Halide Crystals
(b) Vol. **103**, No. 1, p. 337–344 (1981).
See also *4*, p. *57*; 12.2, p. 138.

W. Skrotzki, G. Frommeyer, and P. Haasen
Plasticity of Polycrystalline Ionic Solids
(a) Vol. **66**, No. 1, p. 219–228 (1981).
See also *10.2*, p. *113*; 22.5.2, p. 465.

G. Luther and H. Roemer
Electrical Conductivity of Superionic AgI between 18 and 26.5 GHz
(b) Vol. **106**, No. 2, p. 511–517 (1981).
See also *14.4*, p. *240*.

M. Kobayashi and I. Yokota
Collective Motions in a Metal-Doped Molten Salt
(b) Vol. **108**, No. 2, p. 351–355 (1981).
See also *6*, p. *81*; *13.5.2*, p. *195*.

A. M. Ghodgaonkar and K. Ramani
Evaluation of Bulk Modulus, Pressure Derivative of Bulk Modulus, and Cohesive Energy by Modified Morse-Kratzer Potential
(b) Vol. **108**, No. 2, p. K133–K138 (1981).
See also *4*, p. *61*; *12.2*, p. *141*; 22, p. 400; 22.5.3, p. 472.

22.5.2 Alkali Halides

Review Articles

M. Narayana, V. S. Sivasankar, and S. Radhakrishna
Optical and Magnetic Resonance Studies of Transition Metal Ions and Complexes in Alkali Halides
(b) Vol. **105**, No. 1, p. 11–45 (1981).
See also 10.2, p. 106; 13.4, p. 168; *19*, p. *286*; *20.1*, p. *301*.

P. Müller
Electric Relaxation Currents in High-Resistivity Solids
(a) Vol. **67**, No. 1, p. 11–60 (1981).
See also 14.3.3, p. 225; 14.3.4, p. 232; *14.4*, p. *237*; 22.3, p. 440; 22.5.1, p. 461.

Original Papers and Short Notes

J. Shanker and K. Singh
The Anderson Parameter and Temperature Dependence of Bulk Modulus for Alkali Halide Crystals
(b) Vol. **103**, No. 1, p. 151–157 (1981).
See also 8, p. 84; *12.2*, p. *138*.

S. B. S. Sastry and S. Sapru
Thermoluminescence, Fluorescence, and Optical Absorption Studies on Europium-Doped Rubidium Chloride
(b) Vol. **103**, No. 1, p. 185–192 (1981).
See also 10.2, p. 106; *20.1*, p. *301*; *20.3*, p. *323*.

H. J. Jodl and F. Bolduan
Laser Induced Fluorescence and Raman Studies with $KI:NO_2^-$
(b) Vol. **103**, No. 1, p. 297–310 (1981).
See also *6*, p. *66*; *10.2*, p. *106*; 20.1, p. 302; 20.3, p. 323.

F. van Steen, M. Vanhaelst, P. Matthys, E. Boesman, and D. Schoemaker
Isotropic EPR Spectra of Vanadium in Alkali Chlorides
(b) Vol. **103**, No. 1, p. 345–351 (1981).
See also 11, p. 121; *19*, p. *286*.

M. Moreno
A Study about the Optical Absorption Spectra of $KCl:Hg^+$
(b) Vol. **103**, No. 1, p. K1–K3 (1981).
See also 13.4, p. 169; *20.1*, p. *302*.

L. B. Zuev and V. P. Sergeev
Increase of Dislocation Paths in NaCl:Sr Crystals by Electric Pulses
(a) Vol. **63**, No. 1, p. K119–K121 (1981).
See also *10.2*, p. *106*.

J. Shanker, J. P. Singh, and V. P. Gupta
Evaluation of the Fourth-Order Elastic Constants of NaCl-Structure Alkali Halide Crystals
(b) Vol. **103**, No. 2, p. 573–580 (1981).
See also *12.2*, p. *138*.

A. N. Ermoshkin, E. A. Kotomin, and R. A. Evarestov
Semiempirical Calculations of the Impurity Level Positions with Respect to the Perfect Crystal Bands
(b) Vol. **103**, No. 2, p. 581–587 (1981).
See also *13.1*, p. *152*; *13.4*, p. *170*.

R. S. Lee, J. F. Merklin, C. D. Marrs, and M. H. Richter
Impurity Photoconductivity of γ-Irradiated LiF
(b) Vol. **103**, No. 2, p. 605–611 (1981).
See also 10.2, p. 107; *16*, p. *251*; *20.1*, p. *303*.

B. V. R. Chowdari and S. Selvasekarapandian
Thermoluminescence of Ag^{+}-Doped CsBr Crystals
(b) Vol. **103**, No. 2, p. 623–628 (1981).
See also *10.2*, p. *107*; *20.3*, p. *324*.

A. V. Singh and J. Shanker
Effect of Three-Body Potential on the Dielectric Behaviour and Optic Mode Grüneisen Parameters of Alkali Halides Crystals
(b) Vol. **103**, No. 2, p. 733–738 (1981).
See also *4*, p. *57*; 6, p. 68; 12.2, p. 138; 14.4, p. 238.

T. Suzuki, W. Skrotzki, and P. Haasen
Intrinsic Glide Resistance on {100} Planes in Alkali Halides
(b) Vol. **103**, No. 2, p. 763–768 (1981).
See also *10.2*, p. *107*.

K. Asami, T. Naka, and M. Ishiguro
Assignment of the Optical Absorption Bands in CsI:Tl and CsBr:Tl by Application of High Pressure
(b) Vol. **104**, No. 1, p. 145–149 (1981).
See also 1.2, p. 15; *13.4*, p. *171*; *20.1*, p. *305*.

P. Ketolainen and A. Vaittinen
On the Production of Z_1 Centres in Irradiated Crystals
(b) Vol. **104**, No. 1, p. K53–K56 (1981).
See also *10.2*, p. *108*; *11*, p. *122*; 20.1, p. 305.

L. B. Harris and N. G. Cutmore
Factors Affecting Microhardness Measurements on NaCl Crystals
(a) Vol. **64**, No. 1, p. 105–112 (1981).
See also *12.2*, p. *139*.

P. Sridaran, R. K. Gartia, R. C. Bhuniya, and V. V. Ratnam
Thermoluminescence of Z_2 Centers in X-Irradiated Ca-Doped KCl Crystals
(a) Vol. **64**, No. 1, p. 127–131 (1981).
See also 10.2, p. 108; 20.1, p. 306; *20.3*, p. *325*.

J. C. Lewis, J. McCardle, and R. W. Withworth
Evidence for Glide of Charged Dislocations on (111) Planes in NaCl Crystals
(a) Vol. **64**, No. 1, p. K67–K71 (1981).
See also *10.2*, p. *108*.

M. Sharon, R. R. Pradhananga, and C. M. Sunta
Thermoluminescence Glow Curves of Ca^{2+} and Sr^{2+} Doped LiCl Single Crystals
(a) Vol. **65**, No. 1, p. K67–K70 (1981).
See also *10.2*, p. *111*; *20.3*, p. *327*.

S. P. Kuila, R. K. Gartia, and V. V. Ratnam
Emission Processes in Thermoluminescence of Li-Doped KCl Single Crystals
(a) Vol. **65**, No. 1, p. K71–K74 (1981).
See also *10.2*, p. *111*; *20.3*, p. *327*.

A. A. Predvoditelev, G. I. Nichugovskii, and V. I. Veselov
Simulation of Dislocation Motion through a Dislocation Forest
(a) Vol. **65**, No. 2, p. 469–478 (1981).
See also *10.1*, p. *100*; *10.2*, p. *111*; 21.1, p. 359.

O. E. Aksenov, E. D. Aluker, D. J. Lusis, and I. P. Mezina
Luminescence of Activated Alkali Halide Crystals at Excitation in the Fundamental Absorption Region
(a) Vol. **65**, No. 2, p. 617–622 (1981).
See also *10.2*, p. *111*; *20.3*, p. *328*.

Ch. Kokott and F. Fischer
Dielectric Relaxation of Chalcogen–Vacancy Centres in Potassium Halide Crystals
(b) Vol. **106**, No. 1, p. 141–147 (1981).
See also *10.2*, p. *112*; *14.4*, p. *239*.

G. G. Nilsson
Phonon Widths in RbBr at 80, 290, and 370 K from Neutron Measurements
(b) Vol. **106**, No. 1, p. 271–277 (1981).
See also *6*, p. *76*.

S. Sapru and S. B. S. Sastry
Optical and Polarized Luminescence Studies on $RbI:Pb^{2+}$ at Liquid Helium Temperatures
(b) Vol. **106**, No. 1, p. 295–300 (1981).
See also 10.2, p. 112; *20.1*, p. *313*; *20.3*, p. *329*.

T. Tsuboi
Assignment of $(Tl^{+})_2$ Absorption Bands in Alkali Halides by Magnetic Circular Dichroism Spectra
(b) Vol. **106**, No. 1, p. 301–307 (1981).
See also *13.4*, p. *178*; 18, p. 263; *20.1*, p. *313*.

A. V. Gorbunov, E. M. Nadgornyi, and S. N. Valkovskii
Laser Pulse Induced Dislocation Structure in Ionic Crystals (I): Bulk Damage of NaCl
(a) Vol. **66**, No. 1, p. 53–63 (1981).
See also *10.2*, p. *112*; *11*, p. *127*.

W. Skrotzki, G. Frommeyer, and P. Haasen
Plasticity of Polycrystalline Ionic Solids
(a) Vol. **66**, No. 1, p. 219–228 (1981).
See also *10.2*, p. *113*; 22.5.1, p. 461.

H.-J. Kaufmann, S. V. Lubenets, and E. I. Ostapchuk
Influence of Temperature on the Initial State of Plastic Deformation of LiF Crystals
(a) Vol. **66**, No. 1, p. 229–238 (1981).
See also *10.2*, p. *113*.

B. Chandra, A. R. Lakshmanan, and R. C. Bhatt
Effect of Deep Traps on Sensitization in LiF (TLD-100) Phosphor. A Correlation Study between Peak 7 and 10 Centres
(a) Vol. **66**, No. 1, p. 335–340 (1981).
See also *10.2*, p. *113*; *20.3*, p. *330*.

V. K. Jain
Z_3 Centre Thermoluminescence and Sensitization Mechanism in Lithium Fluoride
(a) Vol. **66**, No. 1, p. 341–346 (1981).
See also *10.2*, p. *113*; 20.1, p. 314; *20.3*, p. *330*.

M. Elyas, B. P. Chandra, and S. P. Kathuria
Spectroscopy of Luminescence Produced during the Application and Release of Pressure from Coloured Alkali Halide Crystals
(a) Vol. **66**, No. 1, p. K21–K24 (1981).
See also 12.2, p. 140; *20.3*, p. *330*.

V. N. Bapat and S. P. Kathuria
Correlation of Optical Absorption and Thermoluminescence in LiF: TLD-100
(a) Vol. **66**, No. 1, p. K67–K70 (1981).
See also 10.2, p. 113; *20.1*, p. *314*; *20.3*, p. *330*.

E. E. Penzina, V. M. Metsik, I. A. Parfinovich, and K. A. Makushev
Interaction of Power Laser Radiation with Colour Centres in KCl Crystals
(a) Vol. **66**, No. 1, p. K71–K74 (1981).
See also *10.2*, p. *114*; *11*, p. *127*; 20.1, p. 314.

O. V. Boev and K. P. Arefiev
Localized Positronium-Like States in Ionic Crystals
(b) Vol. **106**, No. 2, p. 481–488 (1981).
See also 10.2, p. 114; *13*, p. *146*.

V. Hizhnyakov, T. Soovik, and S. Zazubovich
Origin of Radiative Transitions from Metastable Minima of the Excited State of $Sn^{2+}v_c^-$ Centres in Alkali Halides
(b) Vol. **106**, No. 2, p. 581–586 (1981).
See also *13.4*, p. *179*; *20.3*, p. *331*.

O. Sild and I. Sildos
Zeeman Effect in Low-Temperature Luminescence Spectra of Molecular Ions O_2^- and S_2^- in Alkali Halide Crystals
(b) Vol. **106**, No. 2, p. 675–682 (1981).
See also *13.4*, p. *179*; *20.3*, p. *331*.

R. V. Jimenez, F. Lüty, M. Siu Li, and M. de Souza
Paraelastic Alignment and Electric Dipole Relaxation Behaviour of Off-Center Ag^+ Defects in RbI
(b) Vol. **106**, No. 2, p. 683–692 (1981).
See also *10.2*, p. *114*; *20.1*, p. *315*.

N. Dass and F. A. Kasir
Temperature Dependence of Compressibility of Alkali Halides
(b) Vol. **106**, No. 2, p. K99–K102 (1981).
See also *12.2*, p. *140*.

A. V. Gorbunov, E. M. Nadgornyi, and S. N. Valkovskii
Laser Pulse Induced Dislocation Structure in Ionic Crystals (II): Surface Damage of NaCl and MgO
(a) Vol. **66**, No. 2, p. 455–462 (1981).
See also *10.2*, p. *114*; *11*, p. *127*; 22.6, p. 478.

S. V. Moharil
A New Method for Obtaining Trapping Parameters from Glow Curves
(a) Vol. **66**, No. 2, p. 767–772 (1981).
See also *13.4*, p. *179*; *20.3*, p. *331*.

B. R. Sood
Purity of Potassium Iodide Necessary for Its Use in Phasers
(a) Vol. **66**, No. 2, p. K145–K147 (1981).
See also *6*, p. *78*; 10.2, p. 115; *20.2*, p. *322*.

Ch. Högfors, J. Kubát, and M. Rigdahl
A Cooperative Model of Stress Relaxation in Solids
(b) Vol. **107**, No. 1, p. 147–155 (1981).
See also *12.1*, p. *136*; *12.2*, p. *140*; 13.5, p. 187; 21, p. 345; 22.9, p. 511.

K. Vogler
Two-Photon Absorption of F-Centres in KCl
(b) Vol. **107**, No. 1, p. 195–199 (1981).
See also *10.2*, p. *115*; 13.4, p. 180; *20.3*, p. *332*.

H. J. Hoffmann
Recombination Kinetics of Electrons with α-Centers in Colored Alkali Halides
(b) Vol. **107**, No. 1, p. 215–224 (1981).
See also *10.2*, p. *115*; 13.4, p. 180; *16*, p. *254*.

K. Toriumi and N. Itoh
Behaviour of Defects Introduced by a Nanosecond Intense Electron Pulse at High Temperatures in Alkali Halides
(b) Vol. **107**, No. 1, p. 375–386 (1981).
See also *10.2*, p. *115*; 20.1, p. *316*.

A. Watterich
Monovalent Metal Ions Investigated by ESR in Ca or Sr Doped LiF
(b) Vol. **107**, No. 1, p. K35–K38 (1981).
See also *10.2*, p. *116*; *19*, p. *292*.

K.-E. Peiponen and T. Jääskeläinen
Simple Formulas for the Complex Refractive Index of Gaussian Type
(b) Vol. **108**, No. 1, p. K71–K73 (1981).
See also *20.1*, p. *319*.

E. A. Radyabov and A. I. Nepomnyachikh
Neutral Magnesium Atoms on Anion Sites in LiF
(b) Vol. **108**, No. 1, p. K75–K78 (1981).
See also *10.2*, p. *118*; 20.3, p. 333.

E. A. Radyabov and A. I. Nepomnyachikh
Magnesium Color Centers at 3.5 and 5.0 eV in Lithium Fluoride
(a) Vol. **68**, No. 1, p. 77–81 (1981).
See also *10.2*, p. *118*; 20.1, p. 319; 20.3, p. 333.

A. Dauletbekova and A. Elango
Conversion of F–H_A Pairs into α–I_A Pairs in KBr:Li and KBr:Na X-Rayed at 80 K
(b) Vol. **108**, No. 2, p. 299–305 (1981).
See also *10.2*, p. *118*; *20.1*, p. *320*.

Y. V. G. S. Murti and T. V. Selvarajan
An Extended Interionic Potential for Alkali Halides
(b) Vol. **108**, No. 2, p. 315–322 (1981).
See also *4*, p. *61*.

K. Imanaka, A.-H. Kayal, A. C. Mezger, and J. Rossel
Self-Trapped Exciton Luminescence after Tunnelling of V_k and Na^0 Centers in CsI:Na Crystals
(b) Vol. **108**, No. 2, p. 449–458 (1981).
See also *13.5.1*, p. *192*; *20.3*, p. *334*.

A. Baldereschi, K. Maschke, A. Milchev, R. Pickenhain, and K. Unger
Thomas-Fermi Approximation for the Valence Electron Densities in Cubic Semiconductors and Insulators
(b) Vol. **108**, No. 2, p. 511–520 (1981).
See also *13.1*, p. *161*; 14.3, p. 218; 21.7, p. 389; 22.1.1, p. 407; 22.1.2, p. 422; 22.2.1, p. 431; 22.4.2, p. 453.

V. V. Hizhnyakov, V. G. Plekhanov, V. V. Shepelev, and G. S. Zavt
Hot Luminescence of Self-Trapped Excitons in Alkali Halide Crystals
(b) Vol. **108**, No. 2, p. 531–540 (1981).
See also *13.5.1*, p. *193*; *20.3*, p. *334*.

A. L. Shluger and E. A. Kotomin
Semiempirical Calculations of Defect Properties in LiF Crystal (I): Perfect Crystal
(b) Vol. **108**, No. 2, p. 673–681 (1981).
See also *10.2*, p. *119*; *13.4*, p. *184*.

N. S. Minaev, A. V. Mudrii, and V. D. Tkachev
Symmetry and Nature of the 1.0186 eV Luminescence Centre in Neutron-Irradiated Silicon
(b) Vol. **108**, No. 2, p. K89–K94 (1981).
See also *10.2*, p. *119*; 11, p. 131; 12.2, p. 141; *20.3*, p. *334*.

S. Shimanuki and N. Nishimaki
Formation of Interstitial Silver Centers by the Optical Destruction of Ag^- Centers in KCl:Li and KCl:Na Crystals
(b) Vol. **108**, No. 2, p. K95–K100 (1981).
See also *10.2*, p. *119*; *20.1*, p. *320*.

S. V. Moharil
On the Variation of Order of Kinetics during Thermoluminescence Run
(a) Vol. **68**, No. 2, p. 413–418 (1981).
See also *20.3*, p. *334*.

H. Koizumi and T. Suzuki
The Influence of Yield Stress on Fracture of NaCl Crystals
(a) Vol. **68**, No. 2, p. 579–588 (1981).
See also *12.2*, p. *141*.

22.5.3 Alkali-Earth Halides

Original Papers and Short Notes

A. Basu, S. Sengupta, and A. N. Basu
Three-Body Interaction in the Fluorite Structure Ionic Crystals
(b) Vol. **103**, No. 2, p. 535–546 (1981).
See also *4*, p. *57*; *12.2*, p. *138*.

N. V. Starostin, M. P. Shepilov, and A. B. Alekseev
Energy Structure of the Alkaline-Earth Fluorides
(b) Vol. **103**, No. 2, p. 717–723 (1981).
See also *13.1*, p. *152*.

E. Anastassakis and M. Cardona
Internal Strains Raman-Active Optical Phonons
(b) Vol. **104**, No. 2, p. 589–600 (1981).
See also *6*, p. *71*; *12.2*, p. *139*; 22.1, p. 402; 22.6, p. 475.

T. Tsuboi
The Parity-Forbidden $4f^5 \rightarrow 4f^5$ Transition of the Sm^{3+} Ion in CaF_2
(b) Vol. **104**, No. 2, p. 697–701 (1981).
See also *13.4*, p. *173*; *20.1*, p. *307*.

R. Rauch and E. Liebold
VUV Absorption Spectra of Oxygenated CaF_2 Crystals
(a) Vol. **64**, No. 2, p. K165–K168 (1981).
See also *10.2*, p. *110*; *20.1*, p. *309*.

R. K. Sinha and M. L. Mukherjee
On the Kinetics of Phosphorescence and High-Temperature Glow Peak in CaF_2:Pr Single Crystals
(b) Vol. **105**, No. 1, p. 69–75 (1981).
See also *13.4*, p. *174*; *20.3*, p. *326*.

V. Tabakova and K. Kynev
Absorption Spectra of Lead Centers in Barium Chloride
(a) Vol. **65**, No. 1, p. K75–K78 (1981).
See also *13.4*, p. *176*; *20.3*, p. *327*.

J. Kuriata, N. Guskos, and T. Rewaj
Temperature Dependence of Second-Order Spin–Lattice Coefficients for Eu^{2+} in MeF_2 Single Crystals
(b) Vol. **106**, No. 2, p. 607–612 (1981).
See also *19*, p. *291*.

J. F. Pouradier and G. Grimouille
Growth of Semiconducting $CdF_2:Ho^{3+}$ Crystals and Study of Their Optical and Transport Properties
(b) Vol. **106**, No. 2, p. 657–665 (1981).
See also *3*, p. *55*; 13.4, p. 179; *14.3*, p. *215*; 20.1, p. 315.

22.5.4 Mixed Crystals

Original Papers and Short Notes

D. N. TALWAR, M. VANDEVYVER, K. KUNC, and M. ZIGONE
IR Absorption and Raman Scattering Spectra in Mixed Cuprous Halides
(b) Vol. **103**, No. 1, p. 381–389 (1981).
See also *6*, p. *67*; *20.1*, p. *302*.

J. SHANKER, G. D. JAIN, and S. K. SHARMA
Evaluation of Third-Order Elastic Constants and Pressure Derivatives of Second-Order Elastic Constants for AgCl–AgBr Mixed Crystals
(b) Vol. **104**, No. 1, p. 75–80 (1981).
See also *12.2*, p. *139*.

B. ARUNA, Y. V. S. RAMAKRISHNA, and P. A. NARAYANA
Electron Paramagnetic Resonance Studies of Cu^{2+} in $NH_4Cl_{1-x}Br_x$ Single Crystals
(b) Vol. **105**, No. 1, p. 347–351 (1981).
See also *19*, p. *289*.

M. M. BEG, S. MAHMOOD, N. AHMAD, J. ASLAM, Q. H. KHAN, and N. M. BUTT
Concentration Dependence of the Debye Temperature for the $K_xRb_{1-x}F$ System
(b) Vol. **106**, No. 1, p. K43–K46 (1981).
See also *6*, p. *77*.

M. L. MUKHERJEE and R. K. SINHA
Z_1 Centers in Cd Doped $KCl_{0.5}Br_{0.5}$ Mixed Crystal
(b) Vol. **106**, No. 1, p. K53–K56 (1981).
See also *10.2*, p. *112*; *20.3*, p. *330*.

P. J. CHANDLER and E. LILLEY
Ionic Conductivity of Single Crystals of 6 NaCl · $CdCl_2$ Suzuki Phase
(a) Vol. **66**, No. 1, p. 183–187 (1981).
See also *14.4*, p. *239*.

K. RECKER, F. WALLRAFEN, R. BÜSCHER, and G. LEHMANN
Crystal Growth, EPR, and Luminescence of $BaZnF_4$:Mn
(b) Vol. **107**, No. 2, p. 699–706 (1981).
See also 3, p. 55; *19*, p. *293*; *20.3*, p. *333*.

22.6 Simple Oxides (CdO, ZnO See 22.4)

Original Papers and Short Notes

M. A. Blokhin, E. G. Orlova, and I. G. Schweitzer
The Influence of the Environment of the Absorbing Atom on the Photoelectric Yield Spectra of Beryllium Compounds in the Soft X-Ray Region
(b) Vol. **103**, No. 1, p. 63–69 (1981).
See also *4*, p. *57*; *17*, p. *256* ; 22.5, p. 458; 22.8, p. 484; 22.9, p. 507.

S. K. Trikha and S. C. Jain
Computer Simulation of the Phase Transition in Ice
(a) Vol. **63**, No. 1, p. 163–168 (1981).
See also *2*, p. *41*.

A. Krell
Alumina Structure with Improved Fracture Properties
(a) Vol. **63**, No. 1, p. 183–192 (1981).
See also *12.2*, p. *138*.

M. S. Abrahams, J. L. Hutchison, and G. R. Booker
Direct Observation of a Silicon–Sapphire Hetero-Epitaxial Interface by High Resolution Transmission Electron Microscopy
(a) Vol. **63**, No. 1, p. K3–K6 (1981).
See also *1.4*, p. *28*; 3, p. 53; 22.1.2, p. 409.

K. B. Sundaram and G. K. Bhagavat
Preparation and Properties of Indium Oxide Films
(a) Vol. **63**, No. 1, p. K15–K18 (1981).
See also 1.5, p. 34; *3*, p. *53*; *14.3.1*, p. *220*; 15, p. 247; 20.1, p. 303.

H.-J. Fitting
Exoelectron Emission Due to Relaxation Processes at Surfaces
(a) Vol. **63**, No. 1, p. K47–K50 (1981).
See also *17*, p. *256*.

G. C. Psaltakis and M. G. Cottam
Theory of Spin–Phonon Interactions in a System of Paramagnetic $S = 3/2$ Ions
(b) Vol. **103**, No. 2, p. 709–716 (1981).
See also *6*, p. *68*; 8, p. 84; *13.4*, p. *170*; 18.1, p. 265.

V. Siegel and H.-H. Kirchner
The Influence of the Ambient Atmosphere on the TSEE of Ceramic BeO
(a) Vol. **63**, No. 2, p. 481–485 (1981).
See also *17*, p. *256*.

M. Bohl
Electronic Structure, UV Absorption, and Dispersion of Optical Glasses
(a) Vol. **63**, No. 2, p. 529–538 (1981).
See also *2*, p. *41*; 13, p. 143; *20.1*, p. *304*.

V. V. MÜRK, A. I. KUZNETSOV, and B. R. NAMOZOV
Kinetics of Intrinsic Luminescence and Energy Transfer in Third Group Metal Oxides
(a) Vol. **68**, No. 2, p. K131–K135 (1981).
See also *20.3*, p. *325*.

M. G. ABRAMISHVILI, V. I. ALTUKHOV, T. L. KALABEGISHVILI, and V. G. KVACHADZE
Low Temperature Heat Conduction of γ-Irradiated MgO Crystals
(b) Vol. **104**, No. 1, p. 49–55 (1981).
See also *8*, p. *84*; *11*, p. *122*; 20.1, p. 304.

G. DRÄGER, O. BRÜMMER, V. HEERA, G. SEIFERT, and P. ZIESCHE
Polarized TiKβ Valence Band Spectra of TiO_2 (Rutile)
(b) Vol. **104**, No. 1, p. 219–226 (1981).
See also *13.1*, p. *154*; *20*, p. *296*.

A. C. VAN DER STEEN and L. T. F. DIJCKS
The Luminescence Properties of Alkaline-Earth Oxides Activated with $6s^2$ Ions
(b) Vol. **104**, No. 1, p. 283–292 (1981).
See also 6, p. 70; 13.4, p. 171; *20.3*, p. *325*.

K. HÜBNER, L. SCHUMANN, A. LEHMANN, H. H. VAJEN, and G. ZUTHER†
Detection of LO and TO Phonons in Amorphous SiO_2 Films by Oblique Incidence of IR Light
(b) Vol. **104**, No. 1, p. K1–K5 (1981).
See also 2, p. 42; *6*, p. *70*; *20.1*, p. *305*.

J. PÖRSCH and N. B. MANSON
Magnetic Circular Dichroism of the $^4A_{2g} \rightarrow {}^4T_{2g}$ Zero-Phonon Line in $MgO:Cr^{3+}$
(b) Vol. **104**, No. 1, p. K7–K10 (1981).
See also *13.4*, p. *171*; *20.1*, p. *305*.

G. BLASSE
Some Remarks on the Ce^{3+} and Bi^{3+} Luminescence in Alkaline-Earth Chalcogenides
(b) Vol. **104**, No. 1, p. K57–K59 (1981).
See also *13.4*, p. *172*; *20.3*, p. *325*; 22.4.1, p. 447.

E. D. ATANASOVA, K. I. KIROV, and E. I. KANTARDJEVA
Formation of Ultrathin Oxide Films on Silicon in RF Oxygen Plasma
(a) Vol. **64**, No. 1, p. 73–80 (1981).
See also *1.5*, p. *34*; *3*, p. *53*.

G. F. GOLOVANOVA, V. F. KISELEV, S. N. KOZLOV, A. S. PETROV, and E. A. SILAEV
On the Effect of Halogens on Proton Processes in the Si–SiO_2 Structures
(a) Vol. **64**, No. 1, p. 177–186 (1981).
See also *13.3*, p. *165*; 19, p. 288; 22.1.2, p. 411.

E. ANASTASSAKIS and M. CARDONA
Internal Strains Raman-Active Optical Phonons
(b) Vol. **104**, No. 2, p. 589–600 (1981).
See also *6*, p. *71*; *12.2*, p. *139*; 22.1, p. 402; 22.5.3, p. 471.

R. ENGELKE, TH. ROY, H.-G. NEUMANN, and K. HÜBNER
Chemical Bond and Related Properties of SiO_2 (VIII): Experimental and Theoretical Investigation of the Structure of SiO_x
(a) Vol. **65**, No. 1, p. 271–280 (1981).
See also *2*, p. *46*; *4*, p. *58*; 13, p. 145.

V. M. ORERA
The Effect of Lithium Impurity on Flow Stress in MgO
(a) Vol. **65**, No. 1, p. 293–300 (1981).
See also *10.2*, p. *110*; 12.2, p. 140.

M. B. DUTT, R. BANERJEE, and A. K. BARUA
Transport Properties of Lithium and Sodium Doped Nickel Oxide
(a) Vol. **65**, No. 1, p. 365–370 (1981).
See also *14.3*, p. *213*; *15*, p. *248*.

YU. A. GOROKHOVATSKII, A. N. GUBKIN, and A. I. MARGULEV
Thermally Stimulated Depolarization for Quasi-Continuous Distribution of Relaxators According to Activation Energy and Frequency Factor
(a) Vol. **65**, No. 2, p. 485–490 (1981).
See also 13.4, p. 177; *14.4*, p. *239*.

R. P. KHARE and M. KHARE
A New Method of Analysis of Traps Taking Part in the Electroluminescence Process
(a) Vol. **65**, No. 2, p. 725–729 (1981).
See also *13.4*, p. *177*; *20.3*, p. *329*.

SH. A. VAKHIDOV, E. M. IBRAGIMOVA, I. NURITDINOV, A. F. RAKOV, and G. I. IKRAMOV
Self-Trapped Particles in Complex Oxide Crystals
(b) Vol. **106**, No. 1, p. 31–35 (1981).
See also *11*, p. *126*; 13.5.1, p. 190; 19, p. 290; 20.1, p. 312; 20.3, p. 329; 22.8, p. 487.

W. GEHLHOFF and K. H. SEGSA
Line Broadening and Line Displacement of EPR Transitions Produced by Inhomogeneous Plane Stresses in a Two-Layer System
(b) Vol. **106**, No. 1, p. 171–182 (1981).
See also *12.2*, p. *140*; *19*, p. *291*; 22.1.2, p. 415.

D. D. RICHARDSON
A Calculation of Vacancy Formation Energies in Strontium Oxide
(b) Vol. **106**, No. 1, p. 223–227 (1981).
See also *10.2*, p. *112*.

V. K. JAIN and J. SHANKER
Analysis of Interionic Potentials in II–VI Chalcogenide Crystals with NaCl Structure
(b) Vol. **106**, No. 1, p. 287–293 (1981).
See also *4*, p. *59*; 22.4.1, p. 448; 22.4.2, p. 453; 22.4.3, p. 454.

P. NOVÁK
Self-Consistently Embeded Ion – A New Approach to the Electronic Structure Calculation
(b) Vol. **106**, No. 1, p. K39–K41 (1981).
See also *13*, p. *146*.

S. Mochizuki
Polarized Infrared Absorption in Thin-Film Cerium Dioxide
(b) Vol. **107**, No. 1, p. K53–K56 (1981).
See also *6*, p. *78*; *20.1*, p. *317*.

E. Iguchi, H. Sawatari, and R. J. D. Tilley
Polarisabilities of Ions in Rutile (TiO_2) Calculated with Madelung Potentials Determined by Ewald's Method
(b) Vol. **107**, No. 1, p. K63–K66 (1981).
See also *14.4*, p. *240*.

R. W. A. Hillhouse and J. Woods
The Evaluation of Thermally Stimulated Current and Thermoluminescence Curves in Rutile
(a) Vol. **67**, No. 1, p. 119–128 (1981).
See also *13.4*, p. *180*; *14.3*, p. *217*; 20.3, p. 332.

K. J. Bachmann, F. S. L. Hsu, and J. P. Remeika
The Low Temperature Heat Capacities of SnO_2 and In_2O_3
(a) Vol. **67**, No. 1, p. K39–K42 (1981).
See also *8*, p. *87*.

H. Bahadur and R. Parshad
A New Electronic Method for Detection of Hydrogen Concentration in Quartz Crystals
(a) Vol. **67**, No. 1, p. K53–K57 (1981).
See also *10.2*, p. *116*; 11, p. 129.

M. Baran, A. A. Mirzakhanyan, E. G. Sharoyan, and H. Szymczak
Zero Field Splitting of the Mo^{3+} Ground State in Corundum
(b) Vol. **107**, No. 2, p. K129–K132 (1981).
See also *13.4*, p. *181*; *19*, p. *293*.

F. Bechstedt and K. Hübner
Mean Free Path of Photoelectrons in Silicon and Silicon Oxides
(a) Vol. **67**, No. 2, p. 517–526 (1981).
See also *13*, p. *148*; *17*, p. *258*; 22.1.2, p. 420.

X. Aymerich-Humet, F. Serra-Mestres, and J. Millán
On the Resonant Broadening of the Trap Assisted Tunneling in MIM Structures
(a) Vol. **67**, No. 2, p. 597–602 (1981).
See also 13.4, p. 182; *14.3.3*, p. *230*; *14.3.4*, p. *235*.

K. Iguchi and H. Hashimoto
The Flux Distribution of Electron Waves Incident on a MgO Crystal in 110 Orientation
(a) Vol. **67**, No. 2, p. 655–667 (1981).
See also *1.1*, p. *12*.

H. Bahadur and R. Parshad
Some New Results on the Frequency Characteristics of Quartz Crystals Irradiated by Ionizing and Particle Radiations
(a) Vol. **67**, No. 2, p. 683–692 (1981).
See also 11, p. 130; *12.2*, p. *141*.

H. Jain and A. S. Nowick
Electrical Conduction in Paratellurite (TeO_2) Crystals
(a) Vol. **67**, No. 2, p. 701–707 (1981).
See also **10.2**, p. **117**; *14.4*, p. *240*.

E. Hanga, E. Tataru, M. Morariu, L. Diamandescu, and N. Popescu-Pogrion
Magnetic Properties of Lithium Silicate Glass Containing Manganese Ferrite
(a) Vol. **67**, No. 2, p. 725–728 (1981).
See also *5*, p. *65*; *18.1*, p. *266*.

J. Pietrzak, J. Kudyńska, and Z. Kruczyński
Phase Transition Study in VO_2 Using the EPR Method
(a) Vol. **67**, No. 2, p. K175–K178 (1981).
See also 14.1, p. 203; 14.3, p. 218; *19*, p. *293*.

E. F. Kustov, L. I. Uskova, and S. A. Klinchikov
Estimation of Covalency Effects in the Optical Spectra of Crystals Containing Iron Group Ions (II)
(b) Vol. **108**, No. 1, p. 53–58 (1981).
See also *13*, p. *149*; *20.1*, p. *318*; 22.5, p. 460.

W. Grevendonk, L. van den Abeele, P. van den Keybus, B. Vanhuyse, and G. Ruymbeek
Faraday Rotation in Quartz and Oxide Glasses
(b) Vol. **108**, No. 1, p. K53–K56 (1981).
See also 2, p. 50; 13.1, p. 160; *18*, p. *264*; *20*, p. *299*; 22.6.1, p. 482.

G. Badics, I. A. Serbinov, Yu. D. Kalafati, and W. F. Fallmann
Time Periodic Dissipative Structures in VO_2 Films
(a) Vol. **68**, No. 1, p. K101–K104 (1981).
See also *14.3.1*, p. *221*.

S. Asano and N. Yamashita
Effect of a Magnetic Field on the Luminescence of the Pb^{2+} Ion in CaO, CaS, CaSe, and MgS Phosphors
(b) Vol. **108**, No. 2, p. 549–558 (1981).
See also *6*, p. *81*; 13.4, p. 184; 18, p. 264; *20.3*, p. *334*; 22.4.1, p. 450; 22.4.2, p. 453.

J. Bullot, P. Cordier, O. Gallais, M. Gauthier, and J. Livage
Experimental Determination of the Disorder Energy in Amorphous V_2O_5 Layers Deposited from Gels
(a) Vol. **68**, No. 2, p. 357–361 (1981).
See also *2*, p. *51*; *14.3.1*, p. *222*.

J. Finster and D. Schulze
Studies of the Si/SiO_2 Interface by Angular Dependent X-Ray Photoelectron Spectroscopy
(a) Vol. **68**, No. 2, p. 505–517 (1981).
See also *1.5*, p. *37*; *17*, p. *258*; 22.1.2, p. 423.

22.6.1 Mixed Crystals

Original Papers and Short Notes

V. Patráš, J. Krempaský, G. Vlasák, and F. Hanic
Superionic Conductivity in the System PrO_x–HfO_2
(a) Vol. **63**, No. 1, p. 79–83 (1981).
See also *14.4*, p. *237*.

T. van Dijk and A. J. Burggraaf
Grain Boundary Effects on Ionic Conductivity in Ceramic $Gd_xZr_{1-x}O_{2-(x/2)}$ Solid Solutions
(a) Vol. **63**, No. 1, p. 229–240 (1981).
See also *1.4*, p. *28*; *14.4*, p. *237*.

H. Reuther and W. Hinz
Erratum to: Thermotransport in Lithium Silicate Glasses
(a) Vol. **63**, No. 2, p. K217 (1981).
See also *2*, p. *42*; *9*, p. *89*.

B. Banerjee, J. Sharma, and S. Lahiry
Thermoremanent Magnetization in $Bi_2O_3 \cdot Fe_2O_3$ System
(b) Vol. **105**, No. 1, p. 275–283 (1981).
See also *18.2*, p. *270*; *18.4*, p. *282*.

P. Kleinert and R. Leihkauf
On the Low-Temperature Specific Heat of Glasses
(b) Vol. **105**, No. 1, p. K67–K70 (1981).
See also *2*, p. *45*; *8*, p. *85*.

L. Kaihola
Anomalous Soft X-Ray Reflectivity of Borosilicate Glass by Successive Grazing Reflections
(a) Vol. **65**, No. 1, p. 163–166 (1981).
See also *2*, p. *45*; *20*, p. *297*.

H. Gruber, E. Krautz, and H. P. Fritzer
Magnetic Susceptibility and Electrical Properties of the ϰ-Phase $Mo_{17}O_{47}$ with Substitutions of Molybdenum
(a) Vol. **65**, No. 2, p. 589–594 (1981).
See also 14.1, p. 201; *14.3*, p. *214*; *18*, p. *262*; 18.4, p. 283.

A. V. Shendrik
ESR of Vitreous Silica Containing Hydroxyl
(a) Vol. **67**, No. 1, p. K25–K27 (1981).
See also *2*, p. *49*; *19*, p. *292*.

J. L. Soubeyroux, D. Fruchart, J. C. Marmeggi, W. J. Fitzgerald, C. Delmas, and G. le Flem
Magnetic Structure of $LiCrO_2$
(a) Vol. **67**, No. 2, p. 633–642 (1981).
See also *18.4*, p. *284*.

J. G. Pepin and E. R. Vance
Non-Confirmation of a $Cr_2Zr_3O_{10}$ Compound
(a) Vol. **67**, No. 2, p. K167 – K169 (1981).
See also *1.3*, p. *26*.

W. Grevendonk, L. van den Abeele, P. van den Keybus, B. Vanhuyse, and G. Ruymbeek
Faraday Rotation in Quartz and Oxide Glasses
(b) Vol. **108**, No. 1, p. K53 – K56 (1981).
See also 2, p. 50; 13.1, p. 160; *18*, p. *264*; *20*, p. *299*; 22.6, p.480.

R. Göring, H. Bürger, H. Nass, and B. Schnabel
^{11}B NMR Study on Glasses of the System TeO_2–B_2O_3
(a) Vol. **68**, No. 1, p. K29 – K32 (1981).
See also *2*, p. *51*; *19*, p. *294*.

C. A. Hogarth and M. Y. Nadeem
The Optical Absorption Edge in Amorphous Thin Films of Germania and of Germania with Barium Oxide
(a) Vol. **68**, No. 2, p. K181 – K184 (1981).
See also *2*, p. *52*; *20.1*, p. *321*.

22.7 Semiconducting Intermetallic Compounds

Original Papers and Short Notes

G. Simon and W. Eichler
Galvanomagnetic and Thermoelectric Transport Investigations on Sb_2Te_3
(b) Vol. **103**, No. 1, p. 289–295 (1981).
See also *14.3*, p. *209*; *15*, p. *247*.

B. Fellmuth, H. Krüger, R. Rudolph, and R. Herrmann
Composition Dependence of the Maximum Cyclotron Mass m_1^c in n-Type Semiconducting $Bi_{1-x}Sb_x$
(b) Vol. **104**, No. 1, p. 209–217 (1981).
See also *13.1*, p. *153*; *19*, p. *288*.

M. Stordeur, H. T. Langhammer, H. Sobotta, and V. Riede
Valence Band Structure of $(Bi_{1-x}Sb_x)_2Te_3$ Single Crystals
(b) Vol. **104**, No. 2, p. 513–522 (1981).
See also *13.1*, p. *154*; 14.3, p. 211; 15, p. 248; 20.1, p. 307.

V. D. Das and N. Meena
Electrical Conduction in $Bi_{40}Sb_{60}$ Alloy Thin Films
(b) Vol. **104**, No. 2, p. K89–K93 (1981).
See also *14.3.1*, p. *220*.

G. Simon and W. Eichler
Investigations on a Two-Valence Band Model for Sb_2Te_3
(b) Vol. **107**, No. 1, p. 201–206 (1981).
See also *13.1*, p. *159*; 14.3, p. 216.

A. Priemuth
Electromigration of Cu in V_2VI_3 Compounds
(a) Vol. **67**, No. 2, p. 505–510 (1981).
See also *1*, p. *10*; *9*, p. *91*; 14.3, p. 217.

22.8 Tertiary and Higher Compounds

Original Papers and Short Notes

J. Hölsä and M. Leskelä
Luminescence Properties of Terbium-Activated Rare-Earth Oxyhalides
(b) Vol. **103**, No. 2, p. 797–801 (1981).
See also *20.3*, p. *324*.

H. Sobotta, H. Neumann, W. Kissinger, G. Kühn, and V. Riede
Two-Phonon Absorption Spectra in $CuInSe_2$
(b) Vol. **103**, No. 2, p. K125–K129 (1981).
See also *6*, p. *68*; *20.1*, p. *304*.

J. Horák, L. Tichý, and P. Lošťák
Electrical Conductivity of BiTeI Crystals
(a) Vol. **63**, No. 2, p. 407–416 (1981).
See also 13.1, p. 153; *14.3*, p. *210*; 15, p. 247.

J. J. M. Binsma, L. J. Giling, and J. Bloem
Order–Disorder Behaviour and Tetragonal Distortion of Chalcopyrite Compounds
(a) Vol. **63**, No. 2, p. 595–603 (1981).
See also *1.3*, p. *22*; 4, p. 57.

M. K. Teng, M. Massot, M. R. Chaves, M. H. Amaral, S. Ziolkiewicz, and W. Young
Tricritical Point Induced by Atomic Substitution in $Bi_xSb_{1-x}SI$
(a) Vol. **63**, No. 2, p. 605–615 (1981).
See also *14.4.1*, p. *242*.

J. Pietrzak and Z. Kruczyński
EPR of Cr^{3+} in γ-Alum $NaAl(SO_4)_2 \cdot 12\,H_2O$
(a) Vol. **63**, No. 2, p. K143–K145 (1981).
See also *19*, p. *287*.

E. R. Vance, D. K. Agrawal, and J. G. Pepin
Powder X-Ray Diffraction Study of Halide Sodalites
(a) Vol. **63**, No. 2, p. K189–K193 (1981).
See also *1.1*, p. *11*; *10.2*, p. *107*.

F. Prokert
Neutron Scattering Studies on Phase Transitions and Phonon Dispersion in $CsSrCl_3$
(b) Vol. **104**, No. 1, p. 261–265 (1981).
See also *1.2*, p. *15*; *6*, p. *69*.

K. K. P. Srivastava and H. C. Singh
Erratum to: Effect of Second-Order Spin–Orbit Coupling $\lambda(\boldsymbol{L}\cdot\boldsymbol{S})^2$ On Electronic Spin Relaxation of Paramagnetic Fe^{2+} in $CaCO_3$ and $CdCO_3$
(b) Vol. **104**, No. 1, p. 393 (1981).
See also *5*, p. *63*; 18.1, p. 265.

T. X. Hoai, K. H. Herrmann, and D. Genzow
A New Attempt for Understanding Temperature-Dependent Thresholds in Lead Salt Injection Lasers
(a) Vol. **64**, No. 1, p. 239–248 (1981).
See also *20.2*, p. *322*.

T. G. Kerimova, Sh. S. Mamedov, and R. Kh. Nani
On the Band Structure of $CdGa_2S_4$
(b) Vol. **105**, No. 1, p. K39–K44 (1981).
See also *13.1*, p. *156*; 20.1, p. 310.

S. B. Patil, D. K. Chakrabarty, M. V. Babu, and S. N. Shringi
Mössbauer Spectroscopic Studies of the $Ba_xLa_{1-x}CoO_3$ System
(a) Vol. **65**, No. 1, p. 65–71 (1981).
See also *5*, p. *64*; *18.2*, p. *270*.

G. Sorge, U. Straube, and N. R. Ivanov
Elastic, Dilatational, and Electrostrictive Properties of TSCC Crystals
(a) Vol. **65**, No. 1, p. 189–196 (1981).
See also 1.2, p. 16; *12.2*, p. *140*; *14.4.1*, p. *243*.

O. Valassiades, E. K. Polychroniadis, J. Stoemenos, and N. A. Economou
On the Disordered Nature of Single Crystalline $TlBiTe_2$ and Its Influence on Its Electrical Properties
(a) Vol. **65**, No. 1, p. 215–223 (1981).
See also 1.2, p. 17; *10.1*, p. *100*; *14.3*, p. *213*.

D. Bonnin and S. Muller
Study of the Electric Field Gradient in Muscovite by Mössbauer Spectroscopy of Iron
(b) Vol. **105**, No. 2, p. 649–657 (1981).
See also *5*, p. *64*.

V. K. Jain
Host Spin-Lattice Relaxation Narrowing in $Sm_2Co_3(NO_3)_{12} \cdot 24\,H_2O:Mn^{2+}$, Gd^{3+} Single Crystals
(b) Vol. **105**, No. 2, p. 723–726 (1981).
See also *19*, p. *290*.

I. V. Bodnar, L. V. Golubev, V. G. Plotnichenko, and E. A. Smolyaninova
Raman Scattering in $CuGaSe_2$
(b) Vol. **105**, No. 2, p. K111–K114 (1981).
See also *6*, p. *75*; *20.1*, p. *311*.

I. V. Bodnar, B. V. Korzun, and A. I. Lukomskii
Composition Dependence of the Band Gap of $CuInS_{2x}Se_{2(1-x)}$
(b) Vol. **105**, No. 2, p. K143–K145 (1981).
See also *13.1*, p. *157*; *20.1*, p. *311*.

K. Bohmhammel, P. Deus, and H. A. Schneider
Specific Heat, Debye Temperature, and Related Properties of Compound Semiconductors $A^{II}B^{IV}C_2^V$
(a) Vol. **65**, No. 2, p. 563–569 (1981).
See also *6*, p. *75*; *8*, p. *85*.

Sh. A. Vakhidov, E. M. Ibragimova, I. Nuritdinov, A F. Rakov, and G. I. Ikramov
Self-Trapped Particles in Complex Oxide Crystals
(b) Vol. **106**, No. 1, p. 31–35 (1981).
See also *11*, p. *126*; 13.5.1, p. 190; 19, p. 290; 20.1, p. 312; 20.3, p. 329; 22.6, p. 477.

G. Ambrazevičius, G. Babonas, Yu. R. Rud, and A. Šileika
The Fundamental Absorption Edge of $MgSiP_2$
(b) Vol. **106**, No. 1, p. 85–89 (1981).
See also *13.1*, p. *157*; *20.1*, p. *312*.

V. I. Chizhikov, V. L. Panyutin, B. E. Ponedelnikov, and A. E. Rozenson
Electronic Energy Band Structure of $CdAl_2Te_4$ Modifications with Space Group D_{2d}^{11}, D_{2d}^{9}, D_{2d}^{1}, and S_4^2
(b) Vol. **106**, No. 1, p. 91–98 (1981).
See also *13.1*, p. *157*.

W. Bażela and A. Szytuła
Crystal and Magnetic Structure of the $NiMn_{1-t}Ti_tGe$ System
(a) Vol. **66**, No. 1, p. 45–52 (1981).
See also 1.1, p. 12; *18.2*, p. *271*; *18.4*, p. *283*.

H. Tietze, M. Müllner, and H. Jex
The Temperature Dependence of Domain Volumes below the Cubic-to-Tetragonal Structural Phase Transition in $KMnF_3$
(a) Vol. **66**, No. 1, p. 239–242 (1981).
See also *1.2*, p. *18*; *1.4*, p. *30*.

I. A. Belobrova, I. P. Aleksandrova, and A. K. Moskalev
^{81}Br NQR in the Incommensurate Phase of Rb_2ZnBr_4
(a) Vol. **66**, No. 1, p. K17–K20 (1981).
See also *1.2*, p. *18*; 14.4.1, p. 244; *19*, p. *291*.

E. P. Domashevskaya, L. N. Marshakova, V. A. Terekhov, A. N. Lukin, Ya. A. Ugai, V. I. Nefedov, and Ya. V. Salyn
Role of Noble Metal d-States in the Formation of the Electron Structure of Ternary Sulphides
(b) Vol. **106**, No. 2, p. 429–435 (1981).
See also *13.1*, p. *158*; 17, p. 257; 20, p. 298.

C. W. M. Timmermans and G. Blasse
On the Luminescence of $Cs_3Bi_2Br_9$ Single Crystals
(b) Vol. **106**, No. 2, p. 647–655 (1981).
See also 13.5.1, p. 191; *20.3*, p. *331*.

C. J. Delbecq, S. A. Marshall, and S. Susman
Dielectric Properties of the Fast-Ion Conductor $Na_3Sc_2P_3O_{12}$
(b) Vol. **106**, No. 2, p. 731–734 (1981).
See also *14.4*, p. *240*.

B. Kubier, H. Boudriot, K. Deus, R. Gründler, H. Zscheile, and E. Buhrig
Plasmon Data, Dielectric Function, and Optical Constants of $CdGeP_2$ from Electron Energy Loss Measurements between 4.5 and 35 eV
(b) Vol. **106**, No. 2, p. K103–K106 (1981).
See also *13.5.2*, p. *195*; *20.1*, p. *315*.

H.-G. Brühl, H. Neumann, T. Pfeiffer, and G. Kühn
Anisotropic Thermal Expansion of Cu–III–VI_2 Compounds
(a) Vol. **66**, No. 2, p. 597–600 (1981).
See also *8*, p. *86*.

J. Klikorka, M. Frumar, V. Černý, K. Vacek, and L. Tický
ESR of Defect Centres in Ge–Sb–S Glasses
(a) Vol. **66**, No. 2, p. 691–696 (1981).
See also *2*, p. *48*; *19*, p. *292*.

C. Bellabarba and S. M. Wasim
Thermal Conductivity of $AgInTe_2$
(a) Vol. **66**, No. 2, p. K105–K108 (1981).
See also *8*, p. *87*.

W. Bogusz
Influence of Thermal Cycling on Properties of NASICON
(a) Vol. **66**, No. 2, p. K109–K112 (1981).
See also *14.3*, p. *216*.

W. Bogusz, F. Krok, and W. Jakubowski
Electrical Conductivity of $Na_{1+x}Zr_2Si_xP_{3-x}O_{12}$
(a) Vol. **66**, No. 2, p. K113–K116 (1981).
See also *14.3*, p. *216*.

A. Anagnostopoulos and I. Spyridelis
Electrical Conductivity of $ZnIn_2S_4$ Modulated by a Perpendicular Applied Electric Field
(a) Vol. **66**, No. 2, p. K127–K130 (1981).
See also *14.3*, p. *216*.

S. Isomura
Impurity Conduction in p-Type $ZnSnAs_2$
(a) Vol. **66**, No. 2, p. K157–K160 (1981).
See also *14.3*, p. *216*.

A. Heinrich, W. Cordts, and J. Monecke
Electronic Band Structure of Pseudodirect Chalcopyrite Semiconductors (II): $ZnGeP_2$ and $ZnSiAs_2$
(b) Vol. **107**, No. 1, p. 319–333 (1981).
See also *13.1*, p. *159*.

V. I. Chizihikov, V. L. Panyutin, B. E. Ponedelnikov, and A. E. Rozenson
Effective Masses of Electrons and $E_g(x)$ in $CdIn_2$ $(Se_{1-x}Te_x)_4$ Solid Solutions
(b) Vol. **107**, No. 1, p. 347–350 (1981).
See also *13.1*, p. *159*.

T. D. Aitikeeva, A. I. Lebedev, A. E. Yunovich, Karin Herrmann, A. W. Jalyschko, and P. Schäfer
Spectra of Photo- and Electroluminescence of Bismuth-Doped $Pb_{1-x}Sn_x$ Te
(a) Vol. **67**, No. 1, p. 171–175 (1981).
See also *20.3*, p. *332*.

A. H. Abou El Ela, H. H. A. Labib, and N. Abdelmohsen
Electrical Conductivity, Thermoelectric Power, and Thermal Conductivity of $CuTlSe_2$ in the Solid and Liquid States
(a) Vol. **67**, No. 1, p. K43 – K48 (1981).
See also **8**, p. 87; ***14.3***, p. ***217***; **15**, p. 249.

M. K. Kostov, N. P. Ivanchev, and E. M. Kostova
Experimental Study of Electrogyration in the Centrosymmetric Crystals $Pb_5GeO_4(VO_4)_2$ and $Pb_5SiO_4(VO_4)_2$
(a) Vol. **67**, No. 1, p. K71 – K74 (1981).
See also ***14***, p. ***198***; ***20***, p. ***298***.

K. Sierański and J. Szatkowski
Disorder Scattering in Mixed $Pb_{1-x}Sn_xTe$-Type Semiconductors
(b) Vol. **107**, No. 2, p. K87 – K90 (1981).
See also **13.4**, p. 181; ***14.3***, p. ***217***.

V. Dvorak and R. Kind
On the Low-Temperature Phase in Rb_2ZnCl_4
(b) Vol. **107**, No. 2, p. K109 – K113 (1981).
See also ***1***, p. ***10***; **1.2**, p. 19.

J. Tröltzsch and Ch. Dippmann
Thermal Measurements on Chalcogenide Glass Threshold Switching Devices
(a) Vol. **67**, No. 2, p. 527 – 531 (1981).
See also **2**, p. 50; ***8***, p. ***88***; ***14.3.3***, p. ***230***.

Y. Uesu, J. Ogawa, N. Hanami, B. A. Strukov, and J. Kobayashi
X-Ray Dilatometric Study of Phase Transitions of Ferroelectric $(NH_4)_2BeF_4$
(a) Vol. **67**, No. 2, p. 693 – 699 (1981).
See also ***1.2***, p. ***19***; ***14.4.1***, p. ***245***.

T. G. Kerimova, F. R. Adzalova, R. Kh. Nani, and V. Ya. Shteinshreiber
Interaction of Free Current Carriers with Longitudinal Optical Phonons in $CdIn_2Se_4$
(a) Vol. **67**, No. 2, p. K143 – K146 (1981).
See also ***6***, p. ***80***; **14.3**, p. 217; ***20.1***, p. ***318***.

M. Corti, A. Rigamonti, and A. Magistris
^{133}Cs Quadrupole Perturbed NMR Study of Jahn-Teller Phase Transitions in $CsCuCl_3$
(b) Vol. **108**, No. 1, p. 29 – 36 (1981).
See also ***1.2***, p. ***20***; ***19***, p. ***293***.

E. Anachkova, H. Ratajczak, and I. Savatinova
Water Vibrations in Ferroelectric $K_4Fe(CN_6) \cdot 3\ H_2O$
(b) Vol. **108**, No. 1, p. 65 – 70 (1981).
See also ***6***, p. ***80***; **14.4.1**, p. 246; ***20.1***, p. ***318***.

C. Rincón, J. González, and G. Sanchez Perez
Optical Absorption of $CuInSe_2$ in Bulk Single Crystal
(b) Vol. **108**, No. 1, p. K19 – K22 (1981).
See also ***20.1***, p. ***319***.

M. WINTENBERGER and J. C. JUMAS
Magnetic Structure of $Fe_2Te_3O_9$ at 4.2 K
(a) Vol. **68**, No. 1, p. 11–18 (1981).
See also *18.4*, p. *284*.

J. M. PÉREZ-MATO, J. L. MAÑES, H. FERNÁNDEZ, J. ZÚÑIGA, M. J. TELLO, C. SOCÍAS, and M. A. ARRIANDIAGA
Structural Phase Transitions in $(CH_3NH_3)_2ZnCl_4$ Crystals
(a) Vol. **68**, No. 1, p. 29–38 (1981).
See also *1.2*, p. *20*.

E. I. TERUKOV, S. ROTH, G. KRABBES, and H. OPPERMANN
The Magnetic Properties of Cobalt-Doped Iron Sulphide $Fe_{1-y}Co_yS$ ($y \leqq 0.13$)
(a) Vol. **68**, No. 1, p. 233–238 (1981).
See also 1.2, p. 20; *18*, p. *264*.

R. G. GOODCHILD, O. H. HUGHES, and J. C. WOOLLEY
Crystal Structure of I III IV Se_4 Compounds
(a) Vol. **68**, No. 1, p. 239–244 (1981).
See also *1.1*, p. *13*.

G. SORGE, A. ALMEIDA, L. A. SHUVALOV, and R. M. FEDOSYUK
Influence of the Mechanical Stress Component T_5 on the Elastic Behaviour of $KH_3(SeO_3)_2$/$KD_3(SeO_3)_2$ Crystals in the Vicinity of Their Ferroelastic Phase Transition
(a) Vol. **68**, No. 1, p. 245–251 (1981).
See also *1.2*, p. *20*; 12.2, p. 141; *14.4.1*, p. *246*.

D. LAPRAZ and A. BAUMER
Chloroapatite, $Ca_5(PO_4)_3Cl$: Thermoluminescent Properties
(a) Vol. **68**, No. 1, p. 309–319 (1981).
See also *10.2*, p. *118*; *20.3*, p. *334*.

R. KESAVAMOORTHY and R. KRISHNASWAMY
Effect of Solid Dilution and Temperature on the Crystal Field Spectra of Chromium Ions in Single Crystals of Chrome Alum
(b) Vol. **108**, No. 2, p. 307–313 (1981).
See also *13.4*, p. *183*; *20.1*, p. *320*.

E. LAZO, J. RÖESSLER, and L. A. MORAGA
Phase Transition and Spatial Correlation of Localized States in the Falicov-Ramírez-Kimball Model
(b) Vol. **108**, No. 2, p. 341–350 (1981).
See also *1.2*, p. *21*; *13.1*, p. *161*.

S. NARA, A. FREUDENHAMMER, and G. HEBER
Dynamics of Two-Spin Cluster in Spin Glass (I): Simple Model Calculation of Remanent Magnetization
(b) Vol. **108**, No. 2, p. 389–398 (1981).
See also *13*, p. *150*; *18*, p. *264*.

22.8.1 Salts of Oxy-Acids (Carbonates, Phosphates, Silicates Including Mica etc.)

Original Papers and Short Notes

YU. N. POLIVANOV and R. SH. SAYAKHOV
Hyper-Raman Scattering from Infrared Active Modes in Calcite Crystals
(b) Vol. **103**, No. 1, p. 89–92 (1981).
See also *6*, p. *66*; *20.1*, p. *301*.

M. D. FONTANA, G. E. KUGEL, G. METRAT, and C. CARABATOS
Long-Wavelength Phonons in the Different Phases of $KNbO_3$
(b) Vol. **103**, No. 1, p. 211–219 (1981).
See also *6*, p. *66*; **14.4.1**, **p. 242**; *20.1*, p. *301*.

YU. P. TSYASHCHENKO and G. E. KRASNYANSKII
Studies on Anharmonic Damping Mechanism of High-Frequency Phonons in Crystals Containing Complex Ions
(b) Vol. **103**, No. 1, p. 391–395 (1981).
See also *6*, p. *67*; *20.1*, p. *302*.

M. J. J. LAMMERS and G. BLASSE
Luminescence of Sodium Molybdate ($Na_2Mo_2O_7$) and Sodium Tungstate ($Na_2W_2O_7$)
(a) Vol. **63**, No. 1, p. 157–161 (1981).
See also *13.4*, p. *169*; *20.3*, p. *324*.

U. A. JAYASOORIYA, G. J. KEARLEY, S. F. A. KETTLE, and H. J. LAUTER
A Comparison of Inelastic Neutron Scattering and Vibrational Spectral Data for Phase V of Ammonium Nitrate
(a) Vol. **63**, No. 1, p. 169–177 (1981).
See also *6*, p. *67*; *20.1*, p. *303*.

F. EL-KABBANY, Y. BADR, and G. SAID
Pyroelectric and Dielectric Behaviour of Ammonium Nitrate at the Phase Transition II-I
(a) Vol. **63**, No. 1, p. 355–362 (1981).
See also *1.2*, p. *14*; *14.4*, p. *237*.

V. N. BLIZNYUK, M. D. GLINCHUK, A. P. PECHENYI, and L. A. LITVINOV
Determination of the Interblock Misalignment of $SrTiO_3$ Single Crystals by the ESR Technique
(a) Vol. **63**, No. 1, p. K43–K46 (1981).
See also *1.4*, p. *28*; *19*, p. *286*.

J. M. ROBERTSON, and M. V. VAN TOL
Luminescent Single Crystal Thin Films of Manganese Doped Barium Hexa-Aluminate
(a) Vol. **63**, No. 1, p. K59–K62 (1981).
See also 3, p. 53; *20.3*, p. *324*.

D. G. Galimov, K. D. Tarzimanov, and N. R. Yafaev
On the Nature of Colour Centres in Ultramarine and Blue Alkali Borate Glasses Containing Sulphur
(a) Vol. **64**, No. 1, p. 65–71 (1981).
See also *10.2*, p. *108*; 13.4, p. 172; 19, p. 288; 20.1, p. 306.

Hj. Bernhardt and R. Schnell
Modulation-Spectroscopic Investigations of Defect Centres in $PbMoO_4$ Crystals
(a) Vol. **64**, No.1 , p. 207–214 (1981).
See also 10.2, p. 108; *16*, p. *252*; *20.3*, p. *325*.

O. A. Grigorev and T. D. Shermergor
The Scattering of Ultrasonic Waves by Polycrystals
(a) Vol. **64**, No. 1, p. 385–394 (1981).
See also *7*, p. *82*.

W. Elflein, W. Käppler, and G. Arlt
Drift Mobility in Insulating Barium Titanate Ceramics
(a) Vol. **64**, No. 1, p. K1–K4 (1981).
See also 13.5.3, p. 196; *14.4.1*, p. *242*.

R. Le Bihan, M. Maussion, F. Lefaucheux, E. Manghi, and M. C. Robert
Observation of Ferroelectric Domains in Gel Grown $PbHPO_4$ Crystals
(a) Vol. **64**, No. 1, p. K5–K6 (1981).
See also *14.4.1*, p. *243*.

J. Kroupa and Z. Czapla
Far Infrared Spectra of NH_4HSeO_4
(a) Vol. **64**, No. 1, p. K17–K18 (1981).
See also *14.4.1*, p. *243*; *20.1*, p. *306*.

J. Kapturczak and Z. Pająk
Paramagnetic Broadening of PMR Line in Ammonium Manganese Sulphate
(b) Vol. **104**, No. 2, p. 479–483 (1981).
See also *18.1*, p. *265*; *19*, p. *288*.

Sh. M. Efendiev, I. I. Khammadov, I. A. Velichko, E. A. Vinogradov, V. A. Yakovlev, A. Kh. Zeinally, and G. N. Zhizhin
High-Frequency Vibrational Spectrum of $Gd_2(MoO_4)_3$
(b) Vol. **104**, No. 2, p. K129–K133 (1981).
See also *6*, p. *72*; *20.1*, p. *308*.

Sh. M. Efendiev, I. I. Khammadov, I. A. Velichko, E. A. Vinogradov, A. Kh. Zeinally, G. N. Zhizhin, and V. A. Yakovlev
High-Frequency Surface Polariton Dispersion in $Gd_2(MoO_4)_3$
(b) Vol. **104**, No. 2, p. K135–K138 (1981).
See also *6*, p. *72*; *20.1*, p. *308*.

R. Clemens, G. Luther, and H. E. Müser
Dielectric Behaviour of Ferroelectric Barium Titanate in the Microwave Region
(a) Vol. **64**, No. 2, p. 637–645 (1981).
See also *14.4.1*, p. *243*.

I. H. Ismailzade, R. M. Ismailov, and T. M. Stolpakova
The Magnetoelectric Effect $(ME)_H$ in the Ferroelastics $NaFe(MoO_4)_2$ and $K_{0.8}Na_{0.2}Fe(MoO_4)_2$
(a) Vol. **65**, No. 1, p. K25–K28 (1981).
See also *14.4.1*, p. *243*; *18.4*, p. *282*.

M. Koralewski
Dispersion of the Faraday Rotation in KDP-Type Crystals by Pulse High Magnetic Field
(a) Vol. **65**, No. 1, p. K49–K53 (1981).
See also 14.4.1, p. 244; *18*, p. *262*; *20.1*, p. *310*.

J. Pietrzak, A. Maryanowska, and J. Leciejewicz
Magnetic Ordering in $Pb(Fe_{0.5}Nb_{0.5})O_3$ at 4.2 K
(a) Vol. **65**, No. 1, p. K79–K82 (1981).
See also *14.4.1*, p. *244*; *18.4*, p. *283*.

W. Wegener, S. Hautecler, and G. Will
Range of Magnetic Interactions in Some Anhydrous Sulphates of the 3d Transition Metals
(b) Vol. **105**, No. 2, p. 491–497 (1981).
See also *18.1*, p. *266*; *18.4*, p. *283*.

V. S. Gorelik, B. S. Umarov, and L. G. Reznik
The Peculiarity of the A_1-Type Polariton in Lithium-Niobate Crystal
(b) Vol. **105**, No. 2, p. K73–K75 (1981).
See also *6*, p. *75*; *20.1*, p. *311*.

V. P. Dudkevich, V. A. Bukreev, Vl. M. Mukhortov, Yu. I. Golovko, Yu. G. Sindeev, V. M. Mukhortov, and E. G. Fesenko
Internal Size Effect in Condensed $BaTiO_3$ Ferroelectric Films
(a) Vol. **65**, No. 2, p. 463–467 (1981).
See also 6, p. 75; *14.4.1*, p. *244*.

I. R. Dorosh, Yu. S. Kuzminov, N. M. Polozkov, A. M. Prokhorov, V. V. Osiko, N. V. Tkachenko, V. V. Voronov, and D. Kh. Nurligareev
Barium–Strontium Niobate Crystals for Optical Information Recording
(a) Vol. **65**, No. 2, p. 513–522 (1981).
See also *14.4.1*, p. *244*; 16, p. 253; *20.1*, p. *312*.

J. Lecomte, J. P. Loup, M. Hervieu, and B. Raveau
Non-Stoichiometry and Electrical Conductivity of Strontium Niobates with Perovskite Structure (I): Defect Structure of $Sr(Sr_{1/3}Nb_{2/3})O_3$
(a) Vol. **65**, No. 2, p. 743–752 (1981).
See also 1.3, p. 24; *10.2*, p. *112*; *14.4*, p. *239*.

U. Knauer
On the Composition and Crystallinity of the Titanium-Rich Second Phase in $BaTiO_3$ Ceramic
(a) Vol. **65**, No. 2, p. K109–K112 (1981).
See also *1.4*, p. *30*; 14.4.1, p. 244.

J. Lecomte, J. P. Loup, M. Hervieu, and B. Raveau
Non-Stoichiometry and Electrical Conductivity of Strontium Niobates with Perovskite Structure (II): Ionic Conductivity of Sr $(Sr_{(1/3)+x}Nb_{(2/3)-x})$ $O_{3-(3/2)x}$
(a) Vol. **66**, No. 2, p. 551–558 (1981).
See also *10.2*, p. *115*; *14.4*, p. *240*.

Fat Duen Ho
Electron Energy Levels in Lithium Niobate Resulting from Oxygen Vacancies
(a) Vol. **66**, No. 2, p. 793–806 (1981).
See also *13.4*, p. *180*.

K. Samudravijaya and B. K. Basu
Dielectric Measurements in Potassium Dihydrogen Arsenate
(a) Vol. **66**, No. 2, p. K121–K124 (1981).
See also *14.4.1*, p. *245*.

G. Srinivasan and Om Prakash
Initial Permeability in Fe–Zn–Cu Ferrites
(a) Vol. **66**, No. 2, p. K149–K151 (1981).
See also *18.3*, p. *280*.

E. Fischer, W. Hässler, E. Hegenbarth and V. I. Fritsberg
Amorphous Behaviour of PLZT Demonstrated by Thermal Conductivity Measurements
(a) Vol. **66**, No. 2, p. K169–K171 (1981).
See also *8*, p. *87*; *14.4.1*, p. *245*.

D. Wagner and D. Bäuerle
Soft Modes in $SrTiO_3$: Zr, Nb, Ta
(b) Vol. **107**, No. 1, p. K39–K41 (1981).
See also *6*, p. *78*; *14.4.1*, p. *245*.

J. P. Eymery, S. B. Raju and J. F. Dinhut
Mössbauer Study of the Substitution of Iron by Chromium in Mn–Zn Ferrites
(a) Vol. **67**, No. 1, p. 89–95 (1981).
See also *5*, p. *65*; *18.3*, p. *280*.

F. Wagner, H. R. Wenk, C. Esling, and H. J. Bunge
Importance of Odd Coefficients in Texture Calculations for Trigonal–Triclinic Symmetries
(a) Vol. **67**, No. 1, p. 269–285 (1981).
See also *1.3*, p. *25*.

F. El-Kabbany, G. Said, Y. Badr, and S. Taha
Infrared Investigation of the Phase Transition in K_2SO_4
(a) Vol. **67**, No. 1, p. 339–345 (1981).
See also *1.2*, p. *18*; 6, p. 78; 20.1, p. 317.

V. Ramachandran, M. M. Ibrahim, V. C. Padaki, and E. S. R. Gopal
Elastic Constants of Sodium Nitrate
(a) Vol. **67**, No. 1, p. K49–K52 (1981).
See also *12.2*, p. *141*.

A. DA COSTA LAMAS, S.-L. CHANG, and S. CATICHA-ELLIS
On the Use of Powder Diffractometry in the Study of Phase Transitions. Case of $NaNO_2$
(a) Vol. **68**, No. 1, p. 173–178 (1981).
See also *1.2*, p. *20*; *14.4.1*, p. *246*.

G. BACQUET, VO QUAN TRUONG, M. VIGNOLES, and G. BONEL
ESR of the Fe^+ Centre in B-Type Carbonated Hydroxyapatite
(a) Vol. **68**, No. 1, p. K71–K74 (1981).
See also *13.4*, p. *183*; *19*, p. *294*.

I. H. ISMAILZADE, R. M. ISMAILOV, A. I. ALEKBEROV, and F. M. SALAEV
Investigation of the Magnetoelectric $(ME)_H$ Effect in the Solid Solutions of the Systems $BiFeO_3$–$BaTiO_3$ and $BiFeO_3$–$PbTiO_3$
(a) Vol. **68**, No. 1, p. K81–K85 (1981).
See also *14.4.1*, p. *246*; 18.4, p. 284.

F. KELLENDONK and G. BLASSE
On the Luminescence of Sm^{3+}, Gd^{3+}, and Dy^{3+} in Yttrium Aluminium Borate
(b) Vol. **108**, No. 2, p. 541–548 (1981).
See also *13.4*, p. *183*; *20.3*, p. *344*.

F. BREHAT, M. EL SHERIF, and B. WYNCKE
Overdamped Modes in Ammonium Dihydrogen Phosphate at 300 K
(b) Vol. **108**, No. 2, p. K105–K108 (1981).
See also *6*, p. *81*; *20.1*, p. *320*.

A. S. PRADHAN and R. C. BHATT
Photo-Stimulated Luminescence and Thermoluminescence in $CaSO_4$:Dy
(a) Vol. **68**, No. 2, p. 405–411 (1981).
See also *20.3*, p. *334*.

M. NARAYANA and L. KEVAN
The Use of Electron Spin Echo Spectroscopy to Study the Ferroelectric Phase Transition in Ammonium Sulfate
(a) Vol. **68**, No. 2, p. 703–709 (1981).
See also *14.4.1*, p. *246*.

M. CHYBICKI, J. NOWATKOWSKI, and W. SADOWSKI
Non-Ohmic Conductivity in Iron-Phosphate Glasses
(a) Vol. **68**, No. 2, p. K129–K134 (1981).
See also *2*, p. *52*; *14.3.1*, p. *222*.

A. MARYANOWSKA and J. PETRZAK
Critical-Points Anomalies in EPR Linewidth of $Pb(Fe_{0.5}Nb_{0.5})O_3$
(a) Vol. **68**, No. 2, p. K185–K188 (1981).
See also *14.4.1*, p. *246*; 18.4, p. 285; *19*, p. *295*.

22.8.2 Spinels, Garnets, Ferrites

Original Papers and Short Notes

V. N. Belogurov and V. A. Bilinkin
Dependence of the Dynamic and Hyperfine Characteristics of Fe^{3+} Ions on the Rare-Earth Ion Parameters in Garnets above the Néel Temperature
(a) Vol. **63**, No. 1, p. 45–53 (1981).
See also *5*, p. *62*.

G. Wende
Intensities of Magnetostatic Modes in Circular Disks and Films
(a) Vol. **63**, No. 1, p. 151–155 (1981).
See also *18.3*, p. *278*; *19*, p. *286*.

Om. Prakash and G. Srinivasan
Initial Permeability Components and Anisotropy in YIG
(a) Vol. **63**, No. 1, p. 179–182 (1981).
See also *18.3*, p. *278*.

M. Torfeh and H. Le Gall
Theoretical Analysis of Hybrid Modes of Magnetooptical Wave-Guides
(a) Vol. **63**, No. 1, p. 247–258 (1981).
See also *18*, p. *260*; *20*, p. *296*.

M. Wurlitzer and J. Franke
Some Experiments on Recombination in the Photomagnetic Effect in Si-Doped YIG
(a) Vol. **63**, No. 1, p. K11–K14 (1981).
See also *18.3*, p. *278*; *20*, p. *296*.

I. Bunget and V. Raeţchi
Magnetoelectric Effect in the Heterogeneous System NiZn Ferrite–PZT Ceramic
(a) Vol. **63**, No. 1, p. K55–K57 (1981).
See also *18.3*, p. *278*.

V. I. Nikitenko, V. S. Gornakov, L. M. Dedukh, and Yu. P. Kabanov
Free and Forced Oscillations of Bloch Lines in YIG
(a) Vol. **63**, No. 1, p. K63–K65 (1981).
See also *18.3*, p. *278*.

G. Srinivasan and C. M. Srivastava
Electrical Conductivity Mechanism in Zinc and Copper Substituted Magnetite
(b) Vol. **103**, No. 2, p. 665–671 (1981).
See also *14.3*, p. *210*; 15, p. 247.

K. Seshan, D. K. Chakrabarty, and A. B. Biswas†
A New Method for the Preparation of Acicular γ-Fe_2O_3
(a) Vol. **63**, No. 2, p. K123–K125 (1981).
See also *3*, p. *53*; *18.3*, p. *278*.

A. Ślawska-Waniewska
Crystal Field of Cr^{3+} Ion in Spinel Lattice: The Effect of High Hydrostatic Pressure
(b) Vol. **104**, No. 2, p. 679–684 (1981).
See also *1*, p. *9*; *13*, p. *144*; 18.3, p. 279.

M. Wurlitzer and J. Franke
A Model of the Photomagnetic Effect in YIG(Si)
(a) Vol. **64**, No. 2, p. 539–547 (1981).
See also *18.3*, p. *279*; *20*, p. *297*.

A. Selmi, P. Gibart, and G. Weill
Pressure Dependence of the Galvanomagnetic Properties of n- and p-Type Ferromagnetic Semiconductors $HgCr_2Se_4$
(a) Vol. **64**, No. 2, p. 665–668 (1981).
See also 12.2, p. 139; *14.3*, p. *212*; *18.2*, p. *270*.

P. Muthukumarasamy, T. Nagarajan, and A. Narayanasamy
Mössbauer Study of the Ni–Cd Ferrite System
(a) Vol. **64**, No. 2, p. 747–754 (1981).
See also *5*, p. *63*; 18.3, p. 279.

C. M. Srivastava, O. Prakash, and R. Aiyar
Magnetic Relaxation in Polycrystalline Garnets
(a) Vol. **64**, No. 2, p. 787–796 (1981).
See also *18.3*, p. *279*; 19, p. 289.

A. G. Shishkov, E. N. Ilyicheva, N. G. Kanavina, A. N. Ivanov, and A. M. Pirogova
Magnetic and Electrical Properties of the Thin Film Magnetic Semiconductor $CdCr_2Se_4$
(a) Vol. **64**, No. 2, p. K93–K97 (1981).
See also *14.3.1*, p. *220*; *18.2.1*, p. *276*.

D. Kostopoulos and J. Grammatikakis
Disturbance in the State of Magnetite Due to Magnetic Prehistory at 78 K
(a) Vol. **64**, No. 2, p. K155–K159 (1981).
See also *14.3*, p. *212*; *18*, p. *261*.

S. Ligenza, Z. Kucharski, and J. Suwalski
Mössbauer Study of the $3d^5$ Shell Deformation of Fe^{3+} Ions in $MnFe_2O_4$
(b) Vol. **105**, No. 1, p. 353–358 (1981).
See also *5*, p. *64*; 18.3, p. 279.

V. K. Singh, R. Chandra, and S. Lokanathan
Mössbauer Study of Jahn-Teller Distortion in Systems $Co_xMn_{2.9-x}Fe_{0.1}O_4$ and $Ni_xMn_{2.9-x}Fe_{0.1}O_4$
(b) Vol. **105**, No. 1, p. K13–K17 (1981).
See also *5*, p. *64*.

Sh. Sh. Bashkirov and V. V. Parfenov
Conduction Band Width of Mn–Zn Ferrites
(b) Vol. **105**, No. 1, p. K27–K29 (1981).
See also *13.1*, p. *155*; 16, p. 252; 20.1, p. 310.

K. Čermák
Colour Centres in YAG
(a) Vol. **66**, No. 1, p. K37–K39 (1981).
See also *10.2*, p. *113*; *20.1*, p. *314*.

I. Mayer and S. Angelov
The Mechanism of Participation of Non-Magnetic B-Site Cations in the Superexchange in Spinels
(b) Vol. **106**, No. 2, p. 467–471 (1981).
See also *18.4*, p. *284*.

V. P. Miroshkin, Ya. I. Panova, and T. V. Stakhieva
Electrical Conductivity of Manganese–Zinc Ferrospinels
(a) Vol. **66**, No. 2, p. 503–507 (1981).
See also *14.3*, p. *216*.

R. Grössinger
A Critical Examination of the Law of Approach to Saturation (I): Fit Procedure
(a) Vol. **66**, No. 2, p. 665–674 (1981).
See also *18.2*, p. *273*; 18.3, p. 279; 21.1, p. 362.

H. V. Kiran, A. L. Shashimohan, D. K. Chakrabarty, and A. B. Biswas
Structural and Magnetic Properties of Cooper–Nickel Ferrites
(a) Vol. **66**, No. 2, p. 743–747 (1981).
See also *1.3*, p. *25*; *18.3*, p. *279*.

V. P. Miroshkin, Ya. I. Panova, and V. V. Passynkov
Dielectric Relaxation in Polycrystalline Ferrites
(a) Vol. **66**, No. 2, p. 779–782 (1981).
See also *14.4.1*, p. *245*.

J. Kubiak and J. Pietrzak
Magnetocrystalline Anisotropy of Cubic Copper Ferrite
(a) Vol. **67**, No. 1, p. 103–108 (1981).
See also *18.3*, p. *280*.

R. Hergt and H. Pfeiffer
Investigation of Growth Steps Due to Misorientation of Garnet LPE Films
(a) Vol. **67**, No. 1, p. 205–215 (1981).
See also *1.5*, p. *36*; *3*, p. *55*.

I. Tomáš, B. N. D. Shija, and V. K. Srivastava
To the Theory of 90° Stripe Domains in Films with Tetragonal Magnetic Anisotropy
(a) Vol. **67**, No. 1, p. 325–332 (1981).
See also *18.2.1*, p. *277*.

V. G. Kleparskii and I. Pintér
The Domain Wall Widening in High Drive Fields
(a) Vol. **67**, No. 1, p. K29–K32 (1981).
See also *18.3*, p. *280*.

N. N. Sirota, L. A. Bashkirov, E. Z. Katsnelson, and E. F. Shapovalova
The Effect of Small Quantities of Group IV and V Transition Elements on Dielectric Properties of Nickel-Zinc Ferrites
(a) Vol. **67**, No. 1, p. K67–K69 (1981).
See also *14.4*, p. *240*.

Yu. G. Chukalkin, V. V. Petrov, and B. N. Goshchitskii
Radiation Effects in Hexagonal Ferrite $BaFe_{12}O_{19}$
(a) Vol. **67**, No. 2, p. 421–426 (1981).
See also *11*, p. *129*; *18.3*, p. *280*.

V. K. Vlasko-Vlasov, L. M. Dedukh, V. I. Nikitenko, and L. S. Uspenskaya
Dynamic Resonant Domain Structure in YIG Single Crystals
(a) Vol. **67**, No. 2, p. K107–K110 (1981).
See also *18.3*, p. *280*.

K. Seshan, A. L. Shashimohan, D. K. Chakrabarty, and A. B. Biswas†
Effect of Cation Distribution on the Properties of Some Magnesium-Nickel Ferrites
(a) Vol. **68**, No. 1, p. 97–101 (1981).
See also *18.3*, p. *280*.

L. M. Dedukh, V. S. Gornakov, Yu. P. Kabanov, and V. I. Nikitenko
Motion of a Bloch Line in YIG under the Action of a Magnetic Field Normal to It
(a) Vol. **68**, No. 1, p. K1–K4 (1981).
See also *18.3*, p. *280*.

O. A. C. Nunes, M. A. Amato, and K. Skeff Neto
Microwave Resonance Absorption in Ferrites by the Photoacoustic Effect
(a) Vol. **68**, No. 1, p. K25–K27 (1981).
See also *19*, p. *294*.

V. V. Tsurkan, S. I. Radautsan, and V. E. Tezlevan
Influence of Anion Substitution on the Conductivity in Magnetic $Cu_yCr_2Se_{4-x}Br_x$ Semiconductors
(a) Vol. **68**, No. 1, p. K93–K96 (1981).
See also *14.3*, p. *218*; *18.2*, p. *274*.

22.9 Organic Semiconductors

Review Articles

C. A. HOGARTH and T. IQBAL
The Electrical and Structural Properties of Thin Films of Polymers Deposited from the Vapour Phase, with Special Reference to Polypropylene
(a) Vol. **65**, No. 1, p. 11–37 (1981).
See also *1.5*, p. *34*; 13.4, p. 168; *14.3.1*, p. *220*; 14.3.4, p. 232.

H. BÄSSLER
Localized States and Electronic Transport in Single Component Organic Solids with Diagonal Disorder
(b) Vol. **107**, No. 1, p. 9–54 (1981).
See also 2, p. 40; *13.4*, p. *168*; 13.5.1, p. 188; *14.3*, p. *209*; 14.3.4, p. 232; 20.1, p. 301; 20.3, p. 323.

Original Papers and Short Notes

M. A. BLOKHIN, E. G. ORLOVA, and I. G. SCHWEITZER
The Influence of the Environment of the Absorbing Atom on the Photoelectric Yield Spectra of Beryllium Compounds in the Soft X-Ray Region
(b) Vol. **103**, No. 1, p. 63–69 (1981).
See also *4*, p. *57*; *17*, p. *256*; 22.5, p. 458; 22.6, p. 474; 22.8, p. 484.

Y. MARUYAMA and H. BÄSSLER
Mobility and Recombination Coefficient of Charge Carriers in Disordered Pentacene
(b) Vol. **103**, No. 1, p. 263–268 (1981).
See also 13.4, p. 168; *16*, p. *251*.

J. P. SANCHEZ, J. M. FRIEDT, and M. F. DE JESUS FILHO
^{57}Fe Absorption and Emission Mössbauer Spectroscopy in Co(III) Acetylacetonate
(b) Vol. **103**, No. 1, p. 311–318 (1981).
See also *5*, p. *62*.

R. M. HILL
Evaluation of Susceptibility Functions
(b) Vol. **103**, No. 1, p. 319–328 (1981).
See also *14.4*, p. *237*.

J. SINGH
Formation of a Complex Charge Carrier Due to the Interaction between a Free Frenkel Exciton and Excess Electron in Molecular Crystals
(b) Vol. **103**, No. 1, p. 423–428 (1981).
See also *13.5.1*, p. *188*.

H. SCHÄFER and H. W. HELBERG
Dielectric Properties of the One-Dimensional Conductors $(UDMTTF)_2BF_4$ and $(TMTTF)_2$-BF_4
(a) Vol. **63**, No. 1, p. 203–207 (1981).
See also *14.4*, p. *237*.

P. Petelenz
Charge Carrier Detrapping by Triplet Excitons
(b) Vol. **104**, No. 1, p. 133–144 (1981).
See also *13.5.1*, p. *189*.

T. Kröber, S. Wartewig, and W. Windsch
Electron Paramagnetic Resonance Studies of Ferroelectric Fe^{3+}-Doped Tris-Sarcosine Calcium Chloride Monocrystals
(b) Vol. **104**, No. 1, p. 177–184 (1981).
See also *14.4.1*, p. *242*; *19*, p. *287*.

I. M. Krigin, S. N. Lukin, G. N. Neilo, and A. D. Prokohrov
Ground State and Spin–Lattice Relaxation of Ce^{3+} Ions in Lanthan Ethyl Sulphate
(b) Vol. **104**, No. 1, p. K21–K25 (1981).
See also *19*, p. *288*.

T. Kawakubo
Color Centers in Irradiated Durene Crystals
(a) Vol. **64**, No. 1, p. 303–310 (1981).
See also *10.2*, p. *108*; 11, p. 123; *20.1*, p. *306*.

B. P. Chandra
Mechanoluminescence and Piezoelectric Behavior of Molecular Crystals
(a) Vol. **64**, No. 1, p. 395–405 (1981).
See also *12.2*, p. *139*; 14.3, p. 211; *20.3*, p. *325*.

M. A. Shaikh
Electrical Properties of Quinolinium $(TCNQ)_2$
(a) Vol. **64**, No. 1, p. K23–K25 (1981).
See also *14.3*, p. *211*; *15*, p. *247*.

J. Godlewski
Electrode-Limited Photoinjection in Aqueous Electrode–Organic Crystal Systems
(a) Vol. **64**, No. 2, p. 499–507 (1981).
See also *14.3.4*, p. *233*; *16*, p. *252*.

D. K. De
Electron Spin Resonance Study on Copper Salicylate Tetrahydrate (Blue Green) and Anhydrous Copper Salicylate (Brown)
(a) Vol. **64**, No. 2, p. 761–769 (1981).
See also 18.1, p. 265; *19*, p. *289*.

K. Kojima
Erratum to: Slip Systems in Deformed Anthracene Crystals
(a) Vol. **64**, No. 2, p. 797 (1981).
See also *10.2*, p. *109*.

H. Kresse and B. Gajewska
Dielectric Relaxation of the Liquid Crystalline S_E Phase
(a) Vol. **64**, No. 2, p. K161–K163 (1981).
See also *2*, p. *45*; *14.4*, p. *239*.

I. Natkaniec, A. V. Bielushkin, and T. Wasiutyński
Calculations of Phonon Dynamics in Solid Biphenyl. Mixing between Interphenyl and Lattice Vibrations
(b) Vol. **105**, No. 1, p. 413–423 (1981).
See also *6*, p. *74*.

A. Rajewska, A. Jabłonka, D. Sikorska, and A. Sobaszek
Neutron Critical Scattering in the Nematic Phase and at the Nematic–Isotropic Liquid Transition in PAA
(b) Vol. **105**, No. 1, p. K35–K37 (1981).
See also *1.2*, p. *16*; *2*, p. *45*.

J. P. Blanc, M. Dugay, H. Robert, and C. Thibaud
EPR Study of a Triplet Fine Structure in a 1d Alternating System: $Cs_2(TCNQ)_3$
(b) Vol. **105**, No. 2, p. 659–675 (1981).
See also *19*, p. *290*.

D. D. Eley, S. Kinnear, and M. R. Willis
Real and Apparent Photoconduction in Single Crystals of Methyltriphenylphosphonium $(TCNQ)_2^-$
(b) Vol. **105**, No. 2, p. 677–683 (1981).
See also *16*, p. *253*.

R. O. Loufty
Effect of Temperature on the Dark and Photoconductivity of Metal-Free Phthalocyanine
(a) Vol. **65**, No. 2, p. 659–668 (1981).
See also *14.3.1*, p. *221*; *16*, p. *253*.

A. K. Ghoshal, D. K. Dey, J. Sthanapati, D. Pal, and A. K. Pal
Single Crystal Magnetic Studies on the γ-Form of Bis N-Methylsalicylaldimine Copper (II)
(a) Vol. **65**, No. 2, p. K157–K161 (1981).
See also *19*, p. *290*.

R. Świetlik and A. Graja
On the Dielectric Properties of Tetracyanoquinodimethane (TCNQ) Salts with N-Methyl Derivatives of Pyridine
(a) Vol. **65**, No. 2, p. K165–K168 (1981).
See also *14.4*, p. *239*.

I. I. Ukrainskii
Effective Electron–Electron Interaction in Conjugated Polymers
(b) Vol. **106**, No. 1, p. 55–62 (1981).
See also *13.5.1*, p. *190*.

V. A. Benderskii, V. Kh. Brikenshtein, P. G. Filippov, and I. A. Shamovskii
Non-Linear Fluorescence Quenching in Molecular Crystals (III): Recombination of Static Excitations
(b) Vol. **106**, No. 1, p. 407–418 (1981).
See also *13.5.1*, p. *191*; *20.3*, p. *330*.

J. M. Viana and A. S. T. Pires
Proton Spin–Lattice Relaxation in TMMC
(b) Vol. **106**, No. 1, p. K1 – K4 (1981).
See also 18.4, p. 283; *19*, p. *291*.

F. C. Montenegro and A. S. T. Pires
Effects of the Anisotropy in the Positions of the Resonance Lines in EPR in TMMC
(b) Vol. **106**, No. 1, p. K13 – K15 (1981).
See also *19*, p. *291*.

A. Montaner, M. Galtier, C. Benoit, and M. Aldissi
Far Infrared Optical Properties of Pure and SbF_5 Doped Trans Polyacetylene
(a) Vol. **66**, No. 1, p. 267 – 270 (1981).
See also *20.1*, p. *314*.

V. Madhusudan, S. G. Sathyanarayan, and G. Sivarama Sastry
Electrical Conductivity and Stimulated Thermocurrent Studies in Dicalcium Barium Propionate Single Crystals
(a) Vol. **66**, No. 1, p. 371 – 376 (1981).
See also *14*, p. *198*; 15, p. 249.

R. Cach
The Measurement of the Gap Exponent Δ_4 for TGS Single Crystal
(b) Vol. **106**, No. 2, p. K137 – K140 (1981).
See also *14.4.1*, p. *244*.

A. Polyakov, Yu. Ryabikin, B. Alimbek, and S. Pivovarov
Relaxation Spectra and Their Use in the Structural Studies
(b) Vol. **106**, No. 2, p. K149 – K152 (1981).
See also 11, p. 127; *19*, p. *292*.

P. Kindl and H. Sormann
Investigations on the Phase Transition of PTFE at the Melting Point by Positron Annihilation
(a) Vol. **66**, No. 2, p. 627 – 633 (1981).
See also *1.2*, p. *18*.

A. A. Maksimov and I. I. Tartakovskii
Stimulated Resonant Raman Scattering of Light in Anthracene Crystals
(b) Vol. **107**, No. 1, p. 55 – 60 (1981).
See also *6*, p. *78*; *20.1*, p. *316*.

Ch. Högfors, J. Kubát, and M. Rigdahl
A Cooperative Model of Stress Relaxation in Solids
(b) Vol. **107**, No. 1, p. 147 – 155 (1981).
See also *12.1*, p. *136*; *12.2*, p. *140*; 13.5, p. 187; 21, p. 345; 22.5.2, p. 467.

B. R. Bułka
Stability of the Peierls State Due to Impurities
(b) Vol. **107**, No. 1, p. 359 – 365 (1981).
See also *10.2*, p. *115*; *13.4*, p. *180*.

H. KRESSE, S. DIELE, A. WIEGELEBEN, D. DEMUS, and B. GAJEWSKA
A New Liquid Crystalline Low Temperature Phase
(a) Vol. **67**, No. 1, p. K21 – K23 (1981).
See also *2*, p. *49*; 14.4, p. 240.

A. N. VTYURIN, V. P. ERMAKOV, B. I. OSTROVSKII, and V. F. SHABANOV
Study of Optical Second Harmonic Generation in Ferroelectric Liquid Crystal
(b) Vol. **107**, No. 2, p. 397 – 402 (1981).
See also *2*, p. *49*; *14.4.1*, p. *245*; 20, p. 299.

E. L. FRANKEVICH, M. M. TRIBEL, I. A. SOKOLIK, L. I. KOLESNIKOVA, and YU. M. STOLOVITSKII
Magnetic Spin Phenomena in the Chlorophyll "a" Photoconductivity
(b) Vol. **107**, No. 2, p. 423 – 434 (1981).
See also 10.2, p. 116; *16*, p. *255*; 18, p. 263.

V. E. SCHNEIDER and E. E. TORNAU
The Model of the Phase Transition in Squaric Acid
(b) Vol. **107**, No. 2, p. 491 – 496 (1981).
See also *1.2*, p. *19*; *14.4.1*, p. *245*.

P. K. C. PILLAI, G. K. NARULA, A. K. TRIPATHI, and R. G. MENDIRATTA
Thermally Stimulated Discharge Current Studies in Polarized Polypropylene
(a) Vol. **67**, No. 2, p. 649 – 654 (1981).
See also *14.4*, p. *240*.

T. E. JENKINS and P. O'BRIEN
A Room Temperature Specific Heat Anomaly in Triamantane ($C_{18}H_{24}$)
(a) Vol. **67**, No. 2, p. K161 – K162 (1981).
See also *2*, p. *50*; *8*, p. *88*.

S. RADHAKRISHNAN, M. N. KAMALASANAN, and P. C. MEHENDRU
Surface Charge Decay Characteristics in Polyvinyl Acetate Films
(a) Vol. **68**, No. 1, p. 117 – 121 (1981).
See also *14.4*, p. *241*.

A. A. BEDNYAKOV, YU. V. BULGAKOV, V. S. NIKOLAEV, and V. L. CHERNOV
Energy Straggling of Hydrogen and Helium Ions in Al, C, and Polystyrene at Energies of Tens and Hundreds keV per Nucleon
(a) Vol. **68**, No. 1, p. 187 – 192 (1981).
See also *11*, p. *131*; 21, p. 348; 22.1, p. 403.

C. BENOIT, M. ROLLAND, M. ALDISSI, A. ROSSI, M. CADENE, and P. BERNIER
Diffusion Study of Iodine and Antimony Pentafluoride in Polyacetylene
(a) Vol. **68**, No. 1, p. 209 – 220 (1981).
See also *9*, p. *92*.

D. D. ELEY, S. KINNEAR, and M. R. WILLIS
Lifetime of Photogenerated Carriers in TCNQ Single Crystals
(b) Vol. **108**, No. 2, p. 373 – 376 (1981).
See also *16*, p. *255*.

G. Völkel, H.-E. Müller, and W. Flohrer
Critical Anomaly of the Electron Spin–Lattice Relaxation in X-Irradiated Rochelle Salt
(b) Vol. **108**, No. 2, p. 501–506 (1981).
See also 1.2, p. 21; 6, p. 81; 11, p. 131; *14.4.1*, p. *246*.

D. K. De, A. K. Ghoshal, D. Pal, and A. K. Pal
EPR and Magnetic Studies on γ:bis(N-Methyl Salicylaldimine) Copper (II)
(b) Vol. **108**, No. 2, p. 645–652 (1981).
See also *18.2*, p. *275*; *19*, p. *294*.

V. V. Kind, V. P. Fedorov, and A. V. Korshunov
The Manifestation of Intermolecular Hydrogen Bond in a Binary Naphthalene-β-Naphthol Mixture
(b) Vol. **108**, No. 2, p. K143–K144 (1981).
See also *4*, p. *61*; *20.1*, p. *321*.

A. S. N. Rao, P. N. Murty, C. R. K. Murty, and T. Rs. Reddy
EPR Study of Molecular Ordering in Liquid Crystal: NOBA
(a) Vol. **68**, No. 2, p. 373–377 (1981).
See also *2*, p. *51*; *19*, p. *295*.

23. Solidified Gases

Original Papers and Short Notes

R. J. Sayer, R. H. Prince, and W. W. Duley
Site Effects on N and O Luminescence in Electron-Irradiated Neon Films
(b) Vol. **105**, No. 2, p. 727–738 (1981).
See also 10.2, p. 111; *20.3*, p. *328*.

V. B. Magalinskii and V. I. Zubov
On the Constancy of the Configurational Entropy and Related Properties of Solids along the Melting Curve
(b) Vol. **105**, No. 2, p. K139–K142 (1981).
See also *1.2*, p. *17*; *8*, p. *85*.

N. Schwentner, G. Martens, and H. W. Rudolf
Free Exciton Energy Transfer in Kr–Au Sandwiches
(b) Vol. **106**, No. 1, p. 183–192 (1981).
See also *13.5.1*, p. *190*; 17, p. 257.

R. J. Sayer, R. H. Prince, and W. W. Duley
Emission Spectra of O and N Atoms and the Structure of Solid N_2/Ne Mixtures
(b) Vol. **106**, No. 1, p. 249–258 (1981).
See also *2*, p. *47*; *20.3*, p. *329*.

M. D. Tiwari
Specific Heat of Ar–Xe Solid Solutions at Low Temperatures
(b) Vol. **106**, No. 2, p. 569–573 (1981).
See also *6*, p. *77*; *8*, p. *86*.

S. B. Trickey, A. K. Ray, and J. P. Worth
Adequacy of Local-Exchange Excitation Hamiltonians in Insulators
(b) Vol. **106**, No. 2, p. 613–620 (1981).
See also *13.1*, p. *159*.

V. I. Zubov
On the Statistical Theory of Point Defects in Pure Crystals with Strong Anharmonicity
(b) Vol. **106**, No. 2, p. K145–K148 (1981).
See also *10.2*, p. *114*.

R. K. Singh and D. K. Neb
On Cohesion and Many-Body Forces in Rare Gas Solids
(b) Vol. **107**, No. 1, p. K47–K52 (1981).
See also *4*, p. *60*; *12.2*, p. *140*.

R. Kink, A. Lõhmus, and M. Selg
Self-Trapping and Hot Luminescence of Excitons in Rare Gas Solids
(b) Vol. **107**, No. 2, p. 479–490 (1981).
See also *13.5.1*, p. *192*; 20.1, p. 317; *20.3*, p. *332*.

Author Index

19. Magnetic Resonance

20. Optical Properties
    20.1 Spectra. Optical Constants (X-Ray Spectra See 20)
    20.2 Lasers
    20.3 Luminescence (X-Ray Spectra See 20)

21. Metals. Alloys
    21.1 Transition Metals (Group Sc to Zn) and Their Alloys
        21.1.1 Fe and Fe Alloys
    21.2 Alkali Metals
    21.3 Alkali-Earth Metals
    21.4 Rare-Earth Metals
    21.5 Actinides
    21.6 Noble Metals
    21.7 Semi-Metals (Semiconducting Alloys See 22.7)

22. Semiconductors and Ionic Crystals
    22.1 Elements
        22.1.1 Germanium
        22.1.2 Silicon
        22.1.3 Group VI Elements
    22.2 III–V Compounds
        22.2.1 Arsenides
        22.2.2 Phosphides
        22.2.3 Antimonides
        22.2.4 Mixed Crystals
    22.3 IV–IV, II–V, and III–VI Compounds
        22.3.1 Mixed Crystals
    22.4 II–VI Compounds
        22.4.1 Sulphides
        22.4.2 Selenides
        22.4.3 Tellurides
        22.4.4 Mixed Crystals
    22.5 Halides
        22.5.1 Silver Halides
        22.5.2 Alkali Halides
        22.5.3 Alkali-Earth Halides
        22.5.4 Mixed Crystals
    22.6 Simple Oxides (CdO, ZnO See 22.4)
        22.6.1 Mixed Crystals
    22.7 Semiconducting Intermetallic Compounds
        22.7.1 Mixed Crystals
    22.8 Tertiary and Higher Compounds
        22.8.1 Salts of Oxy-Acids (Carbonates, Phosphates, Silicates Including Mica etc.)
        22.8.2 Spinels, Garnets, Ferrites
    22.9 Organic Semiconductors

23. Solidified Gases

For comments on the use of the classification scheme see p. 6.

32913/32914

Manuscripts and letters for physica status solidi (b) — basic research and physica status solidi (a) — applied research

should be sent to

DDR-1020 Berlin, Neue Schönhauser Str. 20

or

Internationaler Buchversand GmbH, 1000 Berlin 19, Eschenallee 21.

The object of "physica status solidi (b)" and "physica status solidi (a)" is to publish rapidly important and topical results concerning fundamental research in the field of solid state physics and aspects of application and experimental techniques of fundamental solid state research, respectively. The issues of "physica status solidi (b)" and "physica status solidi (a)" will be published on the 1st and the 16th day of each month, respectively. Either series will contain the following categories:

1. Review articles written by arrangement with the editors. They will be remunerated.
2. Original papers selected from those which have not been submitted for publication elsewhere after being reviewed. (The latest date for acceptance of contributions is the 8th and the 23rd day of each month for "physica status solidi (b)" and "physica status solidi (a)", respectively 50 days before the delivery of the journal.)
3. Short notes published in the off-set part of the journal. They are, if necessary after being reviewed, selected by the editorial staff. (The latest date for acceptance of manuscripts for this part is the 12th and the 27th day of each month for "physica status solidi (b)" and "physica status solidi (a)", respectively, 16 days before the delivery of the journal.)
4. Pre-printed titles of papers to be published in the next issue of both series.

Editorial Information

Manuscripts should be submitted preferentially in English. Besides, papers in German, French, and Russian are admitted. The final decision on the incorporation of the papers in the series (b) or (a) is taken by the editorial staff after consultation of the referee. Authors are asked to observe the following rules:

1. Manuscripts have to be written as concisely as possible. As most, 20 and 4 pages (28 lines per page; abstracts, tables, references, figures, and captions of figures included, where 3 figures are counted as about one page) are admitted for original papers and short notes, respectively. All manuscripts should be type-written on one side only. Two copies are required. Badly arranged manuscripts are returned.
2. The author's name, initials, and institution should be given.
3. One abstract of about ten lines in English and one in German, French, or Russian are required for original papers.
4. Because of lack of space conclusions (if any) must not have the character of a summary (i. e. substantially a repetition of the contents of the abstract).
5. Formulae should not be written too small and not with pencil. Separate lines for formulae are desirable. Indistinct writing will lead to a return of the manuscript.
6. Drawings should be enclosed on separate sheets of paper, not exceeding the maximum and minimum sizes of 20×30 and $10 \times 15\ cm^2$, respectively. Photographs should be submitted, on white paper, only for review articles and original papers. Duplicates are required.
7. Captions for the figures should be submitted on separate sheets of paper, and special instructions with regard to the size of illustrations should be given on these.
8. References should be numbered in order of appearance and should be given, on a separate sheet, in the form used in physica status solidi.
9. SI-units should be used.
10. Each paper should contain a proposal for the subject classification according to the classification scheme appearing on the second and third cover of each issue.

The authors will be sent galley proofs of review articles and original papers. Proof reading of short notes and pre-printed titles will be done by the editorial staff.
For each original paper, and short note, a total of 75 reprints is available free of charge. The number of reprints free of charge per review article is 25. Requests for additional reprints (against payment) should be given together with the submitted manuscript. **The deadline for orders of additional reprints is the 15th and the 30th day of each month** for "physica status solidi (b)" and "physica status solidi (a)", respectively, two weeks before the delivery of the journal.

Kolibri 360 GmbH
Ferdinand-Jühlke-Straße 7
99095 Erfurt, Deutschland
produktsicherheit@kolibri360.de